Student's Solutions Manual

to accompany

Basic College Mathematics

Second Edition

Julie Miller
Daytona State College

Molly O'Neill
Daytona State College

Nancy Hyde
Broward Community College

Prepared by
Jon Weerts

 **Higher Education**

Boston   Burr Ridge, IL   Dubuque, IA   New York   San Francisco   St. Louis
Bangkok   Bogotá   Caracas   Kuala Lumpur   Lisbon   London   Madrid   Mexico City
Milan   Montreal   New Delhi   Santiago   Seoul   Singapore   Sydney   Taipei   Toronto

The **McGraw·Hill** Companies

Higher Education

STUDENT'S SOLUTIONS MANUAL to accompany BASIC COLLEGE MATHEMATICS, SECOND EDITION
JULIE MILLER, MOLLY O'NEILL, AND NANCY HYDE

Published by McGraw-Hill Higher Education, an imprint of The McGraw-Hill Companies, Inc., 1221 Avenue of the Americas, New York, NY 10020. Copyright © 2009 and 2007 by The McGraw-Hill Companies, Inc. All rights reserved.

This book is printed on recycled, acid-free paper containing 10% post consumer waste.

1 2 3 4 5 6 7 8 9 0 OPD/QPD 0 9 8

ISBN: 978-0-07-335812-3
MHID: 0-07-335812-6

www.mhhe.com

Contents

Chapter 4   Decimals

Chapter 5   Ratio and Proportion

Chapter 6   Percents

Chapter 7   Measurement

Chapter 8   Geometry

Chapter 9   Introduction to Statistics

Chapter 10   Real Numbers

Chapter 11   Solving Equations

Additional Topics Appendix

Preface

The *Student's Solutions Manual* to accompany *Basic College Mathematics*, Second Edition by Julie Miller, Molly O'Neill, and Nancy Hyde contains detailed solutions of the odd-numbered practice exercises, the odd-numbered problem recognition exercises, the odd-numbered review exercises, the odd-numbered chapter test exercises, and the odd-numbered cumulative review exercises. I have attempted to provide solutions consistent with the procedures introduced in the textbook. Every attempt has been made to make this manual as error free as possible.

A number of people need to be recognized for their contributions in the preparation of this manual. Thanks go to David Millage, Emilie Berglund, and Adam Fischer of McGraw-Hill Higher Education for giving me the opportunity to author this manual. I am grateful for all the responses to my questions and their guidance throughout the development process of this manual. I wish to express my appreciation to Julie Miller, Molly O'Neill, and Nancy Hyde for writing a Basic College Mathematics text that is well organized and assists students in their understanding of the basic topics of mathematics.

A special word of appreciation goes to my wife, Jan, for her support and understanding during the many hours that went into the preparation of this manual.

Jon D. Weerts
Triton College
2000 Fifth Avenue
River Grove, IL 60171
e-mail: jweerts@triton.edu

Chapter 1   Whole Numbers

Chapter Opener Puzzle

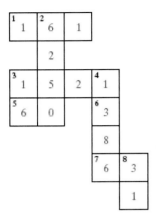

Section 1.1   Introduction to Whole Numbers

Section 1.1   Practice Exercises

1. Answers will vary.

3. 8, 213,457
   7: ones
   5: tens
   4: hundreds
   3: thousands
   1: ten-thousands
   2: hundred-thousands
   8: millions

5. 3<u>2</u>1   tens

7. 21<u>4</u>   ones

9. 8,<u>7</u>10   hundreds

11. <u>1</u>,430   thousands

13. <u>4</u>52,723   hundred-thousands

15. <u>1</u>,023,676,207   billions

17. <u>2</u>2,422   ten-thousands

19. 5<u>1</u>,033,201   millions

21. <u>1</u>0,677,881   ten-millions

23. <u>7</u>,653,468,440   billions

25. 5 tens + 8 ones

27. 5 hundreds + 3 tens + 9 ones

29. 5 hundreds + 3 ones

31. 1 ten-thousand + 2 hundreds + 4 tens + 1 one

33. 524

35. 150

37. 1,906

39. 85,007

41. Ones, thousands, millions, billions

43. Two hundred forty-one

45. Six hundred three

47. Thirty-one thousand, five hundred thirty

49. One hundred thousand, two hundred thirty-four

51. Nine thousand, five hundred thirty-five

1

53. Twenty thousand, three hundred twenty

55. One thousand, three hundred seventy-seven

57. 6,005

59. 672,000

61. 1,484,250

63.

65. Counting on a number line, 10 is 4 units to the right of 6.

67. Counting on a number line, 4 is 3 units to the left of 7.

69. $8 > 2$
8 is greater than 2, or 2 is less than 8.

71. $3 < 7$
3 is less than 7, or 7 is greater than 3.

73. $6 < 11$

75. $21 > 18$

77. $3 < 7$

79. $95 > 89$

81. $0 < 3$

83. $90 < 91$

85. False; 12 is made up of the digits 1 and 2.

87. 99

89. There is no greatest whole number.

91. 10,000,000   7 zeros

93. 964

Section 1.2   Addition of Whole Numbers and Perimeter

Section 1.2 Practice Exercises

1. Answers will vary.

3. 3 hundreds + 5 tens + 1 one

5. 1 hundred + 7 ones

7. 4012

9. Fill in the table. Use the number line if necessary.

+	0	1	2	3	4	5	6	7	8	9
0	0	1	2	3	4	5	6	7	8	9
1	1	2	3	4	5	6	7	8	9	10
2	2	3	4	5	6	7	8	9	10	11
3	3	4	5	6	7	8	9	10	11	12
4	4	5	6	7	8	9	10	11	12	13
5	5	6	7	8	9	10	11	12	13	14
6	6	7	8	9	10	11	12	13	14	15
7	7	8	9	10	11	12	13	14	15	16
8	8	9	10	11	12	13	14	15	16	17
9	9	10	11	12	13	14	15	16	17	18

11. $2 + 8 = 10$
Addends: 2, 8
Sum: 10

13. $11 + 10 = 21$
Addends: 11, 10
Sum: 21

15. $5 + 8 + 2 = 15$
Addends: 5, 8, 2
Sum: 15

17.
$$\begin{array}{l} 21 = 2 \text{ tens} + 1 \text{ one} \\ + 53 = 5 \text{ tens} + 3 \text{ ones} \\ \hline 74 = 7 \text{ tens} + 4 \text{ ones} \end{array}$$

19.
$$\begin{array}{l} 15 = 1 \text{ ten } + 5 \text{ ones} \\ + 43 = 4 \text{ tens} + 3 \text{ ones} \\ \hline 58 = 5 \text{ tens} + 8 \text{ ones} \end{array}$$

21.
$$\begin{array}{l} 10 = 1 \text{ ten } + 0 \text{ ones} \\ 8 = 0 \text{ tens} + 8 \text{ ones} \\ 30 = 3 \text{ tens} + 0 \text{ ones} \\ \hline 48 = 4 \text{ tens} + 8 \text{ ones} \end{array}$$

23.
$$\begin{array}{l} 6 = 0 \text{ tens} + 6 \text{ ones} \\ 11 = 1 \text{ ten } + 1 \text{ one} \\ + 2 = 0 \text{ tens} + 2 \text{ ones} \\ \hline 19 = 1 \text{ ten } + 9 \text{ ones} \end{array}$$

25.
$$\begin{array}{r} 407 \\ + 181 \\ \hline 588 \end{array}$$

27.
$$\begin{array}{r} 444 \\ + 354 \\ \hline 798 \end{array}$$

29.
$$\begin{array}{r} 11 \\ 221 \\ + 5 \\ \hline 237 \end{array}$$

31.
$$\begin{array}{r} 24 \\ 14 \\ + 160 \\ \hline 198 \end{array}$$

33.
$$\begin{array}{r} 1 \\ 25 \\ + 59 \\ \hline 84 \end{array}$$

35.
$$\begin{array}{r} 1 \\ 38 \\ + 77 \\ \hline 115 \end{array}$$

37.
$$\begin{array}{r} 1 \\ 642 \\ + 295 \\ \hline 937 \end{array}$$

39.
$$\begin{array}{r} 11 \\ 462 \\ + 388 \\ \hline 850 \end{array}$$

41.
$$\begin{array}{r} 1 \\ 2 \\ 31 \\ + 8 \\ \hline 41 \end{array}$$

43.
$$\begin{array}{r} 1 \\ 7 \\ 18 \\ + 4 \\ \hline 29 \end{array}$$

45.
$$\begin{array}{r} 1\,1 \\ 62 \\ 907 \\ + 34 \\ \hline 1003 \end{array}$$

47.
$$\begin{array}{r} 1\,1 \\ 87 \\ 119 \\ + 630 \\ \hline 836 \end{array}$$

49.
$$\begin{array}{r} 11 \\ 23112 \\ 892 \\ \hline 24,004 \end{array}$$

51.
$$\begin{array}{r} \overset{11\ \ 11}{92\ 377} \\ 5\ 622 \\ 34\ 659 \\ \hline 132,658 \end{array}$$

53. $30 + 21 = 21 + 30$

55. $8 + 13 = 13 + 8$

57. $(23 + 9) + 10 = 23 + (9 + 10)$

59. $41 + (3 + 22) = (41 + 3) + 22$

61. The sum of any number and 0 is that number.

    (a) $423 + 0 = 423$

    (b) $0 + 25 = 25$

    (c) $67 + 0 = 67$

63. $100 + 42$
$$\begin{array}{r} 100 \\ + 42 \\ \hline 142 \end{array}$$

65. $23 + 81$
$$\begin{array}{r} 23 \\ + 81 \\ \hline 104 \end{array}$$

67. $76 + 2$
$$\begin{array}{r} 76 \\ + 2 \\ \hline 78 \end{array}$$

69. $1320 + 448$
$$\begin{array}{r} 1\ 320 \\ + 448 \\ \hline 1,768 \end{array}$$

71. For example: The sum of 54 and 24

73. For example: 88 added to 12

75. For example: The total of 4, 23, and 77

77. For example: 10 increased by 8

79.
$$\begin{array}{r} 103 \\ 112 \\ + 61 \\ \hline 276 \end{array}$$
276 people attended the play.

81.
$$\begin{array}{r} \overset{1\,2\ \ \ 1}{26{,}548{,}000} \\ 26{,}930{,}000 \\ + 20{,}805{,}000 \\ \hline 74{,}283{,}000 \end{array}$$
The shows had a total of 74,283,000 viewers.

83.
$$\begin{array}{r} \$43{,}000 \\ + \ 2{,}500 \\ \hline \$45{,}500 \end{array}$$
Nora earns $45,500.

85.
$$\begin{array}{r} \overset{1\,1}{115} \\ 104 \\ 93 \\ + 111 \\ \hline 423 \end{array}$$
423 desks were delivered.

87.
$$\begin{array}{r} \overset{5\,3\,3}{2\,787} \\ 1\,956 \\ 991 \\ 1\,817 \\ 1\,567 \\ 715 \\ + 3\,705 \\ \hline 13{,}538 \end{array}$$
There are 13,538 participants.

89.
$$\begin{array}{r} \overset{111\ 11}{100{,}052} \\ 675{,}038 \\ + 45{,}934 \\ \hline 821{,}024 \end{array}$$
There are 821,024 nonteachers.

91.
$$\begin{array}{r} \overset{1}{35}\ \text{cm} \\ 35\ \text{cm} \\ + 34\ \text{cm} \\ \hline 104\ \text{cm} \end{array}$$

93.
$$\begin{array}{r} 2 \\ 21 \text{ m} \\ 20 \text{ m} \\ 18 \text{ m} \\ 19 \text{ m} \\ 11 \text{ m} \\ + 21 \text{ m} \\ \hline 110 \text{ m} \end{array}$$

95.
$$\begin{array}{r} 2 \\ 6 \text{ yd} \\ 10 \text{ yd} \\ 11 \text{ yd} \\ 3 \text{ yd} \\ 5 \text{ yd} \\ + 7 \text{ yd} \\ \hline 42 \text{ yd} \end{array}$$

97.
$$\begin{array}{r} 94 \text{ ft} \\ 94 \text{ ft} \\ 50 \text{ ft} \\ + 50 \text{ ft} \\ \hline 288 \text{ ft} \end{array}$$

99. $9,084,037 + 452,903 = 9,536,940$

101. $7,201,529 + 962,411 = 8,163,940$

103.
$$\begin{array}{r} 9,300,050 \\ 7,803,513 \\ 3,480,009 \\ + 907,822 \\ \hline 21,491,394 \end{array}$$

105.
$$\begin{array}{r} 17,457,000 \\ 17,164,000 \\ 17,004,000 \\ + 15,717,000 \\ \hline 67,342,000 \text{ viewers} \end{array}$$

Section 1.3   Subtraction of Whole Numbers

Section 1.3   Practice Exercises

1. Answers will vary.

3.
$$\begin{array}{r} 330 \\ + 821 \\ \hline 1151 \end{array}$$

5.
$$\begin{array}{r} 1 \\ 46 \\ 804 \\ + 49 \\ \hline 899 \end{array}$$

7. $0 < 10$

9. $12 - 8 = 4$
minuend: 12
subtrahend: 8
difference: 4

11. $21 - 12 = 9$
minuend: 21
subtrahend: 12
difference: 9

13.
$$\begin{array}{r} 9 \\ - 6 \\ \hline 3 \end{array}$$
minuend: 9
subtrahend: 6
difference: 3

15. $27 - 9 = 18$ because $18 + 9 = 27$.

17. $102 - 75 = 27$ because $27 + 75 = 102$.

19. $8 - 3 = 5$ Check: $5 + 3 = 8$

21. $4 - 1 = 3$ Check: $\underline{3} + 1 = 4$

23. $6 - 0 = 6$ Check: $\underline{6} + 0 = 6$

25.
$$\begin{array}{r} 68 \\ - 23 \\ \hline 45 \end{array} \quad \text{Check:} \quad \begin{array}{r} 45 \\ + 23 \\ \hline 68 \checkmark \end{array}$$

27.
```
   88     Check:   61
 − 27            + 27
 ─────          ─────
   61              88 ✓
```

29.
```
 1347    Check:  1126
 − 221          + 221
 ─────          ─────
 1126            1347 ✓
```

31.
```
  1525   Check:  1204
 −1204          + 321
 ─────          ─────
  321            1525 ✓
```

33.
```
 12 806   Check:  10 004
 − 2 802          + 2 802
 ──────           ───────
 10,004            12,806 ✓
```

35.
```
  14,356  Check:    1 103
 −13,253          + 13 253
 ──────           ────────
  1,103             14,356 ✓
```

37.
```
   6 16
   7̸6̸     Check:    1
  − 59            + 59
  ─────           ────
    17              76 ✓
```

39.
```
   7 17
   8̸7̸     Check:    1
  − 38            + 38
  ─────           ────
    49              87 ✓
```

41.
```
   3 10
   2̸4̸0̸    Check:    1
  −136            + 136
  ─────           ─────
   104             240 ✓
```

43.
```
       10
    6 0̸ 10
    7̸ 1̸ 0̸   Check:    11
  − 1 8 9           521
  ───────          + 189
    5 2 1          ─────
                    710 ✓
```

45.
```
      4 10
   43̸5̸0̸    Check:     1
  − 432 7            23
  ───────          + 4327
     2 3           ─────
                    4350 ✓
```

47.
```
    9  9
  5 1̸0 1̸0 12
   6̸ 0̸ 0̸ 2̸   Check:   1 11
  −1 2 3 8            4764
  ─────────          +1238
   4 7 6 4           ─────
                      6002 ✓
```

49.
```
   0 10
   1̸0̸,425  Check:   1 403
  − 9 022          + 9 022
  ───────          ───────
   1, 403           10,425 ✓
```

51.
```
       11
    5 1̸ 10
    6̸2̸ 0̸88   Check:       1
  − 59 871          1 2 217
  ───────          + 59 871
    2,217          ────────
                     62,088 ✓
```

53.
```
       16
    3 6̸ 10
    4̸7̸ 0̸   Check:   378
  − 9 2            + 92
  ──────           ────
   37 8             470 ✓
```

55.
```
        16
    2 6̸ 10 10
    3̸7̸ 0̸ 0   Check:     1 1
  − 29 87           1 713
  ────────          + 2987
    7 13           ──────
                     3700 ✓
```

57.
```
          13
    1  3̸ 13
   32̸,4̸ 3̸9   Check:    1 1
  − 1  4 98          30 941
  ─────────          + 1 498
   30 ,9 41          ──────
                      32,439 ✓
```

59.
```
        9
    7 1̸0 10    2 14
   8̸,0̸ 0̸7,23̸ 4̸   Check:   1 1     1
  − 2,3 45,11 5            5 662 119
  ──────────────          + 2 345 115
   5, 6 62,11 9           ───────────
                           8,007,234 ✓
```

61.
```
   78
 − 23
 ────
   55
```

63.
```
   78
 −  6
 ────
   72
```

6

65.
$$\begin{array}{r} 422 \\ -100 \\ \hline 322 \end{array}$$

67.
$$\begin{array}{r} {}^{8\ 10} \\ 10\not{9}\not{0} \\ -7\ 2 \\ \hline 101\ 8 \end{array}$$

69.
$$\begin{array}{r} {}^{4\ 10} \\ \not{5}\not{0} \\ -1\ 3 \\ \hline 3\ 7 \end{array}$$

71.
$$\begin{array}{r} {}^{9\ 13} \\ 10\not{3} \\ -3\ 5 \\ \hline 6\ 8 \end{array}$$

73. For example: 93 minus 27

75. For example: Subtract 85 from 165.

77. The expression 7 − 4 means 7 minus 4, yielding a difference of 3. The expression 4 − 7 means 4 minus 7 which results in a difference of −3.

79.
$$\begin{array}{r} {}^{4\ 10} \\ \$\not{5}\not{0} \\ -1\ 7 \\ \hline \$3\ 3 \end{array}$$
$33 change was received.

81.
$$\begin{array}{r} {}^{0\ 11} \\ \not{1}\not{1}8 \\ -6\ 3 \\ \hline 5\ 5 \end{array}$$
Lennon and McCartney had 55 more hits.

83.
$$\begin{array}{r} {}^{1\ 16} \\ \not{2}\not{6} \\ -1\ 8 \\ \hline 8 \end{array}$$
Lily needs 8 more plants.

85.
$$\begin{array}{r} {}^{9\ 13} \\ {}^{3\ \not{10}\ \not{14}\ 13} \\ \not{4}\not{0}\not{4}\not{3} \\ -2\ 0\ 6\ 4 \\ \hline 1\ 9\ 7\ 9 \end{array}$$
The Lion King had been performed 1979 more times.

87.
$$\begin{array}{r} 14\ \text{m} \\ +12\ \text{m} \\ \hline 26\ \text{m} \end{array} \qquad \begin{array}{r} 39\ \text{m} \\ -26\ \text{m} \\ \hline 13\ \text{m} \end{array}$$
The missing length is 13 m.

89.
$$\begin{array}{r} 4 \\ 14 \\ 14 \\ 14 \\ +10 \\ \hline 46\ \text{yd} \end{array} \qquad \begin{array}{r} 56\ \text{yd} \\ -46\ \text{yd} \\ \hline 10\ \text{yd} \end{array}$$
The missing side is 10 yd long.

91.
$$\begin{array}{r} {}^{9\quad 9} \\ {}^{7\ \not{10}\ \not{10}\ 10} \\ 2\ 39\not{8}\ 0\ 0\ 0 \\ -2\ 390\ 2\ 5\ 2 \\ \hline 7\ 7\ 4\ 8 \end{array}$$
The difference is 7748 marriages.

93.
$$\begin{array}{r} {}^{13} \\ {}^{3\ \not{3}\ 13} \\ 2\not{4}\not{4}\not{3}\ 489 \\ -2\ 248\ 000 \\ \hline 195{,}489 \end{array}$$
The difference is 195,489 marriages.

95.
$$\begin{array}{r} 4{,}905{,}620 \\ -458{,}318 \\ \hline 4{,}447{,}302 \end{array}$$

97.
$$\begin{array}{r} 82{,}025{,}160 \\ -79{,}118{,}705 \\ \hline 2{,}906{,}455 \end{array}$$

99.
$$\begin{array}{r} 41{,}217\ \text{mi}^2 \\ -24{,}078\ \text{mi}^2 \\ \hline 17{,}139\ \text{mi}^2 \end{array}$$

101.
$$\begin{array}{r} 54{,}310\ \text{mi}^2 \\ -41{,}217\ \text{mi}^2 \\ \hline 13{,}093\ \text{mi}^2 \end{array}$$
Wisconsin has 13,093 mi^2 more than Tennessee.

Section 1.4   Rounding and Estimating

Section 1.4   Practice Exercises

1. Answers will vary

3.
$$
\begin{array}{r}
59 \\
-33 \\
\hline
26
\end{array}
$$

5.
$$
\begin{array}{r}
^{1\ 11}\\
4\,009 \\
+\,998 \\
\hline
5{,}007
\end{array}
$$

7. Ten-thousands

9. If the digit in the tens place is 0, 1, 2, 3, or 4, then change the tens and ones digits to 0. If the digit in the tens place is 5, 6, 7, 8, or 9, increase the digit in the hundreds place by 1 and change the tens and ones digits to 0.

11. $34\underline{2} \approx 340$

13. $72\underline{5} \approx 730$

15. $93\underline{8}4 \approx 9400$

17. $85\underline{3}9 \approx 8500$

19. $34{,}\underline{9}92 \approx 35{,}000$

21. $2\underline{5}78 \approx 3000$

23. $99\underline{8}2 \approx 10000$

25. $109{,}\underline{3}37 \approx 109{,}000$

27. $48\underline{9}{,}090 \approx 490{,}000$

29. $\$148\underline{4}31{,}020 \approx \$148{,}000{,}000$

31. $238{,}\underline{8}63 \text{ mi} \approx 239{,}000 \text{ mi}$

33.
$$
\begin{array}{rcr}
57 & \rightarrow & 60 \\
82 & \rightarrow & 80 \\
+\,21 & \rightarrow & +\,20 \\
\hline
& & 160
\end{array}
$$

35.
$$
\begin{array}{rcr}
41 & \rightarrow & 40 \\
12 & \rightarrow & 10 \\
+\,129 & \rightarrow & +\,130 \\
\hline
& & 180
\end{array}
$$

37.
$$
\begin{array}{rcr}
898 & \rightarrow & 900 \\
-\,422 & \rightarrow & -\,400 \\
\hline
& & 500
\end{array}
$$

39.
$$
\begin{array}{rcr}
3412 & \rightarrow & 3400 \\
-\,1252 & \rightarrow & -\,1300 \\
\hline
& & 2100
\end{array}
$$

41.
$$
\begin{array}{rcr}
& & ^{1}\\
97{,}404{,}576 & \rightarrow & 97{,}000{,}000 \\
+\,53{,}695{,}428 & \rightarrow & +\,54{,}000{,}000 \\
\hline
& & 151{,}000{,}000
\end{array}
$$
$\$151{,}000{,}000$ was brought in by Mars.

43.
$$
\begin{array}{rcr}
71{,}339{,}710 & \rightarrow & 71{,}000{,}000 \\
-\,59{,}684{,}076 & \rightarrow & -\,60{,}000{,}000 \\
\hline
& & 11{,}000{,}000
\end{array}
$$
Neil Diamond earned $\$11{,}000{,}000$ more.

45.
$$
\begin{array}{rcr}
& & ^{1}\\
\$3{,}316{,}897 & \rightarrow & \$3{,}300{,}000 \\
3{,}272{,}028 & \rightarrow & 3{,}300{,}000 \\
+\,3{,}360{,}289 & \rightarrow & +\,3{,}400{,}000 \\
\hline
& & \$10{,}000{,}000
\end{array}
$$

47. (a) 2003; $\$3{,}470{,}295 \rightarrow \$3{,}500{,}000$

(b) 2005; $\$1{,}970{,}380 \rightarrow \$2{,}000{,}000$

49. Massachusetts; $78{,}815 \rightarrow 79{,}000$ students

51.
$$
\begin{array}{r}
79{,}000 \\
-\,8{,}000 \\
\hline
71{,}000
\end{array}
$$
The difference is 71,000 students.

53. Answers may vary.

55.

3045 mm	$\rightarrow$	3000 mm
1892 mm	$\rightarrow$	2000 mm
3045 mm	$\rightarrow$	3000 mm
+ 1892 mm	$\rightarrow$	+ 2000 mm
		10,000 mm

57.

		2
105 in.	$\rightarrow$	110 in.
57 in.	$\rightarrow$	60 in.
57 in.	$\rightarrow$	60 in.
105 in.	$\rightarrow$	110 in.
57 in.	$\rightarrow$	60 in.
+ 57 in.	$\rightarrow$	+ 60 in.
		460 in.

Section 1.5   Multiplication of Whole Numbers and Area

Section 1.5   Practice Exercises

1. Answers will vary.

3.

$$\begin{array}{r} 1 \\ 869,240 \rightarrow 870,000 \\ 34,921 \rightarrow 30,000 \\ + 108,332 \rightarrow + 110,000 \\ \hline 1,010,000 \end{array}$$

5.

$$\begin{array}{r} 8821 \rightarrow 8800 \\ - 3401 \rightarrow - 3400 \\ \hline 5400 \end{array}$$

7. $5 + 5 + 5 + 5 + 5 + 5 = 6 \times 5 = 30$

9. $9 + 9 + 9 = 3 \times 9 = 27$

11. $13 \times 42 = 546$
factors: 13, 42; product: 546

13. $3 \cdot 5 \cdot 2 = 30$
factors: 3, 5, 2; product: 30

15. For example: 5×12; $5 \cdot 12$; $5(12)$

17. d

19. e

21. c

23. $14 \times 8 = 8 \times 14$

25. $6 \times (2 \times 10) = (6 \times 2) \times 10$

27. $5(7 + 4) = (5 \times 7) + (5 \times 4)$

29.

$$\begin{array}{r} 24 \\ \times 6 \\ \hline 24 \\ + 120 \\ \hline 144 \end{array}$$

    Multiply 6×4.
    Multiply 6×20.
    Add.

31.

$$\begin{array}{r} 26 \\ \times 2 \\ \hline 12 \\ + 40 \\ \hline 52 \end{array}$$

    Multiply 2×6.
    Multiply 2×20.
    Add.

33.

$$\begin{array}{r} 131 \\ \times 5 \\ \hline 5 \\ 150 \\ + 500 \\ \hline 655 \end{array}$$

    Multiply 5×1.
    Multiply 5×30.
    Multiply 5×100.
    Add.

35.

$$\begin{array}{r} 344 \\ \times 4 \\ \hline 16 \\ 160 \\ + 1200 \\ \hline 1376 \end{array}$$

    Multiply 4×4.
    Multiply 4×40.
    Multiply 4×300.
    Add.

37.

$$\begin{array}{r} 3 \\ 1410 \\ \times 8 \\ \hline 11,280 \end{array}$$

39.

$$\begin{array}{r} 2 1 \\ 3312 \\ \times 7 \\ \hline 23,184 \end{array}$$

41.
$$\begin{array}{r}
{}^{1}\ \ {}^{13}\ \ \\
42,014 \\
\times\ \ \ \ \ \ 9 \\
\hline
378,126
\end{array}$$

43.
$$\begin{array}{r}
32 \\
\times\ \ 14 \\
\hline
128 \\
+\ 320 \\
\hline
448
\end{array}$$

45.
$$\begin{array}{r}
{}^{1}\ \ \\
{}^{3}\ \ \\
68 \\
\times\ 24 \\
\hline
1 \\
272 \\
+\ 1360 \\
\hline
1632
\end{array}$$

47.
$$\begin{array}{r}
72 \\
\times\ \ 12 \\
\hline
144 \\
+\ 720 \\
\hline
864
\end{array}$$

49.
$$\begin{array}{r}
{}^{3\,2}\ \ \\
143 \\
\times\ \ 17 \\
\hline
1001 \\
+\ 1430 \\
\hline
2431
\end{array}$$

51.
$$\begin{array}{r}
{}^{4\,8}\ \ \\
349 \\
\times\ \ 19 \\
\hline
3141 \\
+\ 3490 \\
\hline
6631
\end{array}$$

53.
$$\begin{array}{r}
{}^{1}\ \ \\
{}^{3}\ \ \\
151 \\
\times\ \ 127 \\
\hline
1\,057 \\
3\,020 \\
+\ 15\,100 \\
\hline
19,177
\end{array}$$

55.
$$\begin{array}{r}
{}^{11}\ \ \\
222 \\
\times\ 841 \\
\hline
1 \\
1\ 1\ 222 \\
8\ 880 \\
+\ 177\ 600 \\
\hline
186,702
\end{array}$$

57.
$$\begin{array}{r}
{}^{3\,1\,1}\ \ \\
{}^{2\,1}\ \ \\
3532 \\
\times\ \ \ \ \ \ 6014 \\
\hline
14\ 128 \\
35\ 320 \\
000\ 000 \\
+\ 21\ 192\ 000 \\
\hline
21,241,448
\end{array}$$

59.
$$\begin{array}{r}
{}^{1\,1\,1}\ \ \\
{}^{1\,1}\ \ \\
4122 \\
\times\ \ \ \ \ 982 \\
\hline
8\ 244 \\
329\ 760 \\
+\ 3\ 709\ 800 \\
\hline
4,047,804
\end{array}$$

61.
$$\begin{array}{r}
600 \\
\times\ 40 \\
\end{array}
\rightarrow
\begin{array}{r|l}
6 & 00 \\
\times\ 4 & 0 \\
\hline
24 & 000
\end{array}
= 24,000$$

63.
$$\begin{array}{r}
3000 \\
\times\ 700 \\
\end{array}
\rightarrow
\begin{array}{r|l}
3 & 000 \\
\times\ 7 & 00 \\
\hline
21 & 00000
\end{array}
= 2,100,000$$

65.
$$\begin{array}{r}
8000 \\
\times\ 9000 \\
\end{array}
\rightarrow
\begin{array}{r|l}
8 & 000 \\
\times\ 9 & 000 \\
\hline
72 & 000000
\end{array}
= 72,000,000$$

67.
$$\begin{array}{r}
90,000 \\
\times\ \ 400 \\
\end{array}
\rightarrow
\begin{array}{r|l}
9 & 0000 \\
\times\ 4 & 00 \\
\hline
36 & 000000
\end{array}
= 36,000,000$$

69.
$$\begin{array}{r}
11,784 \\
\times\ 5\ 201 \\
\end{array}
\rightarrow
\begin{array}{r}
12,000 \\
\times\ \ \ \ \ 5,000 \\
\hline
60,000,000
\end{array}$$

71.
$$\begin{array}{r}
82,941 \\
\times\ 29,740 \\
\end{array}
\rightarrow
\begin{array}{r}
80,000 \\
\times\ \ \ \ \ \ 30,000 \\
\hline
2,400,000,000
\end{array}$$

73. $\begin{array}{r} \$189 \\ \times\ \ 5 \end{array} \rightarrow \begin{array}{r} \$200 \\ \times\ \ \ 5 \\ \hline \$1000 \end{array}$

75. $\begin{array}{r} 10,256 \\ \times\ \ \$137 \end{array} \rightarrow \begin{array}{r} 1\ |\ 0000 \\ \times\ 137\ | \\ \hline 137\ |\ 0000\ = \end{array}$

$\$1,370,000$

77. $\begin{array}{r} 1000 \\ \times\ \ \ 4 \\ \hline 4000 \end{array}$

4000 minutes can be stored.

79. $\begin{array}{r} \overset{1}{\underset{3}{}} \\ \$45 \\ \times\ \ \ 37 \\ \hline 315 \\ +1\,350 \\ \hline \$1,665 \end{array}$

81. $\begin{array}{r} \overset{2}{} \\ 115 \\ \times\ 5 \\ \hline 575 \end{array}$

$\begin{array}{r} 3\,2 \\ 57\,5\ | \\ \times\ \ \ 5\ |\ 00 \\ \hline 287,5\ \ \ 00 \end{array}$

287,500 sheets of paper are delivered.

83. $\begin{array}{r} 31 \\ \times\ 12 \\ \hline 62 \\ +\,310 \\ \hline 372 \end{array}$

He can travel 372 miles.

85. $A = l \times w$
$A = (23\ \text{ft}) \times (12\ \text{ft})$

$\begin{array}{r} 23 \\ \times\ 12 \\ \hline 46 \\ +\,230 \\ \hline 276 \end{array}$

The area is $276\ \text{ft}^2$.

87. $A = l \times w$
$A = (73\ \text{cm}) \times (73\ \text{cm})$

$\begin{array}{r} 2 \\ 73 \\ \times\ \ \ 73 \\ \hline 219 \\ +\,5110 \\ \hline 5329 \end{array}$

The area is $5329\ \text{cm}^2$.

89. $A = l \times w$
$A = (390 \text{ mi}) \times (270 \text{ mi})$

$$
\begin{array}{r}
1 \\
6 \\
390 \\
\times \quad 270 \\
\hline
000 \\
27300 \\
+\ 78000 \\
\hline
105,300
\end{array}
$$

The area is $105,300 \text{ mi}^2$.

91. (a) $A = l \times w$
$A = (40 \text{ in.}) \times (60 \text{ in.})$

$$
\begin{array}{r}
2 \\
3 \\
40 \\
\times \quad 60 \\
\hline
00 \\
+\ 2400 \\
\hline
2400 \text{ in}^2.
\end{array}
$$

(b)
$$
\begin{array}{r}
1 \\
14 \\
\times\ 3 \\
\hline
42
\end{array}
$$

There are 42 windows.

(c)
$$
\begin{array}{r}
1 \\
2400 \\
\times \quad\quad 42 \\
\hline
4\ 800 \\
+\ 96\ 000 \\
\hline
100,800
\end{array}
$$

The total area is $100,800 \text{ in}^2$.

93. $A = l \times w$
$A = (8 \text{ ft}) \times (16 \text{ ft})$

$$
\begin{array}{r}
4 \\
16 \\
\times\ 8 \\
\hline
128
\end{array}
$$

The area is 128 ft^2.

Section 1.6   Division of Whole Numbers

Section 1.6   Practice Exercises

1. Answers will vary.

3.
$$
\begin{array}{r}
\overset{1}{\underset{2}{}} \\
103 \\
\times \quad 48 \\
\hline
824 \\
+\ 4\ 120 \\
\hline
4,944
\end{array}
$$

5.
$$
\begin{array}{r}
1 \\
1008 \\
+\ 245 \\
\hline
1253
\end{array}
$$

7.
$$
\begin{array}{r}
1\ 2 \\
5230 \\
\times \quad 127 \\
\hline
1\ 1 \\
36\ 610 \\
104\ 600 \\
+\ 523\ 000 \\
\hline
664,210
\end{array}
$$

9.
$$
\begin{array}{r}
3\ 18\ 8\ 10 \\
\cancel{4}\ \cancel{89}\ \cancel{0} \\
-\ 3\ 98\ 8 \\
\hline
90\ 2
\end{array}
$$

11. $72 \div 8 = 9$ because $9 \times 8 = 72$.
dividend: 72
divisor: 8
quotient: 9

13. $8\overline{)64} = 8$ because $8 \times 8 = 64$.

dividend: 64
divisor: 8
quotient: 8

15. $\dfrac{45}{9} = 5$ because $5 \times 9 = 45$.

dividend: 45
divisor: 9
quotient: 5

17. You cannot divide a number by zero (the quotient is undefined). If you divide zero by a number (other than zero), the quotient is always zero.

19. $15 \div 1 = 15$ because $15 \times 1 = 15$.

21. $0 \div 10 = 0$ because $0 \times 10 = 0$.

23. $0\overline{)9}$ is undefined because division by zero is undefined.

25. $\dfrac{20}{20} = 1$ because $1 \times 20 = 20$.

27. $\dfrac{16}{0}$ is undefined because division by zero is undefined.

29. $8\overline{)0} = 0$ because $0 \times 8 = 0$.

31. $6 \div 3 = 2$ because $2 \times 3 = 6$.
$3 \div 6 \neq 2$ because $2 \times 6 \neq 3$.

33. To check a division problem without a remainder you should multiply the quotient and the divisor to get the dividend.

35.
$$\begin{array}{r} 13 \\ 6\overline{)78} \\ -6 \\ \hline 18 \\ -18 \\ \hline 0 \end{array} \qquad \begin{array}{r} 1 \\ 13 \\ \times\ 6 \\ \hline 78 \ \checkmark \end{array}$$

37.
$$\begin{array}{r} 41 \\ 5\overline{)205} \\ -20 \\ \hline 05 \\ -5 \\ \hline 0 \end{array} \qquad \begin{array}{r} 41 \\ \times\ 5 \\ \hline 205 \ \checkmark \end{array}$$

39.
$$\begin{array}{r} 486 \\ 2\overline{)972} \\ -8 \\ \hline 17 \\ -16 \\ \hline 12 \\ -12 \\ \hline 0 \end{array} \qquad \begin{array}{r} 1\ 1 \\ 486 \\ \times\ 2 \\ \hline 972 \ \checkmark \end{array}$$

41.
$$\begin{array}{r} 409 \\ 3\overline{)1227} \\ -12 \\ \hline 02 \\ -0 \\ \hline 27 \\ -27 \\ \hline 0 \end{array} \qquad \begin{array}{r} 2 \\ 409 \\ \times\ \ 3 \\ \hline 1227 \ \checkmark \end{array}$$

43.
$$\begin{array}{r} 203 \\ 5\overline{)1015} \\ -10 \\ \hline 01 \\ -0 \\ \hline 15 \\ -15 \\ \hline 0 \end{array} \qquad \begin{array}{r} 1 \\ 203 \\ \times\ \ 5 \\ \hline 1015 \ \checkmark \end{array}$$

45.
$$\begin{array}{r} 822 \\ 6\overline{)4932} \\ -48 \\ \hline 13 \\ -12 \\ \hline 12 \\ -12 \\ \hline 0 \end{array} \qquad \begin{array}{r} 1\ 1 \\ 822 \\ \times\ \ 6 \\ \hline 4932 \ \checkmark \end{array}$$

47.
$$\begin{array}{r} 2 \\ 56 \\ \times\ 4 \\ \hline 224 \end{array} \text{ correct}$$

49.
$$\overset{1}{253}$$
$$\times\ 3$$
$$\overline{759}\ \text{incorrect}$$

$$\begin{array}{r} 253\ \text{R 2} \\ 3\overline{)\ 761} \\ \underline{-6} \\ 16 \\ \underline{-15} \\ 11 \\ \underline{-9} \\ 2 \end{array}$$

51.
$$\overset{1\,2}{113}$$
$$\times\ \ 9$$
$$\overset{1}{}$$
$$1017$$
$$\underline{+\ \ 4}\ \ \text{Add the remainder.}$$
$$1021\ \text{correct}$$

53.
$$\overset{4}{25}$$
$$\times\ 8$$
$$\overline{200}$$
$$\underline{+\ 6}$$
$$206\ \text{incorrect}$$

$$\begin{array}{r} 25\ \text{R 3} \\ 8\overline{)\ 203} \\ \underline{-16} \\ 43 \\ \underline{-40} \\ 3 \end{array}$$

55.
$$\begin{array}{r} 7\ \text{R 5} \\ 8\overline{)\ 61} \\ \underline{-56} \\ 5 \end{array}$$
$$7\times8+5=56+5$$
$$=61\ \checkmark$$

57.
$$\begin{array}{r} 10\ \text{R 2} \\ 9\overline{)\ 92} \\ \underline{-9} \\ 02 \end{array}$$
$$10\times9+2=90+2$$
$$=92\ \checkmark$$

59.
$$\begin{array}{r} 27\ \text{R 1} \\ 2\overline{)\ 55} \\ \underline{-4} \\ 15 \\ \underline{-14} \\ 1 \end{array}$$
$$27\times2+1=54+1$$
$$=55\ \checkmark$$

61.
$$\begin{array}{r} 197\ \text{R 2} \\ 3\overline{)\ 593} \\ \underline{-3} \\ 29 \\ \underline{-27} \\ 23 \\ \underline{-21} \\ 2 \end{array}$$
$$197\times3+2=591+2$$
$$=593\ \checkmark$$

63.
$$\begin{array}{r} 42\ \text{R 4} \\ 9\overline{)\ 382} \\ \underline{-36} \\ 22 \\ \underline{-18} \\ 4 \end{array}$$
$$42\times9+4=378+4$$
$$=382\ \checkmark$$

65.
$$\begin{array}{r} 1557\ \text{R 1} \\ 2\overline{)\ 3115} \\ \underline{-2} \\ 11 \\ \underline{-10} \\ 11 \\ \underline{-10} \\ 15 \\ \underline{-14} \\ 1 \end{array}$$

$$\overset{1\,1\,1}{1557}$$
$$\times\ \ 2$$
$$\overline{3114}$$
$$\underline{+\ \ 1}$$
$$3115\ \checkmark$$

67.
$$\begin{array}{r} 751\ \text{R 6} \\ 8\overline{)\ 6014} \\ \underline{-56} \\ 41 \\ \underline{-40} \\ 14 \\ \underline{-8} \\ 6 \end{array}$$

$$\overset{4}{751}$$
$$\times\ \ 8$$
$$\overline{6008}$$
$$\underline{+\ \ 6}$$
$$6014\ \checkmark$$

69.
$$\begin{array}{r} 835\ \text{R 2} \\ 6\overline{)\ 5012} \\ \underline{-48} \\ 21 \\ \underline{-18} \\ 32 \\ \underline{-30} \\ 2 \end{array}$$

$$\overset{2\,3}{835}$$
$$\times\ \ 6$$
$$\overline{5010}$$
$$\underline{+\ \ 2}$$
$$5012\ \checkmark$$

71.
$$\begin{array}{r} 479\ \text{R 9} \\ 19\overline{)\ 9110} \\ \underline{-76} \\ 151 \\ \underline{-133} \\ 180 \\ \underline{-171} \\ 9 \end{array}$$

73.
```
        43 R 19
   24) 1051
       −96
        91
       −72
        19
```

75.
```
       308
   26) 8008
      −78
       20
       −0
       208
      −208
         0
```

77.
```
       1259 R 26
   54) 68012
      −54
       140
      −108
        321
       −270
         512
        −486
          26
```

79.
```
        229 R 96
   304) 69712
       −608
        891
       −608
        2832
       −2736
          96
```

81.
```
        302
   114) 34428
       −342
        228
       −228
          0
```

83. $497 \div 71 = 7$
```
         7
   71) 497
      −497
        0
```

85. $877 \div 14 = 62 \text{ R } 9$
```
         62 R 9
   14) 877
      −84
       37
      −28
        9
```

87. $42 \div 6 = 7$

89.
```
        14 classrooms
   28) 392
      −28
       112
      −112
         0
```

91.
```
         5 R 8
   32) 168
      −160
         8
```
5 cases; 8 cans left over

93.
```
        52 mph
   6) 312
     −30
      12
     −12
       0
```

95.
```
         22 lb
   100) 2200
       −200
        200
       −200
          0
```

97. $1200 \div 20 = 60$

$$
\begin{array}{r}
60 \\
20{\overline{\smash{\big)}\,1200}} \\
\underline{-120} \\
00 \\
\underline{-0} \\
0
\end{array}
$$

Approximately 60 words per minute

99.
$$
\begin{array}{r}
25 \\
18{\overline{\smash{\big)}\,450}} \\
\underline{-36} \\
90 \\
\underline{-90} \\
0
\end{array}
$$

Yes they can all attend if they sit in the second balcony.

101. (a)
$$
\begin{array}{r}
12\,\text{R}\,2 \\
4{\overline{\smash{\big)}\,50}} \\
\underline{-4} \\
10 \\
\underline{-8} \\
2
\end{array}
$$

12 loads can be done.

(b) 2 ounces of detergent are left over.

103.
$$
\begin{array}{r}
21{,}000{,}000 \\
\times \quad 365 \\
\hline
7{,}665{,}000{,}000 \ \text{bbl}
\end{array}
$$

105. $2532 \div 12 = 211$
$211 billion

Problem Recognition Exercises: Operations on Whole Numbers

1.
$$
\begin{array}{r}
92 \\
+41 \\
\hline
133
\end{array}
$$

3.
$$
\begin{array}{r}
89 \\
-22 \\
\hline
67
\end{array}
$$

5.
$$
\begin{array}{r}
221 \\
\times 14 \\
\hline
884 \\
+2210 \\
\hline
3094
\end{array}
$$

7.
$$
\begin{array}{r}
64 \\
35{\overline{\smash{\big)}\,2240}} \\
\underline{-210} \\
140 \\
\underline{-140} \\
0
\end{array}
$$

9.
$$
\begin{array}{r}
946 \\
-612 \\
\hline
334
\end{array}
$$

11.
$$
\begin{array}{r}
2311 \\
+2652 \\
\hline
4963
\end{array}
$$

13.
$$
\begin{array}{r}
328 \\
4{\overline{\smash{\big)}\,1312}} \\
\underline{-12} \\
11 \\
-8 \\
\hline
32 \\
\underline{32} \\
0
\end{array}
$$

15.
$$
\begin{array}{r}
3000 \\
\times \ 82 \\
\hline
6000 \\
+240000 \\
\hline
246{,}000
\end{array}
$$

17.
$$
\begin{array}{r}
1 \\
113 \\
+59 \\
\hline
172
\end{array}
$$

19.
$$
\begin{array}{r}
621 \\
-539 \\
\hline
82
\end{array}
$$

21.
$$
\begin{array}{r}
230\ \text{R}\,4 \\
22\overline{)\,5064} \\
\underline{-44} \\
66 \\
\underline{-66} \\
04 \\
\underline{-0} \\
4
\end{array}
$$

23.
$$
\begin{array}{r}
400 \\
\times\ \ 50 \\
\hline
000 \\
+\,20\,000 \\
\hline
20{,}000
\end{array}
$$

25.
$$
\begin{array}{r}
{}^{4}\ {}^{15} \\
34{,}8\,\cancel{5}\ \cancel{5} \\
-\,12{,}13\,7 \\
\hline
22{,}71\,8
\end{array}
$$

27.
$$
\begin{array}{r}
8231 \\
+\,3412 \\
\hline
11{,}643
\end{array}
$$

29.
$$
\begin{array}{r}
548 \\
63\overline{)\,34524} \\
\underline{-315} \\
302 \\
\underline{-252} \\
504 \\
\underline{-504} \\
0
\end{array}
$$

31.
$$
\begin{array}{r}
{}^{24} \\
{}^{12} \\
548 \\
\times\ \ 63 \\
\hline
1644 \\
+\,32880 \\
\hline
34{,}524
\end{array}
$$

33. $418 \times 10 = 4180$

35. $418 \times 1000 = 418{,}000$

37. $350{,}000 \div 10 = 35{,}000$

39. $350{,}000 \div 1000 = 350$

41.
$$
\begin{array}{r}
{}^{11} \\
159 \\
224 \\
+\,123 \\
\hline
506
\end{array}
$$

43.
$$
\begin{array}{r}
{}^{11} \\
534 \\
12 \\
+\,66 \\
\hline
612
\end{array}
$$

Section 1.7   Exponents, Square Roots, and the Order of Operations

Section 1.7   Practice Exercises

1. Answers will vary.

3. True: $5 + 3 = 8$ and $3 + 5 = 8$

5. False: $6 \times 0 = 0$

7. True: $0 \times 8 = 0$

9. 9^4

11. 2^7

13. $3 \cdot 3 \cdot 3 \cdot 3 \cdot 3 \cdot 3 = 3^6$

15. $4 \cdot 4 \cdot 4 \cdot 4 \cdot 2 \cdot 2 \cdot 2 = 4^4 \cdot 2^3$

17. $8^4 = 8 \cdot 8 \cdot 8 \cdot 8$

19. $4^8 = 4 \cdot 4 \cdot 4 \cdot 4 \cdot 4 \cdot 4 \cdot 4 \cdot 4$

21. $2^3 = 2 \cdot 2 \cdot 2 = 4 \cdot 2 = 8$

23. $3^2 = 3 \cdot 3 = 9$

25. $3^3 = 3 \cdot 3 \cdot 3 = 9 \cdot 3 = 27$

27. $5^3 = 5 \cdot 5 \cdot 5 = 25 \cdot 5 = 125$

29. $2^5 = 2 \cdot 2 \cdot 2 \cdot 2 \cdot 2 = 4 \cdot 4 \cdot 2 = 16 \cdot 2 = 32$

31. $3^4 = 3 \cdot 3 \cdot 3 \cdot 3 = 9 \cdot 9 = 81$

33. $1^2 = 1 \cdot 1 = 1$

35. $1^4 = 1 \cdot 1 \cdot 1 \cdot 1 = 1$

37. The number 1 raised to any power equals 1.

39. $10^3 = 10 \cdot 10 \cdot 10 = 1000$

41. $10^5 = 10 \cdot 10 \cdot 10 \cdot 10 \cdot 10 = 100,000$

43. $\sqrt{4} = 2$ because $2 \cdot 2 = 4$.

45. $\sqrt{36} = 6$ because $6 \cdot 6 = 36$.

47. $\sqrt{100} = 10$ because $10 \cdot 10 = 100$.

49. $\sqrt{0} = 0$ because $0 \cdot 0 = 0$.

51. No, addition and subtraction should be performed in the order in which they appear from left to right.

53. $6 + 10 \cdot 2 = 6 + 20 = 26$

55. $10 - 3^2 = 10 - 9 = 1$

57. $(10 - 3)^2 = 7^2 = 49$

59. $36 \div 2 \div 6 = 18 \div 6 = 3$

61. $15 - (5 + 8) = 15 - 13 = 2$

63. $(13 - 2) \cdot 5 - 2 = 11 \cdot 5 - 2 = 55 - 2 = 53$

65. $4 + 12 \div 3 = 4 + 4 = 8$

67. $30 \div 2 \cdot \sqrt{9} = 30 \div 2 \cdot 3 = 15 \cdot 3 = 45$

69. $7^2 - 5^2 = 49 - 25 = 24$

71. $(7 - 5)^2 = 2^2 = 4$

73. $100 \div 5 \cdot 2 = 20 \cdot 2 = 40$

75. $90 \div 3 \cdot 3 = 30 \cdot 3 = 90$

77. $\begin{aligned}\sqrt{81} + 2(9 - 1) &= \sqrt{81} + 2 \cdot 8 \\ &= 9 + 2 \cdot 8 \\ &= 9 + 16 \\ &= 25\end{aligned}$

79. $36 \div (2^2 + 5) = 36 \div (4 + 5) = 36 \div 9 = 4$

81. $80 - (20 \div 4) + 6 = 80 - 5 + 6 = 75 + 6 = 81$

83. $\begin{aligned}(43 - 26) \cdot 2 - 4^2 &= 17 \cdot 2 - 4^2 \\ &= 17 \cdot 2 - 16 \\ &= 34 - 16 \\ &= 18\end{aligned}$

85. $\begin{aligned}(18 - 5) - \left(23 - \sqrt{100}\right) &= 13 - (23 - 10) \\ &= 13 - 13 \\ &= 0\end{aligned}$

87. $\begin{aligned}80 \div (9^2 - 7 \cdot 11)^2 &= 80 \div (81 - 7 \cdot 11)^2 \\ &= 80 \div (81 - 77)^2 \\ &= 80 \div 4^2 \\ &= 80 \div 16 \\ &= 5\end{aligned}$

89. $\begin{aligned}22 - 4\left(\sqrt{25} - 3\right)^2 &= 22 - 4(5 - 3)^2 \\ &= 22 - 4(2)^2 \\ &= 22 - 4 \cdot 4 \\ &= 22 - 16 \\ &= 6\end{aligned}$

91. $\begin{aligned}96 - 3(42 \div 7 \cdot 6 - 5) &= 96 - 3(6 \cdot 6 - 5) \\ &= 96 - 3(36 - 5) \\ &= 96 - 3(31) \\ &= 96 - 93 \\ &= 3\end{aligned}$

93. $\begin{aligned}16 + 5(20 \div 4 \cdot 8 - 3) &= 16 + 5(5 \cdot 8 - 3) \\ &= 16 + 5(40 - 3) \\ &= 16 + 5(37) \\ &= 16 + 185 \\ &= 201\end{aligned}$

95. $\text{Mean} = \dfrac{105 + 114 + 123 + 101 + 100 + 111}{6}$

$\qquad = \dfrac{654}{6} = 109$

97. $\text{Average} = \dfrac{19 + 20 + 18 + 19 + 18 + 14}{6}$

$\qquad = \dfrac{108}{6} = 18$

99. $\text{Average} = \dfrac{33 + 39 + 42}{3}$

$\qquad = \dfrac{114}{3} = 38\cent \text{ per pound}$

101. $\text{Average} = \dfrac{118 + 123 + 122}{3}$

$\qquad = \dfrac{363}{3} = 121 \text{ mm per month}$

103. $3[4 + (6 - 3)^2] - 15 = 3[4 + 3^2] - 15$
$\qquad\qquad\qquad\quad = 3[4 + 9] - 15$
$\qquad\qquad\qquad\quad = 3[13] - 15$
$\qquad\qquad\qquad\quad = 39 - 15$
$\qquad\qquad\qquad\quad = 24$

105. $5\{21 - [3^2 - (4 - 2)]\} = 5\{21 - [3^2 - 2]\}$
$\qquad\qquad\qquad\qquad\quad = 5\{21 - [9 - 2]\}$
$\qquad\qquad\qquad\qquad\quad = 5\{21 - 7\}$
$\qquad\qquad\qquad\qquad\quad = 5\{14\}$
$\qquad\qquad\qquad\qquad\quad = 70$

107. $156^2 = 24{,}336$

109. $12^5 = 248{,}832$

111. $43^3 = 79{,}507$

113. $8126 - 54{,}978 \div 561 = 8126 - 98 = 8028$

115. $(3548 - 3291)^2 = 257^2 = 66{,}049$

117. $\dfrac{89{,}880}{384 + 2184} = \dfrac{89{,}880}{2568} = 35$

Section 1.8   Problem-Solving Strategies

Section 1.8   Practice Exercises

1. Answers will vary.

3. $71 + 14 = 85$

5. $2 \cdot 14 = 28$

7. $102 - 32 = 70$

9. $10 \cdot 13 = 130$

11. $24 \div 6 = 4$

13. $5 + 13 + 25 = 43$

15. For example: sum, added to, increased by, more than, total of, plus

17. For example: difference, minus, decreased by, less, subtract

19. *Given*: The height of each mountain
Find: The difference in height
Operation: Subtract

$$\begin{array}{r} \overset{\scriptstyle 1\,10\;\;2\,11\,10}{20{,}320} \\ -\;14{,}246 \\ \hline 6{,}074 \end{array}$$

Mt McKinley is 6,074 ft higher than White Mountain Peak.

21. *Given*: Oil consumption by country
Find: Total oil consumption for 5 countries
Operation: Addition

```
  1 3 2
  308,600,000
  241,500,000
  128,500,000
  123,600,000
+ 119,300,000
  ───────────
  921,500,000
```

The oil consumption of China, Japan, Russia, Germany, and India is 921,500,000 metric tons.

23. *Given*: The number of rows of pixels and the number of pixels in each row
Find: The number of pixels on the whole screen
Operation: Multiply

```
      5
   2 1 3
     126
×     96
   ─────
   1 756
  11 340
  ──────
  12,096
```

There are 12,096 pixels on the whole screen.

25. *Given*: Number of students and the average class size
Find: Number of classes offered
Operation: Division

```
        120
    25) 3000
       −25
       ───
        50
       −50
       ───
         00
```

There will be 120 classes of Beginning Algebra.

27. *Given*: 45 miles per gallon and driving 405 miles
Find: How many gallons used
Operation: Division

```
         9
    45) 405
       −405
       ────
          0
```

There will be 9 gal used.

29. *Given*: Yearly tuition for two schools
Find: Total tuition paid

Operation: Addition

```
       1
    26,960
+    2,600
   ───────
    29,560
```

Jeannette will pay $29,560 for one year.

31. *Given*: Miles per gallon and number of gallons
Find: How many miles
Operation: Multiplication

```
      1
      55
×     20
   ─────
   1,100
```

The Prius can go 1100 mi.

33. *Given*: Number of rows and number of seats in each row
Find: Total number of seats
Operation: Multiplication

```
      3
      45
×     70
   ─────
   3150
```

The maximum capacity is 3150 seats.

35. *Given*: total price: $16,540
down payment: $2500
payment plan: 36 months
Find: Amount of monthly payments
Operations:
(1) Subtract

```
    16,540
−    2 500
   ───────
    14,040
```

(2) Divide

```
          390
    36) 14040
       − 108
       ─────
         324
        −324
        ────
          00
```

Jackson's monthly payments were $390.

37. *Given*: Distance for each route and speed traveled
Find: Time required for each route
Operations:
(1) Watertown to Utica direct
Divide $80 \div 40 = 2$ hr

(2) Watertown to Syracuse to Utica
   Add distances $70 + 50 = 120$ mi
   Divide $120 \div 60 = 2$ hr
Each trip will take 2 hours.

39. The distance around a figure is the perimeter.

41. *Given*: The dimensions of a room and cost per foot of molding
Find: Total cost
Operations:
(1) Add to find the perimeter, subtract doorway.

$$
\begin{array}{r}
11 \\
12 \\
11 \\
+\,12 \\
\hline
46
\end{array}
\qquad
\begin{array}{r}
46 \\
-\,3 \\
\hline
43 \ \text{ft}
\end{array}
$$

(2) Multiply to find the total cost.

$$
\begin{array}{r}
43 \\
\times\ 2 \\
\hline
86
\end{array}
$$

The cost will be $86.

43. *Given*: dimensions of room and cost per square yard
Find: total cost
Operations:
(1) Multiply to find area

$$6 \times 5 = 30 \ \text{yd}^2$$

(2) Multiply to find total cost

$$
\begin{array}{r}
1 \\
34 \\
\times\ 30 \\
\hline
1020
\end{array}
$$

The total cost is $1020.

45. *Given*: Starting balance in account and individual checks written
Find: Remaining balance in account
Operations:
(1) Add the individual checks

$$
\begin{array}{r}
1 \\
82 \\
159 \\
+\,101 \\
\hline
\$242
\end{array}
$$

(2) Subtract $242 from the initial balance

$$
\begin{array}{r}
278 \\
-\,242 \\
\hline
36
\end{array}
$$

There will be $36 left in Gina's account.

47. *Given*: Number of computers and printers purchased and the cost of each
Find: The total bill
Operations:
(1) Multiply to find the amount spent on computers, then printers.

$$
\begin{array}{r}
1\,1\,5 \\
2118 \\
\times\ \ \ 72 \\
\hline
4\,236 \\
148\,260 \\
\hline
\$152{,}496
\end{array}
\qquad
\begin{array}{r}
3\,3 \\
256 \\
\times\ \ 6 \\
\hline
\$1536
\end{array}
$$

(2) Add to find the total bill.

$$
\begin{array}{r}
1\ \ 11 \\
152{,}496 \\
+\,1\,536 \\
\hline
154{,}032
\end{array}
$$

The total bill was $154,032.

49. *Given*: Amount to sell used CDs, amount to buy used CDs and number of CDs sold

    (a) *Find*: Money from selling 16 CDs

        Operation: Multiply

$$\begin{array}{r} 16 \\ \times\ 3 \\ \hline 48 \end{array}$$

        Latayne will receive $48.

    (b) *Find*: Number of used CDs to buy for $48

        Operation: Division

        $48 \div 8 = 6$

        She can buy 6 CDs.

51. *Given*: Number of field goals, three-point shots and free throws and point values

    Find: Total points scored

    Operations:

    (1) Multiply

field goals	three-point shots
1	2
12,192	581
× 2	× 3
24,384	1,743

    (2) Add

$$\begin{array}{r} {}^{1\ 1}\ {}^{11} \\ 24\ 384 \\ 1\ 743 \\ +\ 7\ 327 \\ \hline 33,454 \end{array}$$

    Michael Jordan scored 33,454 points with the Bulls.

53. *Given*: Number of milliliters in the bottle and the dosage

    (a) *Find*: Days the bottle will last

        Operation: Divide

        $60 \div 2 = 30$

        One bottle will last for 30 days.

    (b) *Find*: Date to reorder

        Operation: Subtract

        $30 - 2 = 28$

        The owner should order a refill no later than September 28.

55. *Given*: Scale on a map

    (a) *Find*: Actual distance between Las Vegas and Salt Lake City

        Operation: Multiply

$$\begin{array}{r} 60 \\ \times\ 6 \\ \hline 360 \end{array}$$

        The distance is 360 mi.

    (b) *Find*: Distance on map between Madison and Dallas

        Operation: Divide

$$\begin{array}{r} 14 \\ 60\overline{)\ 840} \\ \underline{-60} \\ 240 \\ \underline{-240} \\ 0 \end{array}$$

    14 in. represents 840 mi.

57. *Given*: Number of books per box and number of books ordered

    Find: Number of boxes completely filled and number of books left over

    Operation: Divide and find remainder

$$\begin{array}{r} 104\ \text{R}\ 2 \\ 12\overline{)\ 1250} \\ \underline{-12} \\ 050 \\ \underline{-48} \\ 2 \end{array}$$

    104 boxes will be filled completely with 2 books left over.

59. *Given*: Total cost of dinner and type of bill used

    (a) *Find*: Number of $20 bills needed

        Operation: Division

$$\begin{array}{r} 4\ \text{R}\ 4 \\ 20\overline{)\ 84} \\ \underline{-80} \\ 4 \end{array}$$

        Four $20 bills are not enough so Marc needs five $20 bills.

    (b) *Find*: How much change

        Operations: Multiply and subtract

20	100
× 5	− 84
100	16

        He will receive $16 in change.

61. *Given*: Hourly wage and number of hours
worked
Find: Amount earned per week
Operations:
(1) Multiply to find amount per job.
$30 \times 4 = 120$
$10 \times 16 = 160$
$8 \times 30 = 240$

(2) Add to find total.
```
      1
    120
    160
 + 240
 ─────
    520
```
He earned $520.

Chapter 1   Review Exercises

Section 1.1

1. 10,024   Ten-thousands

3. 92,046

5. 3 millions + 4 hundred-thousands
+ 8 hundreds + 2 tens

7. Two hundred forty-five

9. 3602

11.
```
 ─┼─◆─┼─┼─┼─┼─┼─┼─┼─┼─┼─┼─┼─►
  1  2  3  4  5  6  7  8  9 10 11 12 13
```

13. $3 < 10$ True

Section 1.2

15. Addends: 105, 119; sum: 224

17.
```
    2
   18
   24
 + 29
 ────
   71
```

19.
```
      1
   8 403
 + 9 007
 ──────
  17,410
```

21. **(a)** The order changed so it is the
commutative property.
(b) The grouping changed so it is the
associative property.
(c) The order changed so it is the
commutative property.

23. $44 + 92 = 136$
```
    92
  + 44
  ────
   136
```

25. $23 + 6 = 29$

27.
```
    35,377
  + 10,420
  ───────
    45,797  thousand seniors
```

Section 1.3

29. minuend: 14
subtrahend: 8
difference: 6

31.
```
    37
  − 11
  ────
    26
```
    $\underline{26} + 11 = 37$

33.
```
        9
    1 10 10
    2 0̸ 0̸5
  − 1 8 84
  ───────
    1  21
```

35.
```
          9 9
      5 10 10 10
    8̸6̸,0̸ 0̸ 0̸
   − 54  9  8 1
   ──────────
    31,0  1 9
```

37. $38 - 31 = 7$

39. $251 - 42 = 209$

$$
\begin{array}{r}
4\,11 \\
2\,\cancel{5}\,\cancel{1} \\
-\,4\,2 \\
\hline
20\,9
\end{array}
$$

41.
$$
\begin{array}{r}
\overset{10\ 18}{} \\
4\ \cancel{\emptyset}\ \cancel{8}\,11\ \ 5\,11 \\
9\cancel{5},\cancel{1}\,9\,\cancel{1},7\,\cancel{6}\,\cancel{1} \\
-\ 23,\ 29\ 9,3\,2\,3 \\
\hline
71,\ 89\ 2,4\,3\,8 \ \text{tons}
\end{array}
$$

43.
$$
\begin{array}{r}
9 \\
7\ \cancel{10}\ 13 \\
4\,\cancel{8}\,\cancel{\emptyset}\,\cancel{3} \\
-\ 24\,6\,7 \\
\hline
2,3\,3\ 6 \ \text{thousand visitors}
\end{array}
$$

Section 1.4

45. $9,33\boxed{2},945$

$9,330,000$

47.
$$
\begin{array}{rcl}
330 & \rightarrow & 300 \\
489 & \rightarrow & 500 \\
123 & \rightarrow & 100 \\
+\ 571 & \rightarrow & \underline{600} \\
& & 1500
\end{array}
$$

49.
$$
\begin{array}{rcl}
& & \overset{1}{} \\
96,050 & \rightarrow & 96,000 \\
+\ 66,517 & \rightarrow & \underline{+\ 67,000} \\
& & 163,000 \ \text{m}^3
\end{array}
$$

Section 1.5

51. Factors: 33, 40

Product: 1320

53. c

55. d

57. b

59.
$$
\begin{array}{r}
1\,2 \\
1024 \\
\times\ \ \ \ \ 51 \\
\hline
1\ 024 \\
+\ 51\ 200 \\
\hline
52,224
\end{array}
$$

61.
$$
\begin{array}{r}
26 \\
+\ 13 \\
\hline
39
\end{array}
\qquad
\begin{array}{r}
39 \\
\times\ 11 \\
\hline
39 \\
390 \\
\hline
\$429
\end{array}
$$

Section 1.6

63. $42 \div 6 = 7$

divisor: 6, dividend: 42, quotient: 7

65. $3 \div 1 = 3$ because $1 \times 3 = 3$.

67. $3 \div 0$ is undefined.

69. To check a division problem with no remainder you multiply the quotient by the divisor to get the dividend.

71.
$$
\begin{array}{r}
58 \\
6\,\overline{)\,348} \\
\underline{-30} \\
48 \\
\underline{-48} \\
0
\end{array}
\qquad
\begin{array}{r}
3 \\
58 \\
\times\ 6 \\
\hline
348 \ \checkmark
\end{array}
$$

73.
$$
\begin{array}{r}
52 \ \text{R } 3 \\
20\,\overline{)\,1043} \\
\underline{-100} \\
43 \\
\underline{-40} \\
3
\end{array}
\qquad
\begin{array}{r}
52 \\
\times\ 20 \\
\hline
1040 \\
+\ \ 3 \\
\hline
1043 \ \checkmark
\end{array}
$$

75.
$$
\begin{array}{r}
12 \\
9\,\overline{)\,108} \\
\underline{-9} \\
18 \\
\underline{-18} \\
0
\end{array}
$$

77. **(a)** Divide 60 by 15.

$60 \div 15 = 4$ T-shirts

(b) Divide 60 by 12.

$60 \div 12 = 5$ hats

Section 1.7

79. $2 \cdot 2 \cdot 2 \cdot 2 \cdot 5 \cdot 5 \cdot 5 = 2^4 \cdot 5^3$

81. $4^4 = 4 \times 4 \times 4 \times 4 = 16 \times 16 = 256$

83. $10^6 = 10 \times 10 \times 10 \times 10 \times 10 \times 10 = 1,000,000$

85. $\sqrt{144} = 12$ because $12 \times 12 = 144$.

87. $10^2 - 5^2 = 100 - 25 = 75$

89. $2 + 3 \cdot 12 \div 2 - \sqrt{25} = 2 + 3 \cdot 12 \div 2 - 5$
$\qquad = 2 + 36 \div 2 - 5$
$\qquad = 2 + 18 - 5$
$\qquad = 20 - 5$
$\qquad = 15$

91. $26 - 2(10 - 1) + (3 + 4 \cdot 11)$
$\qquad = 26 - 2(9) + (3 + 44)$
$\qquad = 26 - 2(9) + 47$
$\qquad = 26 - 18 + 47$
$\qquad = 8 + 47$
$\qquad = 55$

93. Average $= \dfrac{80 + 78 + 101 + 92 + 94}{5}$
$\qquad = \dfrac{445}{5}$
$\qquad = \$89$

Section 1.8

95. *Given*: Number of animals and species at two zoos
   (a) *Find*: Which zoo has more animals and how many more
   Operation: Subtraction
   $\begin{array}{r} 17,000 \\ -\ 4,000 \\ \hline 13,000 \end{array}$
   The Cincinnati Zoo has 13,000 more animals than the San Diego Zoo.

(b) *Find*: Which zoo has the most species, and how many more
   Operation: Subtract
   $\begin{array}{r} {}^{7\ 10} \\ \cancel{8}00 \\ -\ 750 \\ \hline 50 \end{array}$
   The San Diego Zoo has 50 more species than the Cincinnati Zoo.

97. *Given*: Contract: 252,000,000
   time period: 9 years
   taxes: 75,600,000
   Find: Amount per year after taxes
   Operations
   (1) Subtract
   $\begin{array}{r} {}^{14\ 11} \\ {}^{1\cancel{4}\ \cancel{1}\ 10} \\ 2\cancel{5}\cancel{2},\cancel{0}00,000 \\ -\ 75,600,000 \\ \hline 176,400,000 \end{array}$
   (2) Divide
   $\begin{array}{r} 19,600,000 \\ 9\overline{)176,400,000} \\ \underline{-9} \\ 86 \\ \underline{-81} \\ 54 \\ \underline{-54} \\ 0 \end{array}$

   He will receive \$19,600,000 per year.

Chapter 1   Test

1. (a) 4<u>9</u>2   hundreds
   (b) 2<u>3</u>,441   thousands
   (c) <u>2</u>,340,711   millions
   (d) 3<u>4</u>0,592   ten-thousands

3. (a) $14 > 6$
   (b) $72 < 81$

5. $\begin{array}{r} 82 \\ \times\ 4 \\ \hline 328 \end{array}$

7.
$$\begin{array}{r} 227 \\ 4\overline{)\,908} \\ \underline{-8} \\ 10 \\ \underline{-8} \\ 28 \\ \underline{-28} \\ 0 \end{array}$$

9.
$$\begin{array}{r} \overset{1\ 1}{149} \\ +\ 298 \\ \hline 447 \end{array}$$

11.
$$\begin{array}{r} \overset{\ 9\ \ 9}{\overset{2\ \ \cancel{10}\ \cancel{10}\ 12}{\cancel{3}\ \cancel{0}\ \cancel{0}\ \cancel{2}}} \\ -\ 2\ 4\ 5\ 6 \\ \hline 5\ \ 4\ 6 \end{array}$$

13.
$$\begin{array}{r} 20 \\ 42\overline{)\,840} \\ \underline{-84} \\ 00 \end{array}$$

15.
$$\begin{array}{r} \overset{2\ 1}{34} \\ 89 \\ 191 \\ +\ 22 \\ \hline 336 \end{array}$$

17. $0\overline{)16}$ is undefined.

19. (a) $4,8\boxed{5}0 \rightarrow 4,900$
    (b) $12,\boxed{4}93 \rightarrow 12,000$
    (c) $7,9\boxed{6}3,126 \rightarrow 8,000,000$

21. $8^2 \div 2^4 = 64 \div 16 = 4$

23. $36 \div 3(14 - 10) = 36 \div 3(4) = 12(4) = 48$

25. *Given*: Quiz scores and number of quizzes for Brittany and Jennifer
Find: Who has the higher average
Operations: Find the average of each group.
Brittany:
$$\frac{29 + 28 + 24 + 27 + 30 + 30}{6} = \frac{168}{6} = 28$$
Jennifer:
$$\frac{30 + 30 + 29 + 28 + 28}{5} = \frac{145}{5} = 29$$
Jennifer has the higher average of 29. Brittany has an average of 28.

27. Divide the number of calls by the number of weeks.
North: $80 \div 16 = 5$
South: $72 \div 18 = 4$
East: $84 \div 28 = 3$
The North Side Fire Department is the busiest with an average of 5 calls per week.

29. Add to find the perimeter.
$$\begin{array}{r} \overset{1\ 3}{47} \\ 128 \\ 47 \\ +\ 128 \\ \hline 350\ \text{ft} \end{array}$$
Multiply to find the area.
$$\begin{array}{r} 128 \\ \times\ 47 \\ \hline 896 \\ 5120 \\ \hline 6016\ \ \text{ft}^2 \end{array}$$

Chapter 2   Fractions and Mixed Numbers: Multiplication and Division

Chapter Opener Puzzle

		¹D							
²N	U	M	E	R	A	T	O	R	
		N							
		O		³P					
⁴P		M		R					
⁵R	E	C	I	P	R	O	C	A	L
I		N		P					
M		A		E					
E		T		R					
		O							
⁶I	M	P	R	O	P	E	R		

Section 2.1   Introduction to Fractions and Mixed Numbers

Section 2.1   Practice Exercises

1. Answers will vary.

3. Numerator: 2; denominator: 3

5. Numerator: 12; denominator: 11

7. $6 \div 1$; 6

9. $2 \div 2$; 1

11. $0 \div 3$; 0

13. $2 \div 0$; undefined

15. $\dfrac{3}{4}$

17. $\dfrac{5}{9}$

19. $\dfrac{1}{6}$

21. $\dfrac{3}{8}$

23. $\dfrac{3}{4}$

25. $\dfrac{1}{8}$

27. $\dfrac{41}{103}$

29. $\dfrac{10}{21}$

31. Proper

33. Improper

35. Improper

37. Proper

39. $\dfrac{5}{2}$

41. $\dfrac{12}{4}$

43. $\dfrac{7}{4}; 1\dfrac{3}{4}$

45. $\dfrac{13}{8}; 1\dfrac{5}{8}$

47. $1\dfrac{3}{4} = \dfrac{4\times1+3}{4} = \dfrac{7}{4}$

49. $4\dfrac{2}{9} = \dfrac{4\times9+2}{9} = \dfrac{38}{9}$

51. $3\dfrac{3}{7} = \dfrac{3\times7+3}{7} = \dfrac{24}{7}$

53. $7\dfrac{1}{4} = \dfrac{7\times4+1}{4} = \dfrac{29}{4}$

55. $11\dfrac{5}{12} = \dfrac{11\times12+5}{12} = \dfrac{137}{12}$

57. $21\dfrac{3}{8} = \dfrac{21\times8+3}{8} = \dfrac{171}{8}$

59. $2\dfrac{3}{8} = \dfrac{2\times8+3}{8} = \dfrac{19}{8}$

19 eighths

61. $1\dfrac{3}{4} = \dfrac{1\times4+3}{4} = \dfrac{7}{4}$

7 fourths

63.
```
     4
  8) 37
    -32
      5
```
$4\dfrac{5}{8}$

65.
```
     7
  5) 39
    -35
      4
```
$7\dfrac{4}{5}$

67.
```
     2
  10) 27
     -20
       7
```
$2\dfrac{7}{10}$

69.
```
     5
  9) 52
    -45
      7
```
$5\dfrac{7}{9}$

71.
```
      12
  11) 133
     -11
      23
     -22
       1
```
$12\dfrac{1}{11}$

73.
```
     3
  6) 23
    -18
      5
```
$3\dfrac{5}{6}$

75.
```
     44
  7) 309
    -28
     29
    -28
      1
```
$44\dfrac{1}{7}$

77.
```
      1056
  5) 5281
    -5
     2
    -0
     28
    -25
      31
     -30
       1
```
$1056\dfrac{1}{5}$

79.
```
      810
  11) 8913
     -88
      11
     -11
       3
      -0
       3
```
$810\dfrac{3}{11}$

81.
```
      12
  15) 187
     -15
      37
     -30
       7
```
$12\dfrac{7}{15}$

83. Divide the distance between 0 and 1 into 4 equal parts.

85. Divide the distance between 0 and 1 into 3 equal parts.

87. Divide the distance between 0 and 1 into 3 equal parts.

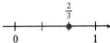

89. $\dfrac{7}{6} = 1\dfrac{1}{6}$

Divide the distance between 1 and 2 into 6 equal parts.

91. $\dfrac{5}{3} = 1\dfrac{2}{3}$

Divide the distance between 1 and 2 into 3 equal parts.

93. False; whole numbers cannot be written as proper fractions.

95. True

Section 2.2   Prime Numbers and Factorization

Section 2.2   Practice Exercises

1. Answers will vary.

3. $\dfrac{8}{12}; \dfrac{4}{12}$

5. $\dfrac{5}{4}; \dfrac{3}{4}$

7. $\dfrac{7}{12}$; proper

9. $5\overline{)23}$ with -20 giving remainder 3 and quotient 4; $4\dfrac{3}{5}$

11. For example: $2 \cdot 4$ and $1 \cdot 8$

13. For example: $4 \cdot 6$ and $2 \cdot 2 \cdot 2 \cdot 3$

15.

Product	42	30	15	81
Factor	7	30	15	27
Factor	6	1	1	3
Sum	13	31	16	30

17. A whole number is divisible by 2 if it is an even number.

19. A whole number is divisible by 3 if the sum of its digits is divisible by 3.

21. 45
  (a) No; 45 is not even.
  (b) Yes; $4 + 5 = 9$ is divisible by 3.
  (c) Yes; the ones-place digit is 5.
  (d) No; the ones-place digit is not 0.

23. 137
  (a) No; 137 is not even.
  (b) No; $1 + 3 + 7 = 11$ is not divisible by 3.
  (c) No; the ones-place digit is not 0 or 5.
  (d) No; the ones-place digit is not 0.

25. 108
  (a) Yes; 108 is even.
  (b) Yes; $1 + 0 + 8 = 9$ is divisible by 3.
  (c) No; the ones-place digit is not 0 or 5.
  (d) No; the ones-place digit is not 0.

27. 3140
  (a) Yes; 3140 is even.
  (b) No; $3 + 1 + 4 + 0 = 8$ is not divisible by 3.
  (c) Yes; the ones-place digit is 0.
  (d) Yes; the ones-place digit is 0.

29.
$$28 \overline{)\ 84} \quad \begin{array}{r} 3 \\ \hline \end{array}$$
$$\underline{-84}$$
$$0$$

Yes, 84 is divisible by 28.

31. Prime

33. Composite $2 \cdot 5 = 10$

35. Composite $3 \cdot 17 = 51$

37. Prime

39. Neither

41. Composite $11 \cdot 11 = 121$

43. Prime

45. Composite $3 \cdot 13 = 39$

47. There are two whole numbers that are neither prime nor composite, 0 and 1.

49. False; 9 is not prime.

51. 2, 3, 5, 7, 11, 13, 17, 19, 23, 29, 31, 37, 41, 43, 47

53. No, 9 is not a prime number.

55. Yes

57.
$$5 \overline{)\ 35} \quad \begin{array}{r} 7 \\ \hline \end{array}$$
$$2 \overline{)\ 70}$$
$$2 \cdot 5 \cdot 7 = 70$$

59.
$$5 \overline{)\ 65} \quad \begin{array}{r} 13 \\ \hline \end{array}$$
$$2 \overline{)\ 130}$$
$$2 \overline{)\ 260}$$
$$2 \cdot 2 \cdot 5 \cdot 13 = 2^2 \cdot 5 \cdot 13 = 260$$

61.
$$7 \overline{)\ 49} \quad \begin{array}{r} 7 \\ \hline \end{array}$$
$$3 \overline{)\ 147}$$
$$3 \cdot 7 \cdot 7 = 3 \cdot 7^2 = 147$$

63.
$$3 \overline{)\ 69} \quad \begin{array}{r} 23 \\ \hline \end{array}$$
$$2 \overline{)\ 138}$$
$$2 \cdot 3 \cdot 23 = 138$$

65.
$$7 \overline{)\ 77} \quad \begin{array}{r} 11 \\ \hline \end{array}$$
$$2 \overline{)\ 154}$$
$$2 \overline{)\ 308}$$
$$2 \overline{)\ 616}$$
$$2 \cdot 2 \cdot 2 \cdot 7 \cdot 11 = 2^3 \cdot 7 \cdot 11 = 616$$

67. 47 is prime.

69. 1, 2, 3, 4, 6, 12

71. 1, 2, 4, 8, 16, 32

73. 1, 3, 9, 27, 81

75. 1, 2, 3, 4, 6, 8, 12, 16, 24, 48

77. No; 30 is not divisible by 4.

79. Yes; 16 is divisible by 4.

81. Yes; 32 is divisible by 8.

83. No; 126 is not divisible by 8.

85. Yes; $3 + 9 + 6 = 18$ is divisible by 9.

87. No; $8 + 4 + 5 + 3 = 20$ is not divisible by 9.

89. Yes; 522 is even and $5 + 2 + 2 = 9$ is divisible by 3.

91. No; 5917 is not even.

Section 2.3   Simplifying Fractions to Lowest Terms

Section 2.3   Practice Exercises

1. Answers will vary.

3. $5\overline{)145}^{\,29}$ $5 \cdot 29 = 145$

5. $\begin{array}{c} 23 \\ 2\overline{)46} \\ 2\overline{)92} \end{array}$ $2 \cdot 2 \cdot 23 = 2^2 \cdot 23 = 92$

7. $5\overline{)85}^{\,17}$ $5 \cdot 17 = 85$

9. $\begin{array}{c} 13 \\ 5\overline{)65} \\ 3\overline{)195} \end{array}$ $3 \cdot 5 \cdot 13 = 195$

11.

13.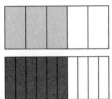

15. False; $5 \times 5 \neq 4 \times 4$

17. $2 \times 5 \blacklozenge 3 \times 3$
$10 \neq 9$
$\dfrac{2}{3} \neq \dfrac{3}{5}$

19. $1 \times 6 \blacklozenge 2 \times 3$
$6 = 6$
$\dfrac{1}{2} = \dfrac{3}{6}$

21. $12 \times 4 \blacklozenge 16 \times 3$
$48 = 48$
$\dfrac{12}{6} = \dfrac{3}{4}$

23. $8 \times 27 \blacklozenge 9 \times 20$
$216 \neq 180$
$\dfrac{8}{9} \neq \dfrac{20}{27}$

25. $\dfrac{12}{24} = \dfrac{\cancel{2} \cdot \cancel{2} \cdot \cancel{3}}{\cancel{2} \cdot \cancel{2} \cdot 2 \cdot \cancel{3}} = \dfrac{1}{2}$

27. $\dfrac{6}{18} = \dfrac{\cancel{2} \cdot \cancel{3}}{\cancel{2} \cdot \cancel{3} \cdot 3} = \dfrac{1}{3}$

29. $\dfrac{36}{20} = \dfrac{\cancel{2} \cdot \cancel{2} \cdot 3 \cdot 3}{\cancel{2} \cdot \cancel{2} \cdot 5} = \dfrac{9}{5}$

31. $\dfrac{15}{12} = \dfrac{\cancel{3} \cdot 5}{2 \cdot 2 \cdot \cancel{3}} = \dfrac{5}{4}$

33. $\dfrac{20}{25} = \dfrac{2 \cdot 2 \cdot \cancel{5}}{\cancel{5} \cdot 5} = \dfrac{4}{5}$

35. $\dfrac{14}{14} = 1$

37. $\dfrac{50}{25} = \dfrac{2 \cdot \cancel{25}}{\cancel{25}} = 2$

39. $\dfrac{9}{9} = 1$

41. $\dfrac{105}{140} = \dfrac{3 \cdot \cancel{5} \cdot \cancel{7}}{2 \cdot 2 \cdot \cancel{5} \cdot \cancel{7}} = \dfrac{3}{4}$

43. $\dfrac{33}{11} = \dfrac{3 \cdot \cancel{11}}{\cancel{11}} = 3$

45. $\dfrac{77}{110} = \dfrac{7 \cdot \cancel{11}}{10 \cdot \cancel{11}} = \dfrac{7}{10}$

47. $\dfrac{130}{150} = \dfrac{\cancel{2} \cdot \cancel{5} \cdot 13}{\cancel{2} \cdot 3 \cdot \cancel{5} \cdot 5} = \dfrac{13}{15}$

49. $\dfrac{385}{195} = \dfrac{\cancel{5} \cdot 7 \cdot 11}{3 \cdot \cancel{5} \cdot 13} = \dfrac{77}{39}$

51. $\dfrac{34}{85} = \dfrac{2 \cdot \cancel{17}}{5 \cdot \cancel{17}} = \dfrac{2}{5}$

53. $\dfrac{6-2}{10+4} = \dfrac{4}{14} = \dfrac{\cancel{2}\cdot 2}{\cancel{2}\cdot 7} = \dfrac{2}{7}$

55. $\dfrac{5-5}{7-2} = \dfrac{0}{5} = 0$

57. $\dfrac{7-2}{5-5} = \dfrac{5}{0} = \text{undefined}$

59. $\dfrac{8-2}{8+2} = \dfrac{6}{10} = \dfrac{\cancel{2}\cdot 3}{\cancel{2}\cdot 5} = \dfrac{3}{5}$

61. $\dfrac{12\cancel{0}}{16\cancel{0}} = \dfrac{12}{16} = \dfrac{\cancel{2}\cdot\cancel{2}\cdot 3}{\cancel{2}\cdot\cancel{2}\cdot 2\cdot 2} = \dfrac{3}{4}$

63. $\dfrac{30\cancel{0}\cancel{0}}{18\cancel{0}\cancel{0}} = \dfrac{30}{18} = \dfrac{\cancel{2}\cdot\cancel{3}\cdot 5}{\cancel{2}\cdot\cancel{3}\cdot 3} = \dfrac{5}{3}$

65. $\dfrac{42,\cancel{0}\cancel{0}\cancel{0}}{22,\cancel{0}\cancel{0}\cancel{0}} = \dfrac{42}{22} = \dfrac{\cancel{2}\cdot 21}{\cancel{2}\cdot 11} = \dfrac{21}{11}$

67. $\dfrac{51\cancel{0}\cancel{0}}{30,0\cancel{0}\cancel{0}} = \dfrac{51}{300} = \dfrac{\cancel{3}\cdot 17}{\cancel{3}\cdot 100} = \dfrac{17}{100}$

69. Heads: $\dfrac{20}{48} = \dfrac{\cancel{2}\cdot\cancel{2}\cdot 5}{\cancel{2}\cdot\cancel{2}\cdot 2\cdot 2\cdot 3} = \dfrac{5}{12}$

Tails: $48 - 20 = 28$

$\dfrac{28}{48} = \dfrac{\cancel{2}\cdot\cancel{2}\cdot 7}{\cancel{2}\cdot\cancel{2}\cdot 2\cdot 2\cdot 3} = \dfrac{7}{12}$

71. (a) $\dfrac{6}{26} = \dfrac{\cancel{2}\cdot 3}{\cancel{2}\cdot 13} = \dfrac{3}{13}$

(b) $26 - 6 = 20$

$\dfrac{20}{26} = \dfrac{\cancel{2}\cdot 2\cdot 5}{\cancel{2}\cdot 13} = \dfrac{10}{13}$

73. (a) Jonathan: $\dfrac{25}{35} = \dfrac{\cancel{5}\cdot 5}{\cancel{5}\cdot 7} = \dfrac{5}{7}$

Jared: $\dfrac{24}{28} = \dfrac{\cancel{2}\cdot\cancel{2}\cdot 2\cdot 3}{\cancel{2}\cdot\cancel{2}\cdot 7} = \dfrac{6}{7}$

(b) Jared sold the greater fractional part because $\dfrac{6}{7} > \dfrac{5}{7}$.

75. (a) Raymond:

$\dfrac{720}{792} = \dfrac{\cancel{2}\cdot\cancel{2}\cdot\cancel{2}\cdot 2\cdot\cancel{3}\cdot\cancel{3}\cdot 5}{\cancel{2}\cdot\cancel{2}\cdot\cancel{2}\cdot\cancel{3}\cdot\cancel{3}\cdot 11} = \dfrac{10}{11}$

Travis: $\dfrac{540}{660} = \dfrac{\cancel{2}\cdot\cancel{2}\cdot\cancel{3}\cdot 3\cdot 3\cdot\cancel{5}}{\cancel{2}\cdot\cancel{2}\cdot\cancel{3}\cdot\cancel{5}\cdot 11} = \dfrac{9}{11}$

(b) Raymond read the greater fractional part because $\dfrac{10}{11} > \dfrac{9}{11}$.

77. (a) 300,000,000

(b) 36,000,000

(c) $\dfrac{36,\cancel{0}\cancel{0}\cancel{0},\cancel{0}\cancel{0}\cancel{0}}{300,\cancel{0}\cancel{0}\cancel{0},\cancel{0}\cancel{0}\cancel{0}} = \dfrac{36}{300}$

$= \dfrac{\cancel{2}\cdot\cancel{3}\cdot 2\cdot 3}{\cancel{2}\cdot 2\cdot\cancel{3}\cdot 5\cdot 5} = \dfrac{3}{25}$

79. For example, $\dfrac{6}{8}, \dfrac{9}{12}, \dfrac{12}{16}$

81. For example, $\dfrac{6}{9}, \dfrac{4}{6}, \dfrac{2}{3}$

83. $\dfrac{792}{891} = \dfrac{8}{9}$

85. $\dfrac{779}{969} = \dfrac{41}{51}$

87. $\dfrac{493}{510} = \dfrac{29}{30}$

89. $\dfrac{969}{646} = \dfrac{3}{2}$

Section 2.4   Multiplication of Fractions and Applications

Section 2.4   Practice Exercises

1. Pages 152–156; answers will vary.

3. Numerator: 10; denominator: 14
$$\frac{10}{14} = \frac{\cancel{2} \cdot 5}{\cancel{2} \cdot 7} = \frac{5}{7}$$

5. Numerator: 25; denominator: 15
$$\frac{25}{15} = \frac{\cancel{5} \cdot 5}{3 \cdot \cancel{5}} = \frac{5}{3}$$

7.

9.

11. $\dfrac{1}{2} \cdot \dfrac{1}{4} = \dfrac{1 \cdot 1}{2 \cdot 4} = \dfrac{1}{8}$

13. $\dfrac{3}{4} \cdot 8 = \dfrac{3}{4} \cdot \dfrac{8}{1} = \dfrac{24}{4} = 6$

15. $\dfrac{1}{2} \times \dfrac{3}{8} = \dfrac{1 \times 3}{2 \times 8} = \dfrac{3}{16}$

17. $\dfrac{14}{9} \cdot \dfrac{1}{9} = \dfrac{14 \cdot 1}{9 \cdot 9} = \dfrac{14}{81}$

19. $\left(\dfrac{12}{7}\right)\left(\dfrac{2}{5}\right) = \dfrac{12 \times 2}{7 \times 5} = \dfrac{24}{35}$

21. $8 \cdot \left(\dfrac{1}{11}\right) = \dfrac{8}{1} \cdot \dfrac{1}{11} = \dfrac{8 \cdot 1}{1 \cdot 11} = \dfrac{8}{11}$

23. $\dfrac{4}{5} \cdot 6 = \dfrac{4}{5} \cdot \dfrac{6}{1} = \dfrac{4 \cdot 6}{5 \cdot 1} = \dfrac{24}{5}$

25. $\dfrac{13}{9} \times \dfrac{5}{4} = \dfrac{13 \times 5}{9 \times 4} = \dfrac{65}{36}$

27. $\dfrac{2}{9} \times \dfrac{3}{5} = \dfrac{2}{\cancel{3} \cdot 3} \times \dfrac{\cancel{3}}{5} = \dfrac{2}{15}$

29. $\dfrac{5}{6} \times \dfrac{3}{4} = \dfrac{5}{2 \cdot \cancel{3}} \times \dfrac{\cancel{3}}{4} = \dfrac{5}{8}$

31. $\dfrac{21}{5} \cdot \dfrac{25}{12} = \dfrac{\cancel{3} \cdot 7}{\cancel{5}} \cdot \dfrac{\cancel{5} \cdot 5}{2 \cdot 2 \cdot \cancel{3}} = \dfrac{35}{4}$

33. $\dfrac{24}{15} \cdot \dfrac{5}{3} = \dfrac{2 \cdot 2 \cdot 2 \cdot \cancel{3}}{\cancel{3} \cdot \cancel{5}} \cdot \dfrac{\cancel{5}}{3} = \dfrac{8}{3}$

35. $\left(\dfrac{6}{11}\right)\left(\dfrac{22}{15}\right) = \dfrac{6 \cdot 22}{11 \cdot 15} = \dfrac{2 \cdot \cancel{3} \cdot 2 \cdot \cancel{11}}{\cancel{11} \cdot \cancel{3} \cdot 5} = \dfrac{4}{5}$

37. $\left(\dfrac{17}{9}\right)\left(\dfrac{72}{17}\right) = \dfrac{17 \cdot 72}{9 \cdot 17} = \dfrac{\cancel{17} \cdot 8 \cdot \cancel{9}}{\cancel{9} \cdot \cancel{17}} = \dfrac{8}{1} = 8$

39. $\dfrac{21}{4} \cdot \dfrac{16}{7} = \dfrac{3 \cdot \cancel{7}}{\cancel{4}} \cdot \dfrac{\cancel{4} \cdot 4}{\cancel{7}} = \dfrac{12}{1} = 12$

41. $12 \times \dfrac{15}{42} = \dfrac{\cancel{2} \cdot 2 \cdot 3}{1} \times \dfrac{\cancel{3} \cdot 5}{\cancel{2} \cdot \cancel{3} \cdot 7} = \dfrac{30}{7}$

43. $\dfrac{9}{15} \times \dfrac{16}{3} \times \dfrac{25}{8}$
$$= \dfrac{\cancel{3} \cdot \cancel{3}}{\cancel{3} \cdot \cancel{5}} \times \dfrac{\cancel{2} \cdot \cancel{2} \cdot \cancel{2} \cdot 2}{\cancel{3}} \times \dfrac{\cancel{5} \cdot 5}{\cancel{2} \cdot \cancel{2} \cdot \cancel{2}}$$
$$= \dfrac{10}{1} = 10$$

45. $\dfrac{5}{2} \times \dfrac{10}{21} \times \dfrac{7}{5} = \dfrac{\cancel{5}}{\cancel{2}} \times \dfrac{\cancel{2} \cdot 5}{3 \cdot \cancel{7}} \times \dfrac{\cancel{7}}{\cancel{5}} = \dfrac{5}{3}$

47. $\dfrac{7}{10} \cdot \dfrac{3}{28} \cdot 5 = \dfrac{\cancel{7}}{2 \cdot \cancel{5}} \cdot \dfrac{3}{2 \cdot 2 \cdot \cancel{7}} \cdot \dfrac{\cancel{5}}{1} = \dfrac{3}{8}$

49. $\dfrac{100}{49} \times 21 \times \dfrac{14}{25} = \dfrac{2 \cdot 2 \cdot \cancel{5} \cdot \cancel{5}}{\cancel{7} \cdot \cancel{7}} \times \dfrac{3 \cdot \cancel{7}}{1} \times \dfrac{2 \cdot \cancel{7}}{\cancel{5} \cdot \cancel{5}}$
$$= \dfrac{24}{1} = 24$$

51. $\left(\dfrac{1}{10}\right)^3 = \dfrac{1}{10} \cdot \dfrac{1}{10} \cdot \dfrac{1}{10} = \dfrac{1}{1000}$

53. $\left(\dfrac{1}{10}\right)^6 = \dfrac{1}{10} \cdot \dfrac{1}{10} \cdot \dfrac{1}{10} \cdot \dfrac{1}{10} \cdot \dfrac{1}{10} \cdot \dfrac{1}{10}$

$\qquad = \dfrac{1}{1,000,000}$

55. $\left(\dfrac{1}{9}\right)^2 = \dfrac{1}{9} \cdot \dfrac{1}{9} = \dfrac{1}{81}$

57. $\left(\dfrac{3}{2}\right)^3 = \dfrac{3}{2} \cdot \dfrac{3}{2} \cdot \dfrac{3}{2} = \dfrac{27}{8}$

59. $\left(4 \cdot \dfrac{3}{4}\right)^3 = \left(\dfrac{\cancel{4}}{1} \cdot \dfrac{3}{\cancel{4}}\right)^3 = 3^3 = 27$

61. $\left(\dfrac{1}{\cancel{9}} \cdot \dfrac{\overset{1}{\cancel{3}}}{5}\right)^2 = \left(\dfrac{1}{15}\right)^2 = \dfrac{1}{15} \cdot \dfrac{1}{15} = \dfrac{1}{225}$
(denominator 9 reduces to 3)

63. $\dfrac{1}{3} \cdot \left(\dfrac{\overset{3}{\cancel{21}}}{\cancel{4}} \cdot \dfrac{\overset{2}{\cancel{8}}}{\cancel{7}}\right) = \dfrac{1}{\cancel{3}} \cdot \dfrac{\overset{2}{\cancel{6}}}{1} = 2$

65. $\dfrac{16}{9} \cdot \left(\dfrac{1}{2}\right)^3 = \dfrac{\overset{2}{\cancel{16}}}{9} \cdot \dfrac{1}{\cancel{8}} = \dfrac{2}{9}$

67.

69.

71. $A = \dfrac{1}{2}bh = \dfrac{1}{2}(11)(8) = \dfrac{1}{\cancel{2}} \cdot \dfrac{11}{1} \cdot \dfrac{\overset{4}{\cancel{8}}}{1} = 44 \text{ cm}^2$

73. $A = \dfrac{1}{2}bh = \dfrac{1}{2}(8)(8) = \dfrac{1}{\cancel{2}} \cdot \dfrac{\overset{4}{\cancel{8}}}{1} \cdot \dfrac{8}{1} = 32 \text{ m}^2$

75. $A = \dfrac{1}{2}bh = \dfrac{1}{2}(5)\left(\dfrac{8}{5}\right) = \dfrac{1}{\cancel{2}} \cdot \dfrac{\overset{1}{\cancel{5}}}{1} \cdot \dfrac{\overset{4}{\cancel{8}}}{\cancel{5}} = 4 \text{ yd}^2$

77. $A = l \times w = \dfrac{\overset{1}{\cancel{3}}}{4} \cdot \dfrac{1}{\cancel{3}} = \dfrac{1}{4} \text{ cm}^2$

79. $A = l \times w = \dfrac{13}{16} \cdot \dfrac{15}{16} = \dfrac{195}{256} \text{ in.}^2$

81. $A = (8)(4) + \dfrac{1}{2}(8)(4) = 32 + 4 \cdot 4 = 32 + 16$

$\qquad = 48 \text{ yd}^2$

83. $A = \dfrac{1}{2}(6)\left(\dfrac{7}{3}\right) + \dfrac{1}{2}(6)\left(\dfrac{2}{3}\right) = 3 \cdot \dfrac{7}{3} + 3 \cdot \dfrac{2}{3}$

$\qquad = \dfrac{\cancel{3}}{1} \cdot \dfrac{7}{\cancel{3}} + \dfrac{\cancel{3}}{1} \cdot \dfrac{2}{\cancel{3}} = 7 + 2 = 9 \text{ cm}^2$

85. $\dfrac{5}{8} \cdot 16 = \dfrac{5}{\cancel{8}} \cdot \dfrac{\overset{2}{\cancel{16}}}{1} = 10$

The amount left is 10 gal.

87. $\dfrac{1}{4} \cdot \dfrac{1}{2} = \dfrac{1}{8}$

Jim ate $\dfrac{1}{8}$ of the pizza for breakfast.

89. $\dfrac{2}{3} \cdot 9,825,000 = \dfrac{2}{\cancel{3}} \cdot \dfrac{\overset{3,275,000}{\cancel{9,825,000}}}{1}$

$\qquad = 6,550,000$

There are 6,550,000 viewers.

91. First place: $\dfrac{2}{3} \cdot 1200 = \dfrac{2}{\cancel{3}} \cdot \dfrac{\overset{400}{\cancel{1200}}}{1} = \$800$

Second place: $\dfrac{1}{4} \cdot 1200 = \dfrac{1}{\cancel{4}} \cdot \dfrac{\overset{300}{\cancel{1200}}}{1} = \$300$

Third place: $\dfrac{1}{12} \cdot 1200 = \dfrac{1}{\cancel{12}} \cdot \dfrac{\overset{100}{\cancel{1200}}}{1} = \$100$

93. (a) $\left(\dfrac{1}{6}\right)^2 = \dfrac{1}{6} \cdot \dfrac{1}{6} = \dfrac{1}{36}$

(b) $\sqrt{\dfrac{1}{36}} = \sqrt{\dfrac{1}{6} \cdot \dfrac{1}{6}} = \dfrac{1}{6}$

95. $\sqrt{\dfrac{1}{25}} = \sqrt{\dfrac{1}{5} \cdot \dfrac{1}{5}} = \dfrac{1}{5}$

97. $\sqrt{\dfrac{64}{81}} = \sqrt{\dfrac{8}{9} \cdot \dfrac{8}{9}} = \dfrac{8}{9}$

99. $\dfrac{1}{2}, \dfrac{1}{4} = \dfrac{1}{2 \cdot 2}, \dfrac{1}{8} = \dfrac{1}{4 \cdot 2}, \dfrac{1}{16} = \dfrac{1}{8 \cdot 2}$

The next number is $\dfrac{1}{16 \cdot 2} = \dfrac{1}{32}$.

101. $\dfrac{1}{2}\left(\dfrac{1}{8}\right) = \dfrac{1}{16}$

$\dfrac{1}{8}\left(\dfrac{1}{2}\right) = \dfrac{1}{16}$

They are the same.

102. $\dfrac{2}{3}\left(\dfrac{1}{4}\right) = \dfrac{2}{12} = \dfrac{1}{6}$

$\dfrac{1}{4}\left(\dfrac{2}{3}\right) = \dfrac{2}{12} = \dfrac{1}{6}$

They are the same.

Section 2.5   Division of Fractions and Applications

Section 2.5   Practice Exercises

1. Page 143
Answers will vary.

3. $\dfrac{9}{\cancel{11}_{1}} \times \dfrac{\overset{2}{\cancel{22}}}{5} = \dfrac{18}{5}$

5. $\dfrac{\overset{2}{\cancel{34}}}{\cancel{8}_{1}} \cdot \dfrac{\overset{1}{\cancel{8}}}{\cancel{11}_{1}} = 2$

7. $8 \cdot \left(\dfrac{5}{24}\right) = \dfrac{\overset{1}{\cancel{8}}}{1} \cdot \dfrac{5}{\cancel{24}_{3}} = \dfrac{5}{3}$

9. $\left(\dfrac{9}{5}\right)\left(\dfrac{5}{9}\right) = \dfrac{45}{45} = 1$

11. $\dfrac{1}{3} \times 3 = \dfrac{1}{3} \cdot \dfrac{3}{1} = \dfrac{3}{3} = 1$

13. $\dfrac{8}{7}$

15. $\dfrac{9}{10}$

17. $\dfrac{1}{4}$

19. No reciprocal exists.

21. $\dfrac{1}{3}$

23. multiplying

25. $\dfrac{2}{15} \div \dfrac{5}{12} = \dfrac{2}{15} \cdot \dfrac{12}{5} = \dfrac{2}{\cancel{3} \cdot 5} \cdot \dfrac{2 \cdot 2 \cdot \cancel{3}}{5} = \dfrac{8}{25}$

27. $\dfrac{7}{13} \div \dfrac{2}{5} = \dfrac{7}{13} \cdot \dfrac{5}{2} = \dfrac{35}{26}$

29. $\dfrac{14}{3} \div \dfrac{6}{5} = \dfrac{14}{3} \cdot \dfrac{5}{\overset{}{\underset{3}{\cancel{6}}}} = \dfrac{35}{9}$ where $\cancel{14}$ has 7 above

31. $\dfrac{15}{2} \div \dfrac{3}{2} = \dfrac{\overset{5}{\cancel{15}}}{\underset{1}{\cancel{2}}} \cdot \dfrac{\overset{1}{\cancel{2}}}{\underset{1}{\cancel{3}}} = 5$

33. $\dfrac{3}{4} \div \dfrac{3}{4} = \dfrac{3}{4} \cdot \dfrac{4}{3} = \dfrac{12}{12} = 1$

35. $7 \div \dfrac{2}{3} = \dfrac{7}{1} \cdot \dfrac{3}{2} = \dfrac{21}{2}$

37. $\dfrac{10}{9} \div \dfrac{1}{18} = \dfrac{10}{\underset{1}{\cancel{9}}} \cdot \dfrac{\overset{2}{\cancel{18}}}{1} = 20$

39. $12 \div \dfrac{3}{4} = \dfrac{\overset{4}{\cancel{12}}}{1} \cdot \dfrac{4}{\underset{1}{\cancel{3}}} = 16$

41. $\dfrac{12}{5} \div 4 = \dfrac{\overset{3}{\cancel{12}}}{5} \cdot \dfrac{1}{\underset{1}{\cancel{4}}} = \dfrac{3}{5}$

43. $\dfrac{9}{50} \div \dfrac{18}{25} = \dfrac{\overset{1}{\cancel{9}}}{\underset{2}{\cancel{50}}} \cdot \dfrac{\overset{1}{\cancel{25}}}{\underset{2}{\cancel{18}}} = \dfrac{1}{4}$

45. $\dfrac{9}{100} \div \dfrac{13}{1000} = \dfrac{9}{\underset{1}{\cancel{100}}} \cdot \dfrac{\overset{10}{\cancel{1000}}}{13} = \dfrac{90}{13}$

47. $\dfrac{36}{5} \div \dfrac{9}{25} = \dfrac{\overset{4}{\cancel{36}}}{\underset{1}{\cancel{5}}} \cdot \dfrac{\overset{5}{\cancel{25}}}{\underset{1}{\cancel{9}}} = 20$

49. $\dfrac{7}{8} \div \dfrac{1}{4} = \dfrac{7}{\underset{2}{\cancel{8}}} \cdot \dfrac{\overset{1}{\cancel{4}}}{1} = \dfrac{7}{2}$

51. $\dfrac{5}{\underset{4}{\cancel{8}}} \cdot \dfrac{\overset{1}{\cancel{2}}}{9} = \dfrac{5}{36}$

53. $6 \cdot \dfrac{4}{3} = \dfrac{\overset{2}{\cancel{6}}}{1} \cdot \dfrac{4}{\underset{1}{\cancel{3}}} = 8$

55. $\dfrac{16}{5} \div 8 = \dfrac{\overset{2}{\cancel{16}}}{5} \cdot \dfrac{1}{\underset{1}{\cancel{8}}} = \dfrac{2}{5}$

57. $\dfrac{16}{3} \div \dfrac{2}{5} = \dfrac{\overset{8}{\cancel{16}}}{3} \cdot \dfrac{5}{\underset{1}{\cancel{2}}} = \dfrac{40}{3}$

59. $\dfrac{1}{8} \cdot 16 = \dfrac{1}{\underset{1}{\cancel{8}}} \cdot \dfrac{\overset{2}{\cancel{16}}}{1} = 2$

61. $\dfrac{22}{7} \cdot \dfrac{5}{16} = \dfrac{\cancel{2} \cdot 11}{7} \cdot \dfrac{5}{\cancel{2} \cdot 8} = \dfrac{55}{56}$

63. $8 \div \dfrac{16}{3} = \dfrac{\overset{1}{\cancel{8}}}{1} \cdot \dfrac{3}{\underset{2}{\cancel{16}}} = \dfrac{3}{2}$

65. $\dfrac{2}{3} \cdot 6$ multiplies $\dfrac{2}{3}$ by $\dfrac{6}{1}$, and $\dfrac{2}{3} \div 6$

multiplies $\dfrac{2}{3}$ by $\dfrac{1}{6}$. So $\dfrac{2}{3} \cdot 6 = \dfrac{2}{\underset{1}{\cancel{3}}} \cdot \dfrac{\overset{2}{\cancel{6}}}{1} = 4$

and $\dfrac{2}{3} \div 6 = \dfrac{\overset{1}{\cancel{2}}}{3} \cdot \dfrac{1}{\underset{3}{\cancel{6}}} = \dfrac{1}{9}$.

67. $\dfrac{54}{21} \div \dfrac{2}{3} \div 9 = \dfrac{\overset{27}{\cancel{54}}}{\underset{7}{\cancel{21}}} \cdot \dfrac{\overset{1}{\cancel{3}}}{\underset{1}{\cancel{2}}} \div 9 = \dfrac{27}{7} \div 9$

$= \dfrac{\overset{3}{\cancel{27}}}{7} \cdot \dfrac{1}{\underset{1}{\cancel{9}}} = \dfrac{3}{7}$

69. $\dfrac{3}{5} \div \dfrac{6}{7} \cdot \dfrac{5}{3} = \dfrac{\overset{1}{\cancel{3}}}{5} \cdot \dfrac{7}{\underset{2}{\cancel{6}}} \cdot \dfrac{5}{3} = \dfrac{7}{\underset{2}{\cancel{10}}} \cdot \dfrac{\overset{1}{\cancel{5}}}{3} = \dfrac{7}{6}$

71. $\left(\dfrac{3}{8}\right)^2 \div \dfrac{9}{14} = \dfrac{3}{8} \cdot \dfrac{3}{8} \div \dfrac{9}{14} = \dfrac{9}{64} \div \dfrac{9}{14}$

$= \dfrac{9}{64} \cdot \dfrac{14}{9} = \dfrac{\cancel{9}}{\cancel{2} \cdot 32} \cdot \dfrac{\cancel{2} \cdot 7}{\cancel{9}} = \dfrac{7}{32}$

73. $\left(\dfrac{2}{5} \div \dfrac{8}{3}\right)^2 = \left(\dfrac{\cancel{2}^1}{5} \cdot \dfrac{3}{\cancel{8}_4}\right)^2 = \left(\dfrac{3}{20}\right)^2 = \dfrac{3}{20} \cdot \dfrac{3}{20}$

$= \dfrac{9}{400}$

75. $\left(\dfrac{63}{8} \div \dfrac{9}{4}\right)^2 \cdot 4 = \left(\dfrac{\cancel{63}^7}{\cancel{8}_2} \cdot \dfrac{\cancel{4}^1}{\cancel{9}_1}\right)^2 \cdot 4 = \left(\dfrac{7}{2}\right)^2 \cdot 4$

$= \dfrac{7}{2} \cdot \dfrac{7}{2} \cdot \dfrac{4}{1} = \dfrac{49}{\cancel{4}} \cdot \dfrac{\cancel{4}^1}{1} = 49$

77. $\dfrac{15}{16} \cdot \left(\dfrac{2}{3}\right)^2 \div \dfrac{20}{21} = \dfrac{15}{16} \cdot \left(\dfrac{2}{3} \cdot \dfrac{2}{3}\right) \div \dfrac{20}{21}$

$= \dfrac{15}{16} \cdot \dfrac{4}{9} \div \dfrac{20}{21} = \dfrac{\cancel{3} \cdot 5}{\cancel{4} \cdot 4} \cdot \dfrac{\cancel{4}}{\cancel{3} \cdot 3} \div \dfrac{20}{21}$

$= \dfrac{5}{12} \div \dfrac{20}{21} = \dfrac{5}{12} \cdot \dfrac{21}{20}$

$= \dfrac{\cancel{5}}{\cancel{3} \cdot 4} \cdot \dfrac{\cancel{3} \cdot 7}{4 \cdot \cancel{5}} = \dfrac{7}{16}$

79. $\dfrac{9}{4} \div \dfrac{1}{8} = \dfrac{9}{\cancel{4}_1} \cdot \dfrac{\cancel{8}^2}{1} = 18$

81. $36 \div \dfrac{2}{3} = \dfrac{\cancel{36}^{18}}{1} \cdot \dfrac{3}{\cancel{2}_1} = 54$

Li wrapped 54 packages.

83. $\dfrac{3}{2} \div \dfrac{1}{16} = \dfrac{3}{\cancel{2}_1} \cdot \dfrac{\cancel{16}^8}{1} = 24$ cups of juice

85. $16 \cdot \dfrac{3}{4} = \dfrac{\cancel{16}^4}{1} \cdot \dfrac{3}{\cancel{4}_1} = 12$

The stack will be 12 in. high.

87. (a) $18 \div \dfrac{2}{3} = \dfrac{\cancel{18}^9}{1} \cdot \dfrac{3}{\cancel{2}_1} = 27$

27 commercials in 1 hr

(b) $27 \times 24 = 648$
648 commercials in 1 day

89. (a) $\dfrac{1}{10} \cdot 240{,}000 = \dfrac{1}{10} \cdot \dfrac{240{,}000}{1}$

$= \dfrac{240{,}00\cancel{0}}{1\cancel{0}}$

$= 24{,}000$

The down payment is \$24,000.

$\dfrac{2}{3} \cdot 24{,}000 = \dfrac{2}{\cancel{3}_1} \cdot \dfrac{\cancel{24{,}000}^{8000}}{1} = 16{,}000$

Ricardo's mother will pay \$16,000.

(b) \$24,000 − \$16,000 = \$8000
Ricardo will have to pay \$8000.

(c) \$240,000 − \$24,000 = \$216,000
He will have to finance \$216,000.

91. (a) $\dfrac{1}{\cancel{3}_1} \cdot \dfrac{\cancel{9}^3}{4} = \dfrac{3}{4}$

She plans to sell $\dfrac{3}{4}$ acre.

(b) She keeps $\dfrac{2}{3}$ of the land.

$\dfrac{\cancel{2}^1}{\cancel{3}_1} \cdot \dfrac{\cancel{9}^3}{\cancel{4}_2} = \dfrac{3}{2}$ or $1\dfrac{1}{2}$ acres

93. $\dfrac{7}{4} \div \dfrac{1}{8} = \dfrac{7}{\cancel{4}_1} \cdot \dfrac{\cancel{8}^2}{1} = 14$

She can prepare 14 samples.

95. The length is 12 ft, because

$$30 \div \frac{5}{2} = \frac{30}{1} \cdot \frac{2}{5} = \frac{\cancel{5} \cdot 6}{1} \cdot \frac{2}{\cancel{5}} = \frac{12}{1} = 12$$

Problem Recognition Exercises: Multiplication and Division of Fractions

1. (a) $\dfrac{8}{3} \cdot \dfrac{6}{5} = \dfrac{8}{\cancel{3}} \cdot \dfrac{\cancel{3} \cdot 2}{5} = \dfrac{16}{5}$

(b) $\dfrac{6}{5} \cdot \dfrac{8}{3} = \dfrac{\cancel{3} \cdot 2}{5} \cdot \dfrac{8}{\cancel{3}} = \dfrac{16}{5}$

(c) $\dfrac{8}{3} \div \dfrac{6}{5} = \dfrac{8}{3} \cdot \dfrac{5}{6} = \dfrac{\cancel{2} \cdot 4}{3} \cdot \dfrac{5}{\cancel{2} \cdot 3} = \dfrac{20}{9}$

(d) $\dfrac{6}{5} \div \dfrac{8}{3} = \dfrac{6}{5} \cdot \dfrac{3}{8} = \dfrac{\cancel{2} \cdot 3}{5} \cdot \dfrac{3}{\cancel{2} \cdot 4} = \dfrac{9}{20}$

3. (a) $12 \cdot \dfrac{9}{8} = \dfrac{12}{1} \cdot \dfrac{9}{8} = \dfrac{3 \cdot \cancel{4}}{1} \cdot \dfrac{9}{2 \cdot \cancel{4}} = \dfrac{27}{2}$

(b) $\dfrac{9}{8} \cdot 12 = \dfrac{9}{8} \cdot \dfrac{12}{1} = \dfrac{9}{2 \cdot \cancel{4}} \cdot \dfrac{3 \cdot \cancel{4}}{1} = \dfrac{27}{2}$

(c) $12 \div \dfrac{9}{8} = \dfrac{12}{1} \cdot \dfrac{8}{9} = \dfrac{\cancel{3} \cdot 4}{1} \cdot \dfrac{8}{\cancel{3} \cdot 3} = \dfrac{32}{3}$

(d) $\dfrac{9}{8} \div 12 = \dfrac{9}{8} \cdot \dfrac{1}{12} = \dfrac{\cancel{3} \cdot 3}{8} \cdot \dfrac{1}{\cancel{3} \cdot 4} = \dfrac{3}{32}$

5. (a) $\dfrac{5}{6} \cdot \dfrac{5}{6} = \dfrac{25}{36}$

(b) $\dfrac{\cancel{5}}{\cancel{6}} \cdot \dfrac{\cancel{6}}{\cancel{5}} = \dfrac{1}{1} = 1$

(c) $\dfrac{5}{6} \div \dfrac{5}{6} = \dfrac{\cancel{5}}{\cancel{6}} \cdot \dfrac{\cancel{6}}{\cancel{5}} = \dfrac{1}{1} = 1$

(d) $\dfrac{5}{6} \div \dfrac{6}{5} = \dfrac{5}{6} \cdot \dfrac{5}{6} = \dfrac{25}{36}$

7. (a) $\dfrac{1}{12} \cdot \dfrac{2}{3} \cdot \dfrac{16}{21} = \dfrac{1}{3 \cdot \cancel{4}} \cdot \dfrac{2}{3} \cdot \dfrac{\cancel{4} \cdot 4}{21} = \dfrac{8}{189}$

(b) $\dfrac{1}{12} \cdot \dfrac{2}{3} \div \dfrac{16}{21} = \dfrac{1}{12} \cdot \dfrac{2}{3} \cdot \dfrac{21}{16}$

$= \dfrac{1}{12} \cdot \dfrac{\cancel{2}}{\cancel{3}} \cdot \dfrac{\cancel{3} \cdot 7}{\cancel{2} \cdot 8} = \dfrac{7}{96}$

(c) $\dfrac{1}{12} \div \dfrac{2}{3} \cdot \dfrac{16}{21} = \dfrac{1}{\cancel{3} \cdot \cancel{4}} \cdot \dfrac{\cancel{3}}{\cancel{2}} \cdot \dfrac{\cancel{4} \cdot \cancel{2} \cdot 2}{21}$

$= \dfrac{2}{21}$

(d) $\dfrac{1}{12} \div \dfrac{2}{3} \div \dfrac{16}{21} = \dfrac{1}{12} \cdot \dfrac{3}{2} \cdot \dfrac{21}{16}$

$= \dfrac{1}{\cancel{3} \cdot 4} \cdot \dfrac{\cancel{3}}{2} \cdot \dfrac{21}{16} = \dfrac{21}{128}$

9. (a) $\dfrac{9}{10} \cdot 6 \cdot \dfrac{1}{4} = \dfrac{9}{10} \cdot \dfrac{6}{1} \cdot \dfrac{1}{4}$

$= \dfrac{9}{10} \cdot \dfrac{\cancel{2} \cdot 3}{1} \cdot \dfrac{1}{\cancel{2} \cdot 2} = \dfrac{27}{20}$

(b) $\dfrac{9}{10} \cdot 6 \div \dfrac{1}{4} = \dfrac{9}{10} \cdot \dfrac{6}{1} \cdot \dfrac{4}{1}$

$= \dfrac{9}{\cancel{2} \cdot 5} \cdot \dfrac{\cancel{2} \cdot 3}{1} \cdot \dfrac{4}{1} = \dfrac{108}{5}$

(c) $\dfrac{9}{10} \div 6 \cdot \dfrac{1}{4} = \dfrac{9}{10} \cdot \dfrac{1}{6} \cdot \dfrac{1}{4}$

$= \dfrac{3 \cdot \cancel{3}}{10} \cdot \dfrac{1}{2 \cdot \cancel{3}} \cdot \dfrac{1}{4} = \dfrac{3}{80}$

(d) $\dfrac{9}{10} \div 6 \div \dfrac{1}{4} = \dfrac{9}{10} \cdot \dfrac{1}{6} \cdot \dfrac{4}{1}$

$= \dfrac{3 \cdot \cancel{3}}{\cancel{2} \cdot 5} \cdot \dfrac{1}{\cancel{2} \cdot \cancel{3}} \cdot \dfrac{\cancel{2} \cdot \cancel{2}}{1} = \dfrac{3}{5}$

11. (a) $\dfrac{2}{3} \cdot 1 = \dfrac{2}{3}$

    (b) $1 \cdot \dfrac{2}{3} = \dfrac{2}{3}$

    (c) $\dfrac{2}{3} \div 1 = \dfrac{2}{3}$

    (d) $1 \div \dfrac{2}{3} = 1 \cdot \dfrac{3}{2} = \dfrac{3}{2}$

13. (a) $8 \div \dfrac{1}{4} = 8 \cdot 4 = 32$

    (b) $8 \cdot \dfrac{1}{4} = \dfrac{8}{4} = 2$

    (c) $8 \div 4 = 2$

    (d) $8 \cdot 4 = 32$

15. (a) $4^2 \cdot \dfrac{1}{6} = 4 \cdot 4 \cdot \dfrac{1}{6} = 16 \cdot \dfrac{1}{6} = \dfrac{\cancel{2} \cdot 8}{1} \cdot \dfrac{1}{\cancel{2} \cdot 3}$

        $= \dfrac{8}{3}$

    (b) $4^2 \div \dfrac{1}{6} = 4 \cdot 4 \div \dfrac{1}{6} = 16 \cdot \dfrac{6}{1} = 16 \cdot 6 = 96$

    (c) $4 \cdot \left(\dfrac{1}{6}\right)^2 = \dfrac{4}{1} \cdot \dfrac{1}{6} \cdot \dfrac{1}{6} = \dfrac{4}{36} = \dfrac{\cancel{4}}{\cancel{4} \cdot 9} = \dfrac{1}{9}$

    (d) $4 \div \left(\dfrac{1}{6}\right)^2 = \dfrac{4}{1} \div \left(\dfrac{1}{6} \cdot \dfrac{1}{6}\right) = \dfrac{4}{1} \div \left(\dfrac{1}{36}\right)$

        $= \dfrac{4}{1} \cdot \dfrac{36}{1} = 144$

Section 2.6   Multiplication and Division of Mixed Numbers

Section 2.6 Practice Exercises

1. Chapter Review Exercises, pages 157–160
Chapter Test, pages 160–161
Cumulative Review Exercises,
pages 161–162
Answers will vary.

3. $\dfrac{13}{\cancel{5}_{1}} \cdot \dfrac{\cancel{10}^{2}}{9} = \dfrac{26}{9}$

5. $\dfrac{42}{11} \div \dfrac{7}{2} = \dfrac{\cancel{42}^{6}}{11} \cdot \dfrac{2}{\cancel{7}_{1}} = \dfrac{12}{11}$

7. $\dfrac{52}{18} \div 13 = \dfrac{\cancel{52}^{4}}{18} \cdot \dfrac{1}{\cancel{13}_{1}} = \dfrac{4}{18} = \dfrac{2}{9}$

9. $3\dfrac{2}{5} = \dfrac{3 \times 5 + 2}{5} = \dfrac{17}{5}$

11. $1\dfrac{4}{7} = \dfrac{1 \times 7 + 4}{7} = \dfrac{11}{7}$

13.
$$
\begin{array}{r}
12 \\
6\,)\overline{\,77} \\
\underline{-6} \\
17 \\
\underline{-12} \\
5
\end{array}
\qquad 12\dfrac{5}{6}
$$

15.
$$
\begin{array}{r}
9 \\
4\,)\overline{\,39} \\
\underline{-36} \\
3
\end{array}
\qquad 9\dfrac{3}{4}
$$

17. $\left(2\dfrac{2}{5}\right)\left(3\dfrac{1}{12}\right) = \dfrac{\cancel{12}^{1}}{5} \cdot \dfrac{37}{\cancel{12}_{1}} = \dfrac{37}{5}$

$$
\begin{array}{r}
7 \\
5\,)\overline{\,37} \\
\underline{-35} \\
2
\end{array}
\qquad = 7\dfrac{2}{5}
$$

19. $2\dfrac{1}{3}\cdot\dfrac{5}{7}=\dfrac{\overset{1}{\cancel{7}}}{3}\cdot\dfrac{5}{\underset{1}{\cancel{7}}}=\dfrac{5}{3}$

$3\overline{)\,5\,}=1\dfrac{2}{3}$
$\quad\underline{-3}$
$\quad\ \ 2$
(quotient 1)

21. $4\dfrac{2}{9}\cdot 9=\dfrac{38}{\underset{1}{\cancel{9}}}\cdot\dfrac{\overset{1}{\cancel{9}}}{1}=38$

23. $\left(5\dfrac{3}{16}\right)\left(5\dfrac{1}{3}\right)=\dfrac{83}{\underset{1}{\cancel{16}}}\cdot\dfrac{\overset{1}{\cancel{16}}}{3}=\dfrac{83}{3}$

$3\overline{)\,83\,}=27\dfrac{2}{3}$
$\quad\underline{-6}$
$\quad\ \ 23$
$\quad\underline{-21}$
$\quad\ \ \ \ 2$
(quotient 27)

25. $\left(7\dfrac{1}{4}\right)\cdot 10=\dfrac{29}{\underset{2}{\cancel{4}}}\cdot\dfrac{\overset{5}{\cancel{10}}}{1}=\dfrac{145}{2}$

$2\overline{)\,145\,}=72\dfrac{1}{2}$
$\quad\underline{-14}$
$\quad\ \ \ 5$
$\quad\ \ \underline{-4}$
$\quad\ \ \ \ 1$
(quotient 72)

27. $4\dfrac{5}{8}\cdot 0=0$

29. $\left(3\dfrac{1}{2}\right)\left(2\dfrac{1}{7}\right)=\dfrac{\overset{1}{\cancel{7}}}{2}\cdot\dfrac{15}{\underset{1}{\cancel{7}}}=\dfrac{15}{2}=7\dfrac{1}{2}$

31. $\left(5\dfrac{2}{5}\right)\left(\dfrac{2}{9}\right)\left(1\dfrac{4}{5}\right)=\dfrac{27}{5}\cdot\dfrac{2}{\underset{1}{\cancel{9}}}\cdot\dfrac{\overset{1}{\cancel{9}}}{5}=\dfrac{54}{25}=2\dfrac{4}{25}$

33. $1\dfrac{7}{10}\div 2\dfrac{3}{4}=\dfrac{17}{10}\div\dfrac{11}{4}=\dfrac{17}{\underset{5}{\cancel{10}}}\cdot\dfrac{\overset{2}{\cancel{4}}}{11}=\dfrac{34}{55}$

35. $5\dfrac{8}{9}\div 1\dfrac{1}{3}=\dfrac{53}{9}\div\dfrac{4}{3}=\dfrac{53}{\underset{3}{\cancel{9}}}\cdot\dfrac{\overset{1}{\cancel{3}}}{4}=\dfrac{53}{12}=4\dfrac{5}{12}$

37. $2\dfrac{1}{2}\div 1\dfrac{1}{16}=\dfrac{5}{2}\div\dfrac{17}{16}=\dfrac{5}{\underset{1}{\cancel{2}}}\cdot\dfrac{\overset{8}{\cancel{16}}}{17}=\dfrac{40}{17}=2\dfrac{6}{17}$

39. $4\dfrac{1}{2}\div 2\dfrac{1}{4}=\dfrac{9}{2}\div\dfrac{9}{4}=\dfrac{\overset{1}{\cancel{9}}}{\underset{1}{\cancel{2}}}\cdot\dfrac{\overset{2}{\cancel{4}}}{\underset{1}{\cancel{9}}}=2$

41. $0\div 6\dfrac{7}{12}=0$

43. $2\dfrac{5}{6}\div\dfrac{1}{6}=\dfrac{17}{6}\div\dfrac{1}{6}=\dfrac{17}{\underset{1}{\cancel{6}}}\cdot\dfrac{\overset{1}{\cancel{6}}}{1}=17$

45. $1\dfrac{1}{3}\div\dfrac{2}{7}=\dfrac{4}{3}\div\dfrac{2}{7}=\dfrac{\overset{2}{\cancel{4}}}{3}\cdot\dfrac{7}{\underset{1}{\cancel{2}}}=\dfrac{14}{3}=4\dfrac{2}{3}$

47. $3\dfrac{1}{2}\div 2=\dfrac{7}{2}\div\dfrac{2}{1}=\dfrac{7}{2}\cdot\dfrac{1}{2}=\dfrac{7}{4}=1\dfrac{3}{4}$

49. $4\dfrac{3}{4}\cdot 8=\dfrac{19}{\underset{1}{\cancel{4}}}\cdot\dfrac{\overset{2}{\cancel{8}}}{1}=38$

Tabitha earned $38.

51. $25\dfrac{7}{10}\cdot 25=\dfrac{257}{\underset{2}{\cancel{10}}}\cdot\dfrac{\overset{5}{\cancel{25}}}{1}=\dfrac{1285}{2}=642\dfrac{1}{2}$

Average Americans consume $642\dfrac{1}{2}$ lb.

53. (a) $1\dfrac{3}{4}\div\dfrac{1}{4}=\dfrac{7}{4}\div\dfrac{1}{4}=\dfrac{7}{\underset{1}{\cancel{4}}}\cdot\dfrac{\overset{1}{\cancel{4}}}{1}=7$ weeks old

(b) $2\dfrac{1}{8}\div\dfrac{1}{4}=\dfrac{17}{8}\div\dfrac{1}{4}$

$=\dfrac{17}{\underset{2}{\cancel{8}}}\cdot\dfrac{\overset{1}{\cancel{4}}}{1}=\dfrac{17}{2}=8\dfrac{1}{2}$ weeks old

55. $28 \div 1\frac{17}{24} = \frac{28}{1} \div \frac{41}{24} = \frac{28}{1} \cdot \frac{24}{41} = \frac{672}{41}$

$$= 16\frac{16}{41}$$

The roll is $16\frac{16}{41}$ ft long.

57. $2\frac{1}{5} \div 1\frac{1}{10} = \frac{11}{5} \div \frac{11}{10} = \frac{\overset{1}{\cancel{11}}}{\cancel{5}} \cdot \frac{\overset{2}{\cancel{10}}}{\cancel{11}} = 2$

59. $6 \div 1\frac{1}{8} = \frac{6}{1} \div \frac{9}{8} = \frac{\overset{2}{\cancel{6}}}{1} \cdot \frac{8}{\underset{3}{\cancel{9}}} = \frac{16}{3} = 5\frac{1}{3}$

61. $\frac{2}{3} \cdot 2\frac{7}{10} = \frac{\overset{1}{\cancel{2}}}{\underset{1}{\cancel{3}}} \cdot \frac{\overset{9}{\cancel{27}}}{\underset{5}{\cancel{10}}} = \frac{9}{5} = 1\frac{4}{5}$

63. $4\frac{1}{12} \cdot 0 = 0$

65. $10\frac{1}{2} \div 9 = \frac{21}{2} \div \frac{9}{1} = \frac{\overset{7}{\cancel{21}}}{2} \cdot \frac{1}{\underset{3}{\cancel{9}}} = \frac{7}{6} = 1\frac{1}{6}$

67. $0 \div 9\frac{2}{3} = 0$

69. $12 \cdot \frac{1}{8} = \frac{\overset{3}{\cancel{12}}}{1} \cdot \frac{1}{\underset{2}{\cancel{8}}} = \frac{3}{2} = 1\frac{1}{2}$

71. $6\frac{8}{9} \div 0$ is undefined.

73. $\left(3\frac{2}{5}\right)\left(\frac{7}{34}\right)\left(3\frac{3}{4}\right) = \frac{\overset{1}{\cancel{17}}}{\underset{1}{\cancel{5}}} \cdot \frac{7}{\underset{2}{\cancel{34}}} \cdot \frac{\overset{3}{\cancel{15}}}{4} = \frac{21}{8}$

$$= 2\frac{5}{8}$$

75. $7\frac{1}{8} \div 1\frac{1}{3} \div 2\frac{1}{4} = \frac{57}{8} \div \frac{4}{3} \div \frac{9}{4}$

$$= \frac{\overset{19}{\cancel{57}}}{8} \cdot \frac{\overset{1}{\cancel{3}}}{\cancel{4}} \cdot \frac{\overset{1}{\cancel{4}}}{\underset{3}{\cancel{9}}} = \frac{19}{8} = 2\frac{3}{8}$$

77. The perimeter of the garden is
$2(20) + 2(15) = 40 + 30 = 70$ ft.

$$70 \div 1\frac{1}{4} = \frac{70}{1} \div \frac{5}{4} = \frac{\overset{14}{\cancel{70}}}{1} \cdot \frac{4}{\underset{1}{\cancel{5}}} = 56$$

56 bricks will be needed.
$56 \times \$3 = \$168$
The total cost is $168.

79. $12\frac{2}{3} \cdot 25\frac{1}{8} = 318\frac{1}{4}$

81. $56\frac{5}{6} \div 3\frac{1}{6} = 17\frac{18}{19}$

83. $32\frac{7}{12} \div 12\frac{1}{6} = 2\frac{99}{146}$

85. $11\frac{1}{2} \cdot 41\frac{3}{4} = 480\frac{1}{8}$

Chapter 2   Review Exercises

Section 2.1

1. $\frac{1}{2}$

3. **(a)** $\frac{5}{3}$

   (b) Improper

5. $\frac{7}{15}$

7. $\frac{7}{6}$ or $1\frac{1}{6}$

9. $11\frac{2}{5} = \frac{11 \times 5 + 2}{5} = \frac{57}{5}$

11. $9{\overline{\smash{\big)}\,47}}$ gives quotient 5, -45, remainder 2 $5\dfrac{2}{9}$

13–15.

Number line from 0 to 4 with points at $\dfrac{13}{8}$ and $\dfrac{10}{5}$

17. $26{\overline{\smash{\big)}\,1582}}$ quotient 60, -156, 22, -0, 22 $60\dfrac{22}{26}=60\dfrac{11}{13}$

Section 2.2

19. 55, 140, 260, 1200

21. Prime

23. Neither

25.
$2{\overline{\smash{\big)}\,4}}$
$2{\overline{\smash{\big)}\,8}}$
$2{\overline{\smash{\big)}\,16}}$
$2{\overline{\smash{\big)}\,32}}$
$2{\overline{\smash{\big)}\,64}}$

$2\cdot2\cdot2\cdot2\cdot2\cdot2=2^6=64$

27.
$3{\overline{\smash{\big)}\,9}}$
$5{\overline{\smash{\big)}\,45}}$
$5{\overline{\smash{\big)}\,225}}$
$2{\overline{\smash{\big)}\,450}}$
$2{\overline{\smash{\big)}\,900}}$

$2\cdot2\cdot3\cdot3\cdot5\cdot5=2^2\cdot3^2\cdot5^2=900$

29. 1, 2, 4, 5, 8, 10, 16, 20, 40, 80

Section 2.3

31. $15\times14 \;\blacklozenge\; 21\times10$
$210=210$
$\dfrac{15}{21}=\dfrac{10}{14}$

33. $\dfrac{14}{49}=\dfrac{2\cdot\cancel{7}}{\cancel{7}\cdot7}=\dfrac{2}{7}$

35. $\dfrac{63}{27}=\dfrac{\cancel{9}\cdot7}{\cancel{9}\cdot3}=\dfrac{7}{3}$

37. $\dfrac{42}{21}=\dfrac{2\cdot\cancel{21}}{\cancel{21}}=2$

39. $\dfrac{14\cancel{00}}{20\cancel{00}}=\dfrac{14}{20}=\dfrac{\cancel{2}\cdot7}{\cancel{2}\cdot10}=\dfrac{7}{10}$

41. (a) $\dfrac{6}{10}=\dfrac{\cancel{2}\cdot3}{\cancel{2}\cdot5}=\dfrac{3}{5}$

(b) $\dfrac{6}{15}=\dfrac{2\cdot\cancel{3}}{\cancel{3}\cdot5}=\dfrac{2}{5}$

Section 2.4

43. $\dfrac{4}{3}\times\dfrac{8}{3}=\dfrac{32}{9}$

45. $33\cdot\dfrac{5}{11}=\dfrac{\overset{3}{\cancel{33}}}{1}\cdot\dfrac{5}{\underset{1}{\cancel{11}}}=15$

47. $\dfrac{\overset{1}{\cancel{45}}}{\underset{1}{\cancel{7}}}\cdot\dfrac{\overset{3}{\cancel{6}}}{\underset{\underset{1}{\cancel{5}}}{\cancel{10}}}\cdot\dfrac{\overset{4}{\cancel{28}}}{\underset{7}{\cancel{63}}}=\dfrac{12}{7}$

49. $\left(\dfrac{2}{5}\right)^2\cdot\left(\dfrac{1}{10}\right)^2=\left(\dfrac{2}{5}\cdot\dfrac{2}{5}\right)\cdot\left(\dfrac{1}{10}\cdot\dfrac{1}{10}\right)$

$=\dfrac{\overset{1}{\cancel{4}}}{25}\cdot\dfrac{1}{\underset{25}{\cancel{100}}}$

$=\dfrac{1}{625}$

51. $\left(\dfrac{1}{10}\right)^3\left(\dfrac{1000}{17}\right)=\dfrac{1}{\underset{1}{\cancel{1000}}}\cdot\dfrac{\overset{1}{\cancel{1000}}}{17}=\dfrac{1}{17}$

53. $A=lw$

55. $A = lw = \dfrac{5}{\cancel{4}_{1}} \cdot \dfrac{\cancel{8}^{2}}{3} = \dfrac{10}{3}$ or $3\dfrac{1}{3}$ m²

57. $4 \cdot \dfrac{7}{8} = \dfrac{\cancel{4}^{1}}{1} \cdot \dfrac{7}{\cancel{8}_{2}} = \dfrac{7}{2}$ or $3\dfrac{1}{2}$

Maximus requires $\dfrac{7}{2}$ or $3\dfrac{1}{2}$ yd of lumber.

59. $\dfrac{1}{12} \cdot 3600 = \dfrac{1}{\cancel{12}_{1}} \cdot \dfrac{\cancel{3600}^{300}}{1} = 300$

There are 300 Asian American students.

61. $\dfrac{1}{2} \cdot \dfrac{5}{12} \cdot 3600 = \dfrac{1}{2} \cdot \dfrac{5}{\cancel{12}_{1}} \cdot \dfrac{\cancel{3600}^{300}}{1} = \dfrac{1500}{2} = 750$

There are 750 Caucasian male students.

Section 2.5

63. $\dfrac{1}{12} \cdot 12 = \dfrac{1}{\cancel{12}_{1}} \cdot \dfrac{\cancel{12}^{1}}{1} = 1$

65. $\dfrac{1}{7}$

67. 6

69. Multiplying

71. $\dfrac{7}{9} \div \dfrac{35}{63} = \dfrac{7}{9} \cdot \dfrac{63}{35} = \dfrac{7}{\cancel{9}} \cdot \dfrac{7 \cdot \cancel{9}}{\cancel{7} \cdot 5} = \dfrac{7}{5}$

73. $\dfrac{3}{10} \div \dfrac{9}{5} = \dfrac{\cancel{3}^{1}}{\cancel{10}_{2}} \cdot \dfrac{\cancel{5}^{1}}{\cancel{9}_{3}} = \dfrac{1}{6}$

75. $12 \div \dfrac{6}{7} = \dfrac{\cancel{12}^{2}}{1} \cdot \dfrac{7}{\cancel{6}_{1}} = 14$

77. $\left(\dfrac{12}{5}\right)^{2} \div \dfrac{36}{5} = \dfrac{144}{25} \div \dfrac{36}{5} = \dfrac{144}{25} \cdot \dfrac{5}{36}$

$\qquad = \dfrac{\cancel{36}^{1} \cdot 4}{\cancel{5}_{1} \cdot 5} \cdot \dfrac{\cancel{5}^{1}}{\cancel{36}_{1}} = \dfrac{4}{5}$

79. $\dfrac{4}{13} \cdot \left(\dfrac{1}{2}\right)^{3} \div 2 = \dfrac{\cancel{4}^{1}}{13} \cdot \dfrac{1}{\cancel{8}_{2}} \div 2 = \dfrac{1}{26} \div 2$

$\qquad = \dfrac{1}{26} \cdot \dfrac{1}{2} = \dfrac{1}{52}$

81. $18 \div \dfrac{2}{3} = \dfrac{\cancel{18}^{9}}{1} \cdot \dfrac{3}{\cancel{2}_{1}} = 27$

83. $\dfrac{4}{5} \cdot 40 = \dfrac{4}{\cancel{5}_{1}} \cdot \dfrac{\cancel{40}^{8}}{1} = 32$ hr

$32 \times \$18 = \$576$
Amelia earned \$576.

85. $9 \div \dfrac{3}{8} = \dfrac{\cancel{9}^{3}}{1} \cdot \dfrac{8}{\cancel{3}_{1}} = 24$

Yes, he will have 24 pieces, which is more than enough for his class.

Section 2.6

87. $\left(11\dfrac{1}{3}\right)\left(2\dfrac{3}{34}\right) = \dfrac{\cancel{34}^{1}}{3} \cdot \dfrac{71}{\cancel{34}_{1}} = \dfrac{71}{3} = 23\dfrac{2}{3}$

89. $4 \cdot \left(5\dfrac{5}{8}\right) = \dfrac{\cancel{4}^{1}}{1} \cdot \dfrac{45}{\cancel{8}_{2}} = \dfrac{45}{2} = 22\dfrac{1}{2}$

91. $4\dfrac{5}{16} \div 2\dfrac{7}{8} = \dfrac{69}{16} \div \dfrac{23}{8} = \dfrac{\cancel{69}^{3}}{\cancel{16}_{2}} \cdot \dfrac{\cancel{8}^{1}}{\cancel{23}_{1}} = \dfrac{3}{2} = 1\dfrac{1}{2}$

93. $7 \div 1\dfrac{5}{9} = \dfrac{7}{1} \div \dfrac{14}{9} = \dfrac{\cancel{7}^{1}}{1} \cdot \dfrac{9}{\cancel{14}_{2}} = \dfrac{9}{2} = 4\dfrac{1}{2}$

95. $10\dfrac{1}{5} \div 17 = \dfrac{51}{5} \div \dfrac{17}{1} = \dfrac{\overset{3}{\cancel{51}}}{5} \cdot \dfrac{1}{\underset{1}{\cancel{17}}} = \dfrac{3}{5}$

97. $2\dfrac{1}{2} \cdot 1\dfrac{1}{4} = \dfrac{5}{2} \cdot \dfrac{5}{4} = \dfrac{25}{8} = 3\dfrac{1}{8}$

It will take $3\dfrac{1}{8}$ gal.

Chapter 2   Test

1. **(a)** $\dfrac{5}{8}$

(b) Proper

3. $\dfrac{11}{2}$; $5\dfrac{1}{2}$

5. **(a)** $12\overline{)44}\;\;\dfrac{-36}{8}$ $\;\;3\dfrac{8}{12} = 3\dfrac{2}{3}$ (quotient 3)

(b) $3\dfrac{7}{9} = \dfrac{3 \times 9 + 7}{9} = \dfrac{34}{9}$

7.

9.

11. **(a)** $1, 3, 5, 9, 15, 45$

(b)
$$3\overline{)9}$$
$$5\overline{)45}$$
$3 \cdot 3 \cdot 5 = 3^2 \cdot 5 = 45$

13. **(a)** No; 1155 is not even.
(b) Yes; $1 + 1 + 5 + 5 = 12$ is divisible by 3.
(c) Yes; the digit in the ones-place is a 5.
(d) No; the digit in the ones-place is not 0.

15. $2 \times 25 \;\blacklozenge\; 5 \times 4$
$50 \neq 20$
$\dfrac{2}{5} \neq \dfrac{4}{25}$

17. $\dfrac{1,2\cancel{0}\cancel{0},\cancel{0}\cancel{0}\cancel{0}}{1,4\cancel{0}\cancel{0},\cancel{0}\cancel{0}\cancel{0}} = \dfrac{12}{14} = \dfrac{\cancel{2} \cdot 6}{\cancel{2} \cdot 7} = \dfrac{6}{7}$

19. $\dfrac{2}{9} \times \dfrac{57}{46} = \dfrac{\cancel{2}}{\cancel{3} \cdot 3} \cdot \dfrac{\cancel{3} \cdot 19}{\cancel{2} \cdot 23} = \dfrac{19}{69}$

21. $\dfrac{28}{24} \div \dfrac{21}{8} = \dfrac{28}{24} \cdot \dfrac{8}{21}$
$\qquad = \dfrac{\cancel{2} \cdot \cancel{2} \cdot 7}{\cancel{2} \cdot \cancel{2} \cdot \cancel{2} \cdot 3} \cdot \dfrac{\cancel{2} \cdot 2 \cdot 2}{3 \cdot \cancel{7}} = \dfrac{4}{9}$

23. $\dfrac{2}{18} \times \dfrac{9}{25} \times \dfrac{40}{6} = \dfrac{\cancel{2}}{\cancel{2} \cdot \cancel{3} \cdot 3} \cdot \dfrac{\cancel{3} \cdot \cancel{3}}{\cancel{5} \cdot 5} \cdot \dfrac{\cancel{2} \cdot 2 \cdot 2 \cdot \cancel{5}}{\cancel{2} \cdot 3}$
$\qquad = \dfrac{4}{15}$

25. $\dfrac{10}{21} \div 4\dfrac{1}{6} = \dfrac{10}{21} \div \dfrac{25}{6}$
$\qquad = \dfrac{10}{21} \cdot \dfrac{6}{25}$
$\qquad = \dfrac{2 \cdot \cancel{5}}{\cancel{3} \cdot 7} \cdot \dfrac{2 \cdot \cancel{3}}{\cancel{5} \cdot 5}$
$\qquad = \dfrac{4}{35}$

27. $\dfrac{52}{72} \div \left[\left(\dfrac{1}{2}\right)^2 \cdot \dfrac{8}{3}\right] = \dfrac{52}{72} \div \left[\dfrac{1}{\underset{1}{\cancel{4}}} \cdot \dfrac{\overset{2}{\cancel{8}}}{3}\right] = \dfrac{52}{72} \div \dfrac{2}{3}$
$\qquad = \dfrac{\overset{26}{\cancel{52}}}{\underset{24}{\cancel{72}}} \cdot \dfrac{\overset{1}{\cancel{3}}}{\cancel{2}} = \dfrac{26}{24} = \dfrac{13}{12}$

29. $20 \cdot \dfrac{1}{4} = \dfrac{\overset{5}{\cancel{20}}}{1} \cdot \dfrac{1}{\underset{1}{\cancel{4}}} = 5$

$20 \div \dfrac{1}{4} = \dfrac{20}{1} \div \dfrac{1}{4} = \dfrac{20}{1} \cdot \dfrac{4}{1} = 80$

$20 \div \dfrac{1}{4}$ is greater.

31. $\dfrac{1}{15} \cdot \dfrac{5}{8} \cdot 120 = \dfrac{1}{\overset{\scriptscriptstyle 3}{\cancel{15}}} \cdot \dfrac{\overset{\scriptscriptstyle 1}{\cancel{5}}}{\overset{\scriptscriptstyle 1}{\cancel{8}}} \cdot \dfrac{\overset{\scriptscriptstyle 15}{\cancel{120}}}{1} = \dfrac{15}{3} = 5$

5 dogs are female pure breeds.

Chapters 1–2   Cumulative Review Exercises

1. 17,000; nineteen thousand, three hundred forty; 22,047; fifteen thousand, seven hundred seventy-one

3.
$$\begin{array}{r} 572 \\ -\ 433 \\ \hline 139 \end{array}$$

5.
$$\begin{array}{r} 24 \\ 16\overline{)384} \\ \underline{-32} \\ 64 \\ \underline{-64} \\ 0 \end{array}$$

7.
$$\begin{array}{r} 18\ \text{R}\ 2 \\ 4\overline{)74} \\ \underline{-4} \\ 34 \\ \underline{-32} \\ 2 \end{array}$$

9.
$$\begin{array}{r} 1007 \\ -\ 823 \\ \hline 184 \end{array}$$

11. $6 + 2 \cdot 8 = 6 + 16 = 22$

13. $(5-3)^2 = 2^2 = 4$

15. c

17. e

19. (a) $\dfrac{4}{7}$

(b) $\dfrac{7}{3}$ or $2\dfrac{1}{3}$

21. (a) 1, 2, 3, 5, 6, 10, 15, 30

(b)
$$\begin{array}{r} 5 \\ 3\overline{)15} \\ 2\overline{)30} \end{array}$$

$2 \cdot 3 \cdot 5 = 30$

23. $\dfrac{35}{27} \cdot \dfrac{51}{95} = \dfrac{\cancel{5} \cdot 7}{\cancel{3} \cdot 3 \cdot 3} \cdot \dfrac{\cancel{3} \cdot 17}{\cancel{5} \cdot 19} = \dfrac{119}{171}$

25. Yes; $\dfrac{\overset{\scriptscriptstyle 1}{\cancel{8}}}{13} \cdot \dfrac{5}{\underset{\scriptscriptstyle 2}{\cancel{16}}} = \dfrac{5}{26}$ and $\dfrac{5}{\underset{\scriptscriptstyle 2}{\cancel{16}}} \cdot \dfrac{\overset{\scriptscriptstyle 1}{\cancel{8}}}{13} = \dfrac{5}{26}.$

27. $\left(\dfrac{\cancel{8}}{\cancel{6}} \cdot \dfrac{\cancel{12}}{\cancel{25}} \right)^2 \div \dfrac{2}{3} = \left(\dfrac{2}{5} \right)^2 \div \dfrac{2}{3} = \dfrac{4}{25} \div \dfrac{2}{3}$

$= \dfrac{\overset{\scriptscriptstyle 2}{\cancel{4}}}{25} \cdot \dfrac{3}{\cancel{2}} = \dfrac{6}{25}$

29. $A = \dfrac{1}{2}bh$

$= \dfrac{1}{2}\left(\dfrac{25}{2} \right)(8) = \dfrac{1}{\cancel{2}} \cdot \dfrac{25}{\cancel{2}} \cdot \dfrac{\cancel{2} \cdot \cancel{2} \cdot 2}{1} = 50 \text{ ft}^2$

Chapter 3   Fractions and Mixed Numbers: Addition and Subtraction

Chapter Opener Puzzle

1. t

2. o

3. p

4. h

5. e

6. a

7. v

8. y

An improper fraction is **top heavy**.

Section 3.1   Addition and Subtraction of Like Fractions

Section 3.1   Practice Exercises

1. Answers will vary.

3. 3 ft + 5 ft = 8 ft

5. 7 m + 13 m = 20 m

7. 1 fourth + 6 fourths = 7 fourths

9.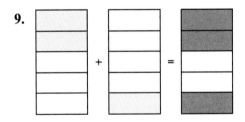

11. $\dfrac{6}{11} + \dfrac{7}{11} = \dfrac{6+7}{11} = \dfrac{13}{11}$

13. $\dfrac{6}{5} + \dfrac{3}{5} = \dfrac{6+3}{5} = \dfrac{9}{5}$

15. $\dfrac{1}{4} + \dfrac{3}{4} = \dfrac{1+3}{4} = \dfrac{4}{4} = 1$

17. $\dfrac{2}{9} + \dfrac{4}{9} = \dfrac{2+4}{9} = \dfrac{6}{9} = \dfrac{2}{3}$

19. $\dfrac{3}{20} + \dfrac{8}{20} + \dfrac{15}{20} = \dfrac{3+8+15}{20} = \dfrac{26}{20}$
$= \dfrac{\cancel{2} \cdot 13}{\cancel{2} \cdot 10} = \dfrac{13}{10}$

21. $\dfrac{18}{14} + \dfrac{11}{14} + \dfrac{6}{14} = \dfrac{18+11+6}{14} = \dfrac{35}{14} = \dfrac{5 \cdot \cancel{7}}{2 \cdot \cancel{7}} = \dfrac{5}{2}$

23. $\dfrac{1}{4} + \dfrac{9}{4} = \dfrac{1+9}{4} = \dfrac{10}{4} = \dfrac{5}{2}$ or $2\dfrac{1}{2}$

   Bethany has $\dfrac{5}{2}$ or $2\dfrac{1}{2}$ cups of bleach and water mixture.

25. 15 baskets – 4 baskets = 11 baskets

27. 7 fifths – 1 fifth = 6 fifths

29.

31. $\dfrac{9}{8} - \dfrac{6}{8} = \dfrac{9-6}{8} = \dfrac{3}{8}$

33. $\dfrac{9}{2} - \dfrac{6}{2} = \dfrac{9-6}{2} = \dfrac{3}{2}$

35. $\dfrac{13}{3} - \dfrac{7}{3} = \dfrac{13-7}{3} = \dfrac{6}{3} = 2$

37. $\dfrac{23}{12} - \dfrac{15}{12} = \dfrac{23-15}{12} = \dfrac{8}{12} = \dfrac{2 \cdot \cancel{4}}{3 \cdot \cancel{4}} = \dfrac{2}{3}$

39. $\dfrac{28}{25} - \dfrac{14}{25} - \dfrac{4}{25} = \dfrac{28-14-4}{25} = \dfrac{10}{25} = \dfrac{2 \cdot \cancel{5}}{5 \cdot \cancel{5}} = \dfrac{2}{5}$

41. $\dfrac{10}{16} - \dfrac{1}{16} - \dfrac{5}{16} = \dfrac{10-1-5}{16} = \dfrac{4}{16} = \dfrac{1}{4}$

43. $\dfrac{5}{8} - \dfrac{3}{8} = \dfrac{5-3}{8} = \dfrac{2}{8} = \dfrac{1}{4}$

$\dfrac{1}{4}$ g is left.

45. $\dfrac{7}{8} + \dfrac{5}{8} = \dfrac{7+5}{8} = \dfrac{12}{8} = \dfrac{\cancel{4} \cdot 3}{\cancel{4} \cdot 2} = \dfrac{3}{2}$

47. $\dfrac{14}{5} - \dfrac{2}{5} = \dfrac{14-2}{5} = \dfrac{12}{5}$

49. $\dfrac{6}{13} + \dfrac{7}{13} = \dfrac{6+7}{13} = \dfrac{13}{13} = 1$

51. $\dfrac{14}{15} + \dfrac{2}{15} - \dfrac{4}{15} = \dfrac{14+2-4}{15} = \dfrac{12}{15} = \dfrac{4 \cdot \cancel{3}}{5 \cdot \cancel{3}} = \dfrac{4}{5}$

53. $\dfrac{7}{2} - \dfrac{3}{2} + \dfrac{1}{2} = \dfrac{7-3+1}{2} = \dfrac{5}{2}$

55. $\dfrac{19}{12} - \dfrac{5}{12} + \dfrac{7}{12} = \dfrac{19-5+7}{12} = \dfrac{21}{12} = \dfrac{\cancel{3} \cdot 7}{\cancel{3} \cdot 4} = \dfrac{7}{4}$

57. $\left(\dfrac{11}{10} - \dfrac{2}{10}\right)^2 = \left(\dfrac{9}{10}\right)^2 = \dfrac{9}{10} \cdot \dfrac{9}{10} = \dfrac{81}{100}$

59. $\dfrac{5}{4} \div \dfrac{3}{2} + \dfrac{5}{6} = \dfrac{5}{\cancel{4}} \cdot \dfrac{\cancel{2}^{1}}{3} + \dfrac{5}{6} = \dfrac{5}{6} + \dfrac{5}{6} = \dfrac{5+5}{6}$

$= \dfrac{10}{6} = \dfrac{5}{3}$

61. $\dfrac{6}{5} + \dfrac{7}{5} - \dfrac{4}{5} = \dfrac{6+7-4}{5} = \dfrac{9}{5}$

63. $\dfrac{3}{7} + \dfrac{13}{14} \cdot 2 = \dfrac{3}{7} + \dfrac{13}{\cancel{14}_{7}} \cdot \dfrac{\cancel{2}^{1}}{1} = \dfrac{3}{7} + \dfrac{13}{7} = \dfrac{16}{7}$

65. $\left(\dfrac{2}{21} + \dfrac{11}{21}\right) \div \dfrac{1}{7} = \dfrac{13}{21} \div \dfrac{1}{7} = \dfrac{13}{\cancel{21}_{3}} \cdot \dfrac{\cancel{7}^{1}}{1} = \dfrac{13}{3}$

67. $\dfrac{17}{30} - \dfrac{1}{2} \cdot \dfrac{7}{15} = \dfrac{17}{30} - \dfrac{7}{30} = \dfrac{10}{30} = \dfrac{\cancel{10}}{3 \cdot \cancel{10}} = \dfrac{1}{3}$

69. Perimeter $= \dfrac{5}{7} + \dfrac{5}{7} + \dfrac{2}{7} = \dfrac{12}{7}$ or $1\dfrac{5}{7}$ m

71. Perimeter $= \dfrac{15}{16} + \dfrac{13}{16} + \dfrac{15}{16} + \dfrac{13}{16} = \dfrac{56}{16}$

$= \dfrac{7 \cdot \cancel{8}}{2 \cdot \cancel{8}} = \dfrac{7}{2}$ or $3\dfrac{1}{2}$ in.

73. $\left(\dfrac{1}{10} + \dfrac{7}{10}\right) - \dfrac{3}{10} = \dfrac{8}{10} - \dfrac{3}{10} = \dfrac{5}{10} = \dfrac{1}{2}$

There was $\dfrac{1}{2}$ gal left over.

75. $\dfrac{1}{4} \cdot \left(\dfrac{5}{8} + \dfrac{7}{8}\right) = \dfrac{1}{\cancel{4}_{1}} \cdot \dfrac{\cancel{12}^{3}}{8} = \dfrac{3}{8}$

He used $\dfrac{3}{8}$ L.

77. (a) $\dfrac{4}{10} + \dfrac{7}{10} + \dfrac{9}{10} + \dfrac{5}{10} + \dfrac{13}{10} + \dfrac{17}{10}$

$= \dfrac{4+7+9+5+13+17}{10} = \dfrac{55}{10}$

$= \dfrac{\cancel{5} \cdot 11}{\cancel{5} \cdot 2} = \dfrac{11}{2} = 5\dfrac{1}{2}$

Thilan walked $5\dfrac{1}{2}$ mi total.

(b) $\dfrac{11}{2} \div 6 = \dfrac{11}{2} \div \dfrac{6}{1} = \dfrac{11}{2} \cdot \dfrac{1}{6} = \dfrac{11}{12}$

He walked an average of $\dfrac{11}{12}$ mi per day.

79. Perimeter $= \dfrac{3}{8} + \dfrac{5}{8} + \dfrac{3}{8} + \dfrac{5}{8} = \dfrac{16}{8} = 2$ ft

Area $= \dfrac{3}{8} \cdot \dfrac{5}{8} = \dfrac{15}{64}$ ft^2

81. Perimeter $= \dfrac{13}{3} + \dfrac{22}{3} + \dfrac{13}{3} + \dfrac{22}{3}$

$= \dfrac{70}{3}$ or $23\dfrac{1}{3}$ yd

Area $= \dfrac{22}{3} \cdot \dfrac{13}{3} = \dfrac{286}{9}$ or $31\dfrac{7}{9}$ yd^2

83. $\dfrac{3}{5} + \dfrac{2}{5}; \dfrac{5}{5} = 1$

85. $\dfrac{11}{15} - \dfrac{8}{15}; \dfrac{3}{15} = \dfrac{1}{5}$

Section 3.2   Least Common Multiple

Section 3.2   Practice Exercises

1. Answers will vary.

3. $\dfrac{19}{6} - \dfrac{16}{6} = \dfrac{3}{6} = \dfrac{1}{2}$

5. $\dfrac{31}{15} + \dfrac{2}{15} - \dfrac{8}{15} = \dfrac{31 + 2 - 8}{15} = \dfrac{\overset{5}{\cancel{25}}}{\underset{3}{\cancel{15}}} = \dfrac{5}{3}$

7. $\dfrac{11}{3} + \dfrac{7}{3} = \dfrac{18}{3} = 6$

9. (a) 48, 72, 240

   (b) 4, 8, 12

11. (a) 72, 360, 108

   (b) 6, 12, 9

13. 10: 10, 20, 30, 40, 50
25: 25, 50, 75
LCM: 50

15. 16: 16, 32, 48, 64
12: 12, 24, 36, 48
LCM: 48

17. 8: 8, 16, 24, 32, 40, 48, 56, 64, 72, 80, 88, 96, 104, 112, 120
10: 10, 20, 30, 40, 50, 60, 70, 80, 90, 100, 110, 120
12: 12, 24, 36, 48, 60, 72, 84, 96, 108, 120
LCM: 120

19. $18 = 2 \cdot 3 \cdot 3$
$24 = 2 \cdot 2 \cdot 2 \cdot 3$
LCM: $2 \cdot 2 \cdot 2 \cdot 3 \cdot 3 = 72$

21. $12 = 2 \cdot 2 \cdot 3$
$15 = 3 \cdot 5$
LCM: $2 \cdot 2 \cdot 3 \cdot 5 = 60$

23. $15 = 3 \cdot 5$
$25 = 5 \cdot 5$
LCM: $3 \cdot 5 \cdot 5 = 75$

25. $24 = 2 \cdot 2 \cdot 2 \cdot 3$
$30 = 2 \cdot 3 \cdot 5$
LCM: $2 \cdot 2 \cdot 2 \cdot 3 \cdot 5 = 120$

27. $42 = 2 \cdot 3 \cdot 7$
$70 = 2 \cdot 5 \cdot 7$
LCM: $2 \cdot 3 \cdot 5 \cdot 7 = 210$

29. $20 = 2 \cdot 2 \cdot 5$
$18 = 2 \cdot 3 \cdot 3$
$27 = 3 \cdot 3 \cdot 3$
LCM: $2 \cdot 2 \cdot 3 \cdot 3 \cdot 3 \cdot 5 = 540$

31. $12 = 2 \cdot 2 \cdot 3$
$15 = 3 \cdot 5$
$20 = 2 \cdot 2 \cdot 5$
LCM: $2 \cdot 2 \cdot 3 \cdot 5 = 60$

33. $16 = 2 \cdot 2 \cdot 2 \cdot 2$
$24 = 2 \cdot 2 \cdot 2 \cdot 3$
$30 = 2 \cdot 3 \cdot 5$
LCM: $2 \cdot 2 \cdot 2 \cdot 2 \cdot 3 \cdot 5 = 240$

35. $6 = 2 \cdot 3$
$12 = 2 \cdot 2 \cdot 3$
$18 = 2 \cdot 3 \cdot 3$
$20 = 2 \cdot 2 \cdot 5$
LCM: $2 \cdot 2 \cdot 3 \cdot 3 \cdot 5 = 180$

37. $5 = 5$
$15 = 3 \cdot 5$
$18 = 2 \cdot 3 \cdot 3$
$20 = 2 \cdot 2 \cdot 5$
LCM: $2 \cdot 2 \cdot 3 \cdot 3 \cdot 5 = 180$

39. Find the LCM of 10, 12, and 15.
$10 = 2 \cdot 5$
$12 = 2 \cdot 2 \cdot 3$
$15 = 3 \cdot 5$
LCM: $2 \cdot 2 \cdot 3 \cdot 5 = 60$
The shortest length of floor space is 60 in.

41. Find the LCM of 6, 8, 10, and 15.
$6 = 2 \cdot 3$
$8 = 2 \cdot 2 \cdot 2$
$10 = 2 \cdot 5$
$15 = 3 \cdot 5$
LCM: $2 \cdot 2 \cdot 2 \cdot 3 \cdot 5 = 120$
It will take 120 hr (5 days) for the satellites to be lined up again.

43. $\dfrac{2}{3} = \dfrac{2 \cdot 7}{3 \cdot 7} = \dfrac{14}{21}$

45. $\dfrac{5}{8} = \dfrac{5 \cdot 2}{8 \cdot 2} = \dfrac{10}{16}$

47. $\dfrac{3}{4} = \dfrac{3 \cdot 4}{4 \cdot 4} = \dfrac{12}{16}$

49. $\dfrac{4}{5} = \dfrac{4 \cdot 3}{5 \cdot 3} = \dfrac{12}{15}$

51. $\dfrac{7}{6} = \dfrac{7 \cdot 7}{6 \cdot 7} = \dfrac{49}{42}$

53. $\dfrac{11}{9} = \dfrac{11 \cdot 11}{9 \cdot 11} = \dfrac{121}{99}$

55. $\dfrac{5}{13} = \dfrac{5 \cdot 3}{13 \cdot 3} = \dfrac{15}{39}$

57. $\dfrac{11}{4} = \dfrac{11 \cdot 1000}{4 \cdot 1000} = \dfrac{11,000}{4000}$

59. $\dfrac{3}{14} = \dfrac{3 \cdot 5}{14 \cdot 5} = \dfrac{15}{70}$

61. $\dfrac{3}{4} = \dfrac{3 \cdot 2}{4 \cdot 2} = \dfrac{6}{8}$
$\dfrac{7}{8} > \dfrac{6}{8}$ so $\dfrac{7}{8} > \dfrac{3}{4}$

63. $\dfrac{13}{10} = \dfrac{13 \cdot 3}{10 \cdot 3} = \dfrac{39}{30}$
$\dfrac{22}{15} = \dfrac{22 \cdot 2}{15 \cdot 2} = \dfrac{44}{30}$
$\dfrac{39}{30} < \dfrac{44}{30}$ so $\dfrac{13}{10} < \dfrac{22}{15}$

65. $\dfrac{3}{12} = \dfrac{3 \cdot 2}{12 \cdot 2} = \dfrac{6}{24}$
$\dfrac{2}{8} = \dfrac{2 \cdot 3}{8 \cdot 3} = \dfrac{6}{24}$
$\dfrac{6}{24} = \dfrac{6}{24}$ so $\dfrac{3}{12} = \dfrac{2}{8}$

67. $\dfrac{5}{18} = \dfrac{5 \cdot 3}{18 \cdot 3} = \dfrac{15}{54}$
$\dfrac{8}{27} = \dfrac{8 \cdot 2}{27 \cdot 2} = \dfrac{16}{54}$
$\dfrac{15}{54} < \dfrac{16}{54}$ so $\dfrac{5}{18} < \dfrac{8}{27}$

69. $\dfrac{2}{3} = \dfrac{2 \cdot 8}{3 \cdot 8} = \dfrac{16}{24}$

$\dfrac{7}{8} = \dfrac{7 \cdot 3}{8 \cdot 3} = \dfrac{21}{24}$

$\dfrac{5}{6} = \dfrac{5 \cdot 4}{6 \cdot 4} = \dfrac{20}{24}$

$\dfrac{1}{2} = \dfrac{1 \cdot 12}{2 \cdot 12} = \dfrac{12}{24}$

$\dfrac{21}{24} = \dfrac{7}{8}$ has the greatest value.

71. $\dfrac{7}{8} = \dfrac{7 \cdot 3}{8 \cdot 3} = \dfrac{21}{24}$

$\dfrac{2}{3} = \dfrac{2 \cdot 8}{3 \cdot 8} = \dfrac{16}{24}$

$\dfrac{3}{4} = \dfrac{3 \cdot 6}{4 \cdot 6} = \dfrac{18}{24}$

$16 < 18 < 21$ so $\dfrac{2}{3} < \dfrac{3}{4} < \dfrac{7}{8}$

73. $\dfrac{5}{16} = \dfrac{5}{16}$

$\dfrac{3}{8} = \dfrac{3 \cdot 2}{8 \cdot 2} = \dfrac{6}{16}$

$\dfrac{1}{4} = \dfrac{1 \cdot 4}{4 \cdot 4} = \dfrac{4}{16}$

$4 < 5 < 6$ so $\dfrac{1}{4} < \dfrac{5}{16} < \dfrac{3}{8}$

75. $\dfrac{4}{3} = \dfrac{4 \cdot 20}{3 \cdot 20} = \dfrac{80}{60}$

$\dfrac{13}{12} = \dfrac{13 \cdot 5}{12 \cdot 5} = \dfrac{65}{60}$

$\dfrac{17}{15} = \dfrac{17 \cdot 4}{15 \cdot 4} = \dfrac{68}{60}$

$65 < 68 < 80$ so $\dfrac{13}{12} < \dfrac{17}{15} < \dfrac{4}{3}$

77. $\dfrac{3}{4} = \dfrac{3 \cdot 4}{4 \cdot 4} = \dfrac{12}{16}$

$\dfrac{11}{16} = \dfrac{11}{16}$

$\dfrac{7}{8} = \dfrac{7 \cdot 2}{8 \cdot 2} = \dfrac{14}{16}$

The longest cut is $\dfrac{14}{16} = \dfrac{7}{8}$ in - above the left eye. The shortest cut is $\dfrac{11}{16}$ in - right hand.

79. $\dfrac{2}{3} = \dfrac{2 \cdot 40}{3 \cdot 40} = \dfrac{80}{120}$

$\dfrac{3}{5} = \dfrac{3 \cdot 24}{5 \cdot 24} = \dfrac{72}{120}$

$\dfrac{5}{8} = \dfrac{5 \cdot 15}{8 \cdot 15} = \dfrac{75}{120}$

The greatest amount is $\dfrac{80}{120} = \dfrac{2}{3}$ lb of turkey. The least amount is $\dfrac{72}{120} = \dfrac{3}{5}$ lb of ham.

81. $\dfrac{1}{4} = \dfrac{1 \cdot 6}{4 \cdot 6} = \dfrac{6}{24}$

$\dfrac{5}{6} = \dfrac{5 \cdot 4}{6 \cdot 4} = \dfrac{20}{24}$

$\dfrac{5}{12} = \dfrac{5 \cdot 2}{12 \cdot 2} = \dfrac{10}{24}$

$\dfrac{2}{3} = \dfrac{2 \cdot 8}{3 \cdot 8} = \dfrac{16}{24}$

$\dfrac{1}{8} = \dfrac{1 \cdot 3}{8 \cdot 3} = \dfrac{3}{24}$

$\dfrac{3}{24} < \dfrac{6}{24} < \dfrac{10}{24} < \dfrac{16}{24} < \dfrac{20}{24}$

$\dfrac{10}{24} = \dfrac{5}{12}$ and $\dfrac{16}{24} = \dfrac{2}{3}$ (a and b) are between $\dfrac{1}{4}$ and $\dfrac{5}{6}$.

Section 3.3   Addition and Subtraction of Unlike Fractions

Section 3.3 Practice Exercises

1. Answers will vary.

3. $\dfrac{6}{7} = \dfrac{6 \cdot 2}{7 \cdot 2} = \dfrac{12}{14}$

5. $\dfrac{2}{3} = \dfrac{2 \cdot 7}{3 \cdot 7} = \dfrac{14}{21}$

7. $\dfrac{5}{1} = \dfrac{5 \cdot 5}{1 \cdot 5} = \dfrac{25}{5}$

9. $\dfrac{2}{1} = \dfrac{2 \cdot 4}{1 \cdot 4} = \dfrac{8}{4}$

11. $\dfrac{4}{5} = \dfrac{4 \cdot 20}{5 \cdot 20} = \dfrac{80}{100}$

13. $\dfrac{1}{8} = \dfrac{1 \cdot 5}{8 \cdot 5} = \dfrac{5}{40}$

15. $\dfrac{7}{8} + \dfrac{5}{16} = \dfrac{7 \cdot 2}{8 \cdot 2} + \dfrac{5}{16} = \dfrac{14}{16} + \dfrac{5}{16} = \dfrac{14+5}{16} = \dfrac{19}{16}$

17. $\dfrac{1}{15} + \dfrac{1}{10} = \dfrac{1 \cdot 2}{15 \cdot 2} + \dfrac{1 \cdot 3}{10 \cdot 3} = \dfrac{2}{30} + \dfrac{3}{30}$

$\qquad = \dfrac{2+3}{30} = \dfrac{5}{30} = \dfrac{1}{6}$

19. $\dfrac{1}{10} + \dfrac{3}{20} = \dfrac{1 \cdot 2}{10 \cdot 2} + \dfrac{3}{20} = \dfrac{2}{20} + \dfrac{3}{20} = \dfrac{2+3}{20}$

$\qquad = \dfrac{5}{20} = \dfrac{\cancel{5}}{4 \cdot \cancel{5}} = \dfrac{1}{4}$

21. $\dfrac{5}{6} + \dfrac{8}{7} = \dfrac{5 \cdot 7}{6 \cdot 7} + \dfrac{8 \cdot 6}{7 \cdot 6} = \dfrac{35}{42} + \dfrac{48}{42}$

$\qquad = \dfrac{35+48}{42} = \dfrac{83}{42}$

23. $\dfrac{7}{8} - \dfrac{1}{2} = \dfrac{7}{8} - \dfrac{1 \cdot 4}{2 \cdot 4} = \dfrac{7}{8} - \dfrac{4}{8} = \dfrac{7-4}{8} = \dfrac{3}{8}$

25. $\dfrac{13}{12} - \dfrac{3}{4} = \dfrac{13}{12} - \dfrac{3 \cdot 3}{4 \cdot 3} = \dfrac{13}{12} - \dfrac{9}{12} = \dfrac{13-9}{12}$

$\qquad = \dfrac{\overset{1}{\cancel{4}}}{\underset{3}{\cancel{12}}} = \dfrac{1}{3}$

27. $\dfrac{10}{9} - \dfrac{5}{12} = \dfrac{10 \cdot 4}{9 \cdot 4} - \dfrac{5 \cdot 3}{12 \cdot 3} = \dfrac{40}{36} - \dfrac{15}{36}$

$\qquad = \dfrac{40-15}{36} = \dfrac{25}{36}$

29. $\dfrac{5}{8} - \dfrac{0}{11} = \dfrac{5}{8} - 0 = \dfrac{5}{8}$

31. $2 + \dfrac{9}{8} = \dfrac{2}{1} \cdot \dfrac{8}{8} + \dfrac{9}{8} = \dfrac{16}{8} + \dfrac{9}{8} = \dfrac{16+9}{8} = \dfrac{25}{8}$

33. $4 - \dfrac{4}{3} = \dfrac{4}{1} \cdot \dfrac{3}{3} - \dfrac{4}{3} = \dfrac{12}{3} - \dfrac{4}{3} = \dfrac{12-4}{3} = \dfrac{8}{3}$

35. $\dfrac{14}{3} + 1 = \dfrac{14}{3} + \dfrac{3}{3} = \dfrac{14+3}{3} = \dfrac{17}{3}$

37. $\dfrac{16}{7} - 2 = \dfrac{16}{7} - \dfrac{2}{1} \cdot \dfrac{7}{7} = \dfrac{16}{7} - \dfrac{14}{7} = \dfrac{16-14}{7} = \dfrac{2}{7}$

39. $\dfrac{7}{10} + \dfrac{19}{100} = \dfrac{7 \cdot 10}{10 \cdot 10} + \dfrac{19}{100} = \dfrac{70}{100} + \dfrac{19}{100}$

$\qquad = \dfrac{70+19}{100} = \dfrac{89}{100}$

41. $\dfrac{1}{10} - \dfrac{9}{100} = \dfrac{1 \cdot 10}{10 \cdot 10} - \dfrac{9}{100} = \dfrac{10}{100} - \dfrac{9}{100}$

$\qquad = \dfrac{10-9}{100} = \dfrac{1}{100}$

43. $\dfrac{3}{10} + \dfrac{9}{100} + \dfrac{1}{1000}$

$$= \dfrac{3 \cdot 100}{10 \cdot 100} + \dfrac{9 \cdot 10}{100 \cdot 10} + \dfrac{1}{1000}$$

$$= \dfrac{300}{1000} + \dfrac{90}{1000} + \dfrac{1}{1000}$$

$$= \dfrac{300 + 90 + 1}{1000} = \dfrac{391}{1000}$$

45. $\dfrac{5}{3} - \dfrac{7}{6} + \dfrac{5}{8} = \dfrac{5 \cdot 8}{3 \cdot 8} - \dfrac{7 \cdot 4}{6 \cdot 4} + \dfrac{5 \cdot 3}{8 \cdot 3}$

$$= \dfrac{40}{24} - \dfrac{28}{24} + \dfrac{15}{24} = \dfrac{40 - 28 + 15}{24}$$

$$= \dfrac{\overset{9}{\cancel{27}}}{\underset{8}{\cancel{24}}} = \dfrac{9}{8}$$

47. $\dfrac{1}{20} + \dfrac{5}{8} - \dfrac{7}{24} = \dfrac{1 \cdot 6}{20 \cdot 6} + \dfrac{5 \cdot 15}{8 \cdot 15} - \dfrac{7 \cdot 5}{24 \cdot 5}$

$$= \dfrac{6}{120} + \dfrac{75}{120} - \dfrac{35}{120}$$

$$= \dfrac{6 + 75 - 35}{120} = \dfrac{46}{120} = \dfrac{23}{60}$$

49. $\dfrac{1}{2} + \dfrac{1}{4} - \dfrac{1}{8} - \dfrac{1}{16} = \dfrac{1 \cdot 8}{2 \cdot 8} + \dfrac{1 \cdot 4}{4 \cdot 4} - \dfrac{1 \cdot 2}{8 \cdot 2} - \dfrac{1}{16}$

$$= \dfrac{8}{16} + \dfrac{4}{16} - \dfrac{2}{16} - \dfrac{1}{16}$$

$$= \dfrac{8 + 4 - 2 - 1}{16} = \dfrac{9}{16}$$

51. $\left(\dfrac{1}{2} - \dfrac{1}{3}\right)^2 = \left(\dfrac{1 \cdot 3}{2 \cdot 3} - \dfrac{1 \cdot 2}{3 \cdot 2}\right)^2 = \left(\dfrac{3}{6} - \dfrac{2}{6}\right)^2$

$$= \left(\dfrac{3 - 2}{6}\right)^2 = \left(\dfrac{1}{6}\right)^2 = \dfrac{1}{6} \cdot \dfrac{1}{6} = \dfrac{1}{36}$$

53. $\dfrac{2}{3} \div \dfrac{1}{2} - \dfrac{3}{4} = \dfrac{2}{3} \cdot \dfrac{2}{1} - \dfrac{3}{4} = \dfrac{4}{3} - \dfrac{3}{4} = \dfrac{4 \cdot 4}{3 \cdot 4} - \dfrac{3 \cdot 3}{4 \cdot 3}$

$$= \dfrac{16}{12} - \dfrac{9}{12} = \dfrac{16 - 9}{12} = \dfrac{7}{12}$$

55. $\dfrac{5}{6} + \dfrac{3}{8} \div \dfrac{1}{4} = \dfrac{5}{6} + \dfrac{3}{8} \cdot \dfrac{4}{1} = \dfrac{5}{6} + \dfrac{3}{2 \cdot \cancel{4}} \cdot \dfrac{\cancel{4}}{1}$

$$= \dfrac{5}{6} + \dfrac{3}{2} = \dfrac{5}{6} + \dfrac{3 \cdot 3}{2 \cdot 3} = \dfrac{5}{6} + \dfrac{9}{6}$$

$$= \dfrac{5 + 9}{6} = \dfrac{14}{6} = \dfrac{7}{3}$$

57. $\left(\dfrac{7}{10} - \dfrac{1}{5}\right) \cdot \dfrac{8}{3} = \left(\dfrac{7}{10} - \dfrac{1 \cdot 2}{5 \cdot 2}\right) \cdot \dfrac{8}{3}$

$$= \left(\dfrac{7}{10} - \dfrac{2}{10}\right) \cdot \dfrac{8}{3} = \left(\dfrac{7 - 2}{10}\right) \cdot \dfrac{8}{3}$$

$$= \dfrac{\overset{}{5}}{\underset{5}{\cancel{10}}} \cdot \dfrac{\overset{4}{\cancel{8}}}{3} = \dfrac{\overset{4}{\cancel{20}}}{\underset{3}{\cancel{15}}} = \dfrac{4}{3}$$

59. $\dfrac{4}{5} + \dfrac{\cancel{8}}{\cancel{8}} \cdot \dfrac{\overset{1}{\cancel{16}}}{\underset{7}{\cancel{35}}} = \dfrac{4}{5} + \dfrac{\overset{2}{2}}{7} = \dfrac{4 \cdot 7}{5 \cdot 7} + \dfrac{2 \cdot 5}{7 \cdot 5}$

$$= \dfrac{28}{35} + \dfrac{10}{35} = \dfrac{28 + 10}{35} = \dfrac{38}{35}$$

61. $\left(\dfrac{2}{5}\right)^3 + \dfrac{1}{25} = \dfrac{2}{5} \cdot \dfrac{2}{5} \cdot \dfrac{2}{5} + \dfrac{1}{25} \cdot \dfrac{5}{5} = \dfrac{8}{125} + \dfrac{5}{125}$

$$= \dfrac{8 + 5}{125} = \dfrac{13}{125}$$

63. $\left(\dfrac{1}{4}\right)^2 \div \left(\dfrac{5}{6} - \dfrac{2}{3}\right) + \dfrac{7}{12}$

$$= \left(\dfrac{1}{4} \cdot \dfrac{1}{4}\right) \div \left(\dfrac{5}{6} - \dfrac{2 \cdot 2}{3 \cdot 2}\right) + \dfrac{7}{12}$$

$$= \dfrac{1}{16} \div \left(\dfrac{5 - 4}{6}\right) + \dfrac{7}{12} = \dfrac{1}{16} \div \dfrac{1}{6} + \dfrac{7}{12}$$

$$= \dfrac{1}{\underset{8}{\cancel{16}}} \cdot \dfrac{\overset{3}{\cancel{6}}}{1} + \dfrac{7}{12} = \dfrac{3}{8} + \dfrac{7}{12}$$

$$= \dfrac{3 \cdot 3}{8 \cdot 3} + \dfrac{7 \cdot 2}{12 \cdot 2} = \dfrac{9}{24} + \dfrac{14}{24}$$

$$= \dfrac{9 + 14}{24} = \dfrac{23}{24}$$

65. $\dfrac{3}{4} + \dfrac{3}{8} = \dfrac{3 \cdot 2}{4 \cdot 2} + \dfrac{3}{8} = \dfrac{6}{8} + \dfrac{3}{8} = \dfrac{6+3}{8} = \dfrac{9}{8}$

Inez added $\dfrac{9}{8}$ or $1\dfrac{1}{8}$ cup.

67. $\dfrac{9}{32} - \dfrac{1}{8} = \dfrac{9}{32} - \dfrac{1 \cdot 4}{8 \cdot 4} = \dfrac{9}{32} - \dfrac{4}{32} = \dfrac{9-4}{32} = \dfrac{5}{32}$

The storm delivered $\dfrac{5}{32}$ in. of rain.

69.

$5 - 4 \cdot \dfrac{3}{8} + \dfrac{3}{2} = \dfrac{5}{1} - \dfrac{4}{1} \cdot \dfrac{3}{4 \cdot 2} + \dfrac{3}{2} = \dfrac{5}{1} - \dfrac{3}{2} + \dfrac{3}{2}$

$= \dfrac{5 \cdot 2}{1 \cdot 2} - \dfrac{3}{2} + \dfrac{3}{2} = \dfrac{10}{2} - \dfrac{3}{2} + \dfrac{3}{2}$

$= \dfrac{10 - 3 + 3}{2} = \dfrac{10}{2} = 5$

The trough now holds the original amount of 5 gal.

71. (a) $\dfrac{7}{36} + \dfrac{1}{6} = \dfrac{7}{36} + \dfrac{1 \cdot 6}{6 \cdot 6} = \dfrac{7}{36} + \dfrac{6}{36} = \dfrac{13}{36}$

(b) $\dfrac{13}{36} + \dfrac{5}{18} = \dfrac{13}{36} + \dfrac{5 \cdot 2}{18 \cdot 2} = \dfrac{13}{36} + \dfrac{10}{36} = \dfrac{23}{36}$

73. $\dfrac{2}{5} + \dfrac{9}{10} + \dfrac{2}{5} + \dfrac{9}{10} = \dfrac{2 \cdot 2}{5 \cdot 2} + \dfrac{9}{10} + \dfrac{2 \cdot 2}{5 \cdot 2} + \dfrac{9}{10}$

$= \dfrac{4}{10} + \dfrac{9}{10} + \dfrac{4}{10} + \dfrac{9}{10}$

$= \dfrac{26}{10} = \dfrac{13}{5}$ or $2\dfrac{3}{5}$ ft

75. First find the missing dimensions.

$\dfrac{5}{8} - \dfrac{1}{4} = \dfrac{5}{8} - \dfrac{1 \cdot 2}{4 \cdot 2} = \dfrac{5}{8} - \dfrac{2}{8} = \dfrac{3}{8}$ ft

$\dfrac{7}{8} - \dfrac{1}{2} = \dfrac{7}{8} - \dfrac{1 \cdot 4}{2 \cdot 4} = \dfrac{7}{8} - \dfrac{4}{8} = \dfrac{3}{8}$ ft

Now find the perimeter.

$\dfrac{1}{4} + \dfrac{1}{2} + \dfrac{3}{8} + \dfrac{3}{8} + \dfrac{5}{8} + \dfrac{7}{8}$

$= \dfrac{1 \cdot 2}{4 \cdot 2} + \dfrac{1 \cdot 4}{2 \cdot 4} + \dfrac{3}{8} + \dfrac{3}{8} + \dfrac{5}{8} + \dfrac{7}{8}$

$= \dfrac{2}{8} + \dfrac{4}{8} + \dfrac{3}{8} + \dfrac{3}{8} + \dfrac{5}{8} + \dfrac{7}{8}$

$= \dfrac{24}{8} = 3$ ft

77. The LCD is 60.

$\dfrac{3}{4} = \dfrac{3 \cdot 15}{4 \cdot 15} = \dfrac{45}{60}$

$\dfrac{7}{10} = \dfrac{7 \cdot 6}{10 \cdot 6} = \dfrac{42}{60}$

$\dfrac{5}{6} = \dfrac{5 \cdot 10}{6 \cdot 10} = \dfrac{50}{60}$

$\dfrac{1}{2} = \dfrac{1 \cdot 30}{2 \cdot 30} = \dfrac{30}{60}$

$\dfrac{42}{60}$ is closest to $\dfrac{30}{60}$, so choice **b** is closest to $\dfrac{1}{2}$.

Section 3.4   Addition and Subtraction of Mixed Numbers

Section 3.4 Practice Exercises

1. Answers will vary.

3. $\dfrac{25}{8} - \dfrac{23}{24} = \dfrac{25 \cdot 3}{8 \cdot 3} - \dfrac{23}{24} = \dfrac{75}{24} - \dfrac{23}{24} = \dfrac{\cancel{52}^{\,13}}{\cancel{24}_{\,6}}$

$= \dfrac{13}{6}$

5. $\dfrac{9}{5} + 3 = \dfrac{9}{5} + \dfrac{3}{1} \cdot \dfrac{5}{5} = \dfrac{9}{5} + \dfrac{15}{5} = \dfrac{24}{5}$

7. $\dfrac{125}{32} - \dfrac{51}{32} - \dfrac{58}{32} = \dfrac{16}{32} = \dfrac{1}{2}$

9.

$$2\frac{1}{11}$$
$$+5\frac{3}{11}$$
$$\overline{7\frac{4}{11}}$$

11.

$$12\frac{1}{14}$$
$$+3\frac{5}{14}$$
$$\overline{15\frac{6}{14}=15\frac{3}{7}}$$

13.

$$4\frac{5}{16} \quad = \quad 4\frac{5}{16}$$
$$+11\frac{1}{4} \quad = \quad +11\frac{4}{16}$$
$$\overline{\qquad\qquad 15\frac{9}{16}}$$

15.

$$6\frac{2}{3} \;=\; 6\frac{2\cdot 5}{3\cdot 5} \;=\; 6\frac{10}{15}$$
$$+4\frac{1}{5} \;=\; +4\frac{1\cdot 3}{5\cdot 3} \;=\; +4\frac{3}{15}$$
$$\overline{\qquad\qquad\qquad\qquad 10\frac{13}{15}}$$

17. 5

19. 2

21. $2\frac{6}{5}=2+1\frac{1}{5}=3\frac{1}{5}$

23. $7\frac{5}{3}=7+1\frac{2}{3}=8\frac{2}{3}$

25. Estimate   Exact

$$7 \qquad\qquad 6\frac{3}{4}$$
$$+8 \qquad\qquad +7\frac{3}{4}$$
$$\overline{15} \qquad\qquad \overline{13\frac{6}{4}=13+1\frac{2}{4}=14\frac{1}{2}}$$

27. Estimate   Exact

$$15 \qquad\qquad 14\frac{7}{8} \;=\; 14\frac{7}{8}$$
$$+8 \qquad\qquad +8\frac{1}{4} \;=\; +8\frac{2}{8}$$
$$\overline{23} \qquad\qquad \overline{\qquad 22\frac{9}{8}=23\frac{1}{8}}$$

29. Estimate   Exact

$$4 \qquad\qquad 3\frac{7}{16} \;=\; 3\frac{21}{48}$$
$$+16 \qquad\qquad +15\frac{11}{12} \;=\; +15\frac{44}{48}$$
$$\overline{20} \qquad\qquad \overline{\qquad 18\frac{65}{48}=19\frac{17}{48}}$$

31. $3+6\frac{7}{8}=9\frac{7}{8}$

33. $32\frac{2}{7}+10=42\frac{2}{7}$

35.

$$21\frac{9}{10}$$
$$-10\frac{3}{10}$$
$$\overline{11\frac{6}{10}=11\frac{3}{5}}$$

37.

$$5\frac{9}{15}$$
$$-3\frac{7}{15}$$
$$\overline{2\frac{2}{15}}$$

39.

$$18\frac{5}{6} \quad = \quad 18\frac{5}{6}$$
$$-6\frac{2}{3} \quad = \quad -6\frac{4}{6}$$
$$\overline{\qquad\qquad 12\frac{1}{6}}$$

41. $\quad 11\dfrac{5}{7} = 11\dfrac{10}{14}$

$\quad\quad -9\dfrac{5}{14} = -9\dfrac{5}{14}$

$\quad\quad\quad\quad\quad\quad\quad\ 2\dfrac{5}{14}$

43. $\dfrac{3}{3}$

45. $\dfrac{12}{12}$

47. Estimate: $\quad\quad\begin{array}{r} 25 \\ -14 \\ \hline 11 \end{array}$

Exact:

$25\dfrac{1}{4} \quad = \quad 24\dfrac{1}{4} + \dfrac{4}{4} \quad = \quad 24\dfrac{5}{4}$

$-13\dfrac{3}{4} \ = \ -13\dfrac{3}{4} \quad\quad = \quad -13\dfrac{3}{4}$

$\quad\quad\quad\quad\quad\quad\quad\quad\quad\quad\quad\quad\quad 11\dfrac{2}{4} = 11\dfrac{1}{2}$

49. Estimate: $\quad\quad\begin{array}{r} 17 \\ -15 \\ \hline 2 \end{array}$

Exact:

$17\dfrac{1}{6} = \ 17\dfrac{2}{12} = \ 16\dfrac{2}{12} + \dfrac{12}{12} = \ 16\dfrac{14}{12}$

$-15\dfrac{5}{12} = -15\dfrac{5}{12} = -15\dfrac{5}{12} \quad\quad = -15\dfrac{5}{12}$

$\quad\quad\quad\quad\quad\quad\quad\quad\quad\quad\quad\quad\quad\quad\quad 1\dfrac{9}{12} = 1\dfrac{3}{4}$

51. Estimate: $\begin{array}{r} 46 \\ -39 \\ \hline 7 \end{array}$

Exact:

$46\dfrac{3}{7} = \ 46\dfrac{6}{14} = \ 45\dfrac{6}{14} + \dfrac{14}{14} = \ 45\dfrac{20}{14}$

$-38\dfrac{1}{2} = -38\dfrac{7}{14} = -38\dfrac{7}{14} \quad\quad = -38\dfrac{7}{14}$

$\quad\quad\quad\quad\quad\quad\quad\quad\quad\quad\quad\quad\quad\quad\quad\quad 7\dfrac{13}{14}$

53. Estimate: Exact:

$\begin{array}{r} 6 \\ -3 \\ \hline 3 \end{array}$ $\quad 6 - 2\dfrac{5}{6} = 5\dfrac{6}{6} - 2\dfrac{5}{6} = 3\dfrac{1}{6}$

55. Estimate: Exact:

$\begin{array}{r} 12 \\ -9 \\ \hline 3 \end{array}$ $\quad 12 - 9\dfrac{2}{9} = 11\dfrac{9}{9} - 9\dfrac{2}{9} = 2\dfrac{7}{9}$

57. Estimate: Exact:

$\begin{array}{r} 5 \\ -3 \\ \hline 2 \end{array}$ $\quad 5\dfrac{3}{17} - 3 = 2\dfrac{3}{17}$

59. Estimate: Exact:

$\begin{array}{r} 23 \\ -17 \\ \hline 6 \end{array}$ $\quad 23\dfrac{5}{14} - 17 = 6\dfrac{5}{14}$

61. $2\dfrac{2}{3} + 4\dfrac{5}{8} = \dfrac{8}{3} + \dfrac{37}{8} = \dfrac{8 \cdot 8}{3 \cdot 8} + \dfrac{37 \cdot 3}{8 \cdot 3}$

$\quad\quad = \dfrac{64}{24} + \dfrac{111}{24} = \dfrac{175}{24} = 7\dfrac{7}{24}$

63. $1\dfrac{11}{15} + 4\dfrac{2}{5} = \dfrac{26}{15} + \dfrac{22}{5} = \dfrac{26}{15} + \dfrac{22 \cdot 3}{5 \cdot 3}$

$\quad\quad = \dfrac{26}{15} + \dfrac{66}{15} = \dfrac{92}{15} = 6\dfrac{2}{15}$

65. $3\dfrac{7}{8} - 3\dfrac{3}{16} = \dfrac{31}{8} - \dfrac{51}{16} = \dfrac{31 \cdot 2}{8 \cdot 2} - \dfrac{51}{16}$

$\quad\quad = \dfrac{62}{16} - \dfrac{51}{16} = \dfrac{11}{16}$

67. $4\dfrac{1}{12} + 5\dfrac{1}{9} = \dfrac{49}{12} + \dfrac{46}{9} = \dfrac{49 \cdot 3}{12 \cdot 3} + \dfrac{46 \cdot 4}{9 \cdot 4}$

$\quad\quad = \dfrac{147}{36} + \dfrac{184}{36} = \dfrac{331}{36} = 9\dfrac{7}{36}$

69. $9\dfrac{5}{32} - 8\dfrac{1}{4} = \dfrac{293}{32} - \dfrac{33}{4} = \dfrac{293}{32} - \dfrac{33 \cdot 8}{4 \cdot 8}$

$\quad\quad = \dfrac{293}{32} - \dfrac{264}{32} = \dfrac{29}{32}$

71. $6\dfrac{11}{14}+4\dfrac{1}{6}=\dfrac{95}{14}+\dfrac{25}{6}=\dfrac{95\cdot3}{14\cdot3}+\dfrac{25\cdot7}{6\cdot7}$

$\qquad=\dfrac{285}{42}+\dfrac{175}{42}=\dfrac{460}{42}=10\dfrac{40}{42}$

$\qquad=10\dfrac{20}{21}$

73. $12\dfrac{1}{5}-11\dfrac{2}{7}=\dfrac{61}{5}-\dfrac{79}{7}=\dfrac{61\cdot7}{5\cdot7}-\dfrac{79\cdot5}{7\cdot5}$

$\qquad=\dfrac{427}{35}-\dfrac{395}{35}=\dfrac{32}{35}$

75. $10\dfrac{1}{8}-2\dfrac{17}{18}=\dfrac{81}{8}-\dfrac{53}{18}=\dfrac{81\cdot9}{8\cdot9}-\dfrac{53\cdot4}{18\cdot4}$

$\qquad=\dfrac{729}{72}-\dfrac{212}{72}=\dfrac{517}{72}=7\dfrac{13}{72}$

77. $11\dfrac{1}{4}-3\dfrac{1}{2}=\dfrac{45}{4}-\dfrac{7}{2}=\dfrac{45}{4}-\dfrac{7\cdot2}{2\cdot2}=\dfrac{45}{4}-\dfrac{14}{4}$

$\qquad=\dfrac{31}{4}=7\dfrac{3}{4}$ in.

79. The index finger is longer.

81.

$\qquad 8\dfrac{2}{3}\quad=\quad 8\dfrac{8}{12}$

$\qquad 4\dfrac{1}{2}\quad=\quad 4\dfrac{6}{12}$

$\qquad +3\dfrac{3}{4}\quad=\quad +3\dfrac{9}{12}$

$\qquad\qquad\qquad\qquad 15\dfrac{23}{12}=16\dfrac{11}{12}$

The total is $16\dfrac{11}{12}$ hr.

83. $5\dfrac{3}{4}-2\dfrac{1}{3}=\dfrac{23}{4}-\dfrac{7}{3}=\dfrac{23\cdot3}{4\cdot3}-\dfrac{7\cdot4}{3\cdot4}$

$\qquad=\dfrac{69}{12}-\dfrac{28}{12}=\dfrac{41}{12}=3\dfrac{5}{12}$ ft.

85. $1\dfrac{3}{16}-1\dfrac{1}{8}=\dfrac{19}{16}-\dfrac{9}{8}=\dfrac{19}{16}-\dfrac{9\cdot2}{8\cdot2}$

$\qquad=\dfrac{19}{16}-\dfrac{18}{16}=\dfrac{1}{16}$ in.

Divide by 2 to get the thickness:

$\dfrac{1}{16}\div2=\dfrac{1}{16}\cdot\dfrac{1}{2}=\dfrac{1}{32}$ in.

87. $8\dfrac{1}{2}-1\dfrac{1}{4}-1\dfrac{1}{4}=\dfrac{17}{2}-\dfrac{5}{4}-\dfrac{5}{4}=\dfrac{17\cdot2}{2\cdot2}-\dfrac{5}{4}-\dfrac{5}{4}$

$\qquad=\dfrac{34}{4}-\dfrac{5}{4}-\dfrac{5}{4}=\dfrac{24}{4}=6$ in.

89.

$\qquad 5\dfrac{1}{3}\quad=\quad 5\dfrac{2}{6}\quad=\quad 4\dfrac{8}{6}$

$\qquad -2\dfrac{1}{2}\quad=\quad -2\dfrac{3}{6}\quad=\quad -2\dfrac{3}{6}$

$\qquad\qquad\qquad\qquad\qquad\qquad\qquad 2\dfrac{5}{6}$

There is $2\dfrac{5}{6}$ hr remaining.

91.

$\qquad 3\dfrac{3}{4}\quad=\quad 3\dfrac{9}{12}$

$\qquad -3\dfrac{5}{12}\quad=\quad -3\dfrac{5}{12}$

$\qquad\qquad\qquad\qquad \dfrac{4}{12}=\dfrac{1}{3}$

The blind will hang $\dfrac{1}{3}$ ft below the window.

93. (a) $1\dfrac{1}{4}+\dfrac{7}{8}+\dfrac{3}{4}+\dfrac{1}{2}=\dfrac{5}{4}+\dfrac{7}{8}+\dfrac{3}{4}+\dfrac{1}{2}$

$\qquad=\dfrac{5\cdot2}{4\cdot2}+\dfrac{7}{8}+\dfrac{3\cdot2}{4\cdot2}+\dfrac{1\cdot4}{2\cdot4}$

$\qquad=\dfrac{10}{8}+\dfrac{7}{8}+\dfrac{6}{8}+\dfrac{4}{8}$

$\qquad=\dfrac{27}{8}=3\dfrac{3}{8}$ L

(b) $4-3\dfrac{3}{8}=3\dfrac{8}{8}-3\dfrac{3}{8}=\dfrac{5}{8}$ L

95. Each number is $\dfrac{3}{4}$ greater than the last;

$3\dfrac{1}{4}+\dfrac{3}{4}=4$

97. Each number is $\dfrac{3}{4}$ greater than the last;

$3\dfrac{1}{2}+\dfrac{3}{4}=3\dfrac{2}{4}+\dfrac{3}{4}=3\dfrac{5}{4}=4\dfrac{1}{4}$

99. $\dfrac{14}{75}+\dfrac{9}{50}=\dfrac{11}{30}$

101. $\dfrac{29}{68} - \dfrac{7}{92} = \dfrac{137}{391}$

103. $21\dfrac{3}{28} + 4\dfrac{31}{42} = \dfrac{2171}{84}$ or $25\dfrac{71}{84}$

105. $5\dfrac{14}{17} - 2\dfrac{47}{68} = \dfrac{213}{68}$ or $3\dfrac{9}{68}$

Problem Recognition Exercises: Operations on Fractions and Mixed Numbers

1. $\dfrac{7}{5} + \dfrac{2}{5} = \dfrac{9}{5}$ or $1\dfrac{4}{5}$

3. $\dfrac{7}{5} \div \dfrac{2}{5} = \dfrac{7}{\cancel{5}} \cdot \dfrac{\cancel{5}}{2} = \dfrac{7}{2}$ or $3\dfrac{1}{2}$

5. $\dfrac{4}{3} \times \dfrac{5}{6} = \dfrac{\cancel{2} \cdot 2}{3} \times \dfrac{5}{\cancel{2} \cdot 3} = \dfrac{10}{9}$ or $1\dfrac{1}{9}$

7. $\dfrac{4}{3} + \dfrac{5}{6} = \dfrac{4 \cdot 2}{3 \cdot 2} + \dfrac{5}{6} = \dfrac{8}{6} + \dfrac{5}{6} = \dfrac{13}{6}$ or $2\dfrac{1}{6}$

9. $2\dfrac{3}{4} + 1\dfrac{1}{2} = \dfrac{11}{4} + \dfrac{3}{2} = \dfrac{11}{4} + \dfrac{3 \cdot 2}{2 \cdot 2} = \dfrac{11}{4} + \dfrac{6}{4}$

$\quad = \dfrac{17}{4}$ or $4\dfrac{1}{4}$

11. $2\dfrac{3}{4} \div 1\dfrac{1}{2} = \dfrac{11}{4} \div \dfrac{3}{2} = \dfrac{11}{4} \cdot \dfrac{2}{3} = \dfrac{11}{2 \cdot \cancel{2}} \cdot \dfrac{\cancel{2}}{3}$

$\quad = \dfrac{11}{6}$ or $1\dfrac{5}{6}$

13. $4\dfrac{1}{3} \times 2\dfrac{5}{6} = \dfrac{13}{3} \cdot \dfrac{17}{6} = \dfrac{221}{18}$ or $12\dfrac{5}{18}$

15. $4\dfrac{1}{3} - 2\dfrac{5}{6} = \dfrac{13}{3} - \dfrac{17}{6} = \dfrac{13 \cdot 2}{3 \cdot 2} - \dfrac{17}{6}$

$\quad = \dfrac{26}{6} - \dfrac{17}{6} = \dfrac{9}{6} = \dfrac{3}{2}$ or $1\dfrac{1}{2}$

17. $4 - \dfrac{3}{8} = \dfrac{4}{1} - \dfrac{3}{8} = \dfrac{4 \cdot 8}{1 \cdot 8} - \dfrac{3}{8} = \dfrac{32}{8} - \dfrac{3}{8}$

$\quad = \dfrac{29}{8}$ or $3\dfrac{5}{8}$

19. $4 \div \dfrac{3}{8} = \dfrac{4}{1} \times \dfrac{8}{3} = \dfrac{32}{3}$ or $10\dfrac{2}{3}$

21. $3\dfrac{2}{3} \div 2 = \dfrac{11}{3} \times \dfrac{1}{2} = \dfrac{11}{6}$ or $1\dfrac{5}{6}$

23. $3\dfrac{2}{3} + 2 = 5\dfrac{2}{3}$ or $\dfrac{17}{3}$

25. $4\dfrac{1}{5} - \dfrac{2}{3} = \dfrac{21}{5} - \dfrac{2}{3} = \dfrac{21 \cdot 3}{5 \cdot 3} - \dfrac{2 \cdot 5}{3 \cdot 5} = \dfrac{63}{15} - \dfrac{10}{15}$

$\quad = \dfrac{53}{15}$ or $3\dfrac{8}{15}$

27. $4\dfrac{1}{5} \cdot \dfrac{2}{3} = \dfrac{21}{5} \cdot \dfrac{2}{3} = \dfrac{7 \cdot \cancel{3}}{5} \cdot \dfrac{2}{\cancel{3}} = \dfrac{14}{5}$ or $2\dfrac{4}{5}$

29. $\dfrac{25}{9} \div 2 = \dfrac{25}{9} \cdot \dfrac{1}{2} = \dfrac{25}{18}$ or $1\dfrac{7}{18}$

31. $\dfrac{25}{9} - 2 = \dfrac{25}{9} - \dfrac{2 \cdot 9}{1 \cdot 9} = \dfrac{25}{9} - \dfrac{18}{9} = \dfrac{7}{9}$

33. $1\dfrac{4}{5} \cdot \dfrac{5}{9} = \dfrac{\cancel{9}}{\cancel{5}} \cdot \dfrac{\cancel{5}}{\cancel{9}} = \dfrac{1}{1} = 1$

35. $1\dfrac{4}{5} \div \dfrac{5}{9} = \dfrac{9}{5} \cdot \dfrac{9}{5} = \dfrac{81}{25}$ or $3\dfrac{6}{25}$

37. $8 \cdot \dfrac{1}{8} = \dfrac{\cancel{8}}{1} \cdot \dfrac{1}{\cancel{8}} = \dfrac{1}{1} = 1$

39. $\dfrac{3}{7} \cdot \dfrac{7}{3} = \dfrac{\cancel{3}}{\cancel{7}} \cdot \dfrac{\cancel{7}}{\cancel{3}} = \dfrac{1}{1} = 1$

Section 3.5   Order of Operations and Applications of Fractions and Mixed Numbers

Section 3.5   Practice Exercises

1. Answers will vary.

3. $16 - 3\dfrac{7}{9} = 15\dfrac{9}{9} - 3\dfrac{7}{9} = 12\dfrac{2}{9}$

5. $7\dfrac{1}{9} \div 2\dfrac{2}{3} = \dfrac{64}{9} \div \dfrac{8}{3} = \dfrac{\overset{8}{\cancel{64}}}{\underset{3}{\cancel{9}}} \cdot \dfrac{\overset{1}{\cancel{3}}}{\cancel{8}} = \dfrac{8}{3} = 2\dfrac{2}{3}$

7. $\left(1\dfrac{5}{6}\right)^2 = \left(\dfrac{11}{6}\right)^2 = \dfrac{121}{36} = 3\dfrac{13}{36}$

9. $5\dfrac{2}{13} = \dfrac{5 \times 13 + 2}{13} = \dfrac{67}{13}$

11. $3\dfrac{9}{10} = \dfrac{3 \times 10 + 9}{10} = \dfrac{39}{10}$

13. $\dfrac{29}{5} = 5\dfrac{4}{5}$

$$5\overline{\smash{\big)}\,29}$$
$$\underline{-25}$$
$$\ \ 4$$

15. $\dfrac{30}{19} = 1\dfrac{11}{19}$

$$19\overline{\smash{\big)}\,30}$$
$$\underline{-19}$$
$$\ \ 11$$

17. $\left(2 - \dfrac{1}{2}\right)^2 = \left(\dfrac{4}{2} - \dfrac{1}{2}\right)^2 = \left(\dfrac{3}{2}\right)^2 = \dfrac{3}{2} \cdot \dfrac{3}{2}$

$= \dfrac{9}{4} = 2\dfrac{1}{4}$

19. $1\dfrac{5}{6} \cdot 2\dfrac{1}{2} \div 1\dfrac{1}{4} = \dfrac{11}{6} \cdot \dfrac{5}{2} \div \dfrac{5}{4}$

$= \dfrac{11}{\underset{3}{\cancel{6}}} \cdot \dfrac{\cancel{5}}{\underset{1}{\cancel{2}}} \cdot \dfrac{\overset{1}{\overset{2}{\cancel{4}}}}{\underset{1}{\cancel{5}}} = \dfrac{11}{3} = 3\dfrac{2}{3}$

21. $6\dfrac{1}{6} + 2\dfrac{1}{3} \div 1\dfrac{3}{4} = \dfrac{37}{6} + \dfrac{7}{3} \div \dfrac{7}{4}$

$= \dfrac{37}{6} + \dfrac{\cancel{7}}{3} \cdot \dfrac{4}{\underset{1}{\cancel{7}}} = \dfrac{37}{6} + \dfrac{4}{3}$

$= \dfrac{37}{6} + \dfrac{8}{6} = \dfrac{45}{6} = 7\dfrac{3}{6} = 7\dfrac{1}{2}$

23. $6 - 5\dfrac{1}{7} \cdot \dfrac{1}{3} = 6 - \dfrac{\overset{12}{\cancel{36}}}{7} \cdot \dfrac{1}{\underset{1}{\cancel{3}}} = 6 - \dfrac{12}{7}$

$= \dfrac{42}{7} - \dfrac{12}{7} = \dfrac{30}{7} = 4\dfrac{2}{7}$

25. $\left(3\dfrac{1}{4} + 1\dfrac{5}{8}\right) \cdot 2\dfrac{2}{3} = \left(3\dfrac{2}{8} + 1\dfrac{5}{8}\right) \cdot 2\dfrac{2}{3}$

$= \left(4\dfrac{7}{8}\right) \cdot 2\dfrac{2}{3}$

$= \dfrac{\overset{13}{\cancel{39}}}{\cancel{8}} \cdot \dfrac{\overset{1}{\cancel{8}}}{\underset{1}{\cancel{3}}} = 13$

27. $\left(1\dfrac{1}{5}\right)^2 \cdot \left(1\dfrac{7}{9} - 1\dfrac{5}{12}\right) = \left(\dfrac{6}{5}\right)^2 \cdot \left(\dfrac{16}{9} - \dfrac{17}{12}\right)$

$= \dfrac{36}{25} \cdot \left(\dfrac{64}{36} - \dfrac{51}{36}\right)$

$= \dfrac{\overset{1}{\cancel{36}}}{25} \cdot \dfrac{13}{\underset{1}{\cancel{36}}} = \dfrac{13}{25}$

29. $\left(6\frac{3}{4} - 2\frac{1}{8}\right) \div \left(1\frac{1}{2}\right)^3 = \left(\frac{27}{4} - \frac{17}{8}\right) \div \left(\frac{3}{2}\right)^3$

$\qquad = \left(\frac{54}{8} - \frac{17}{8}\right) \div \left(\frac{27}{8}\right)$

$\qquad = \frac{37}{\cancel{8}} \cdot \frac{\cancel{8}^1}{27} = \frac{37}{27} = 1\frac{10}{27}$

31. $\left(\frac{1}{2}\right)^2 + \left(1\frac{1}{3}\right) \cdot \frac{9}{4} = \frac{1}{4} + \frac{\cancel{4}}{\cancel{3}} \cdot \frac{\cancel{3} \cdot 3}{\cancel{4}}$

$\qquad = \frac{1}{4} + 3 = 3\frac{1}{4}$

33. $\left(5 - 1\frac{7}{8}\right) \div \left(3 - \frac{13}{16}\right)$

$\qquad = \left(\frac{40}{8} - \frac{15}{8}\right) \div \left(\frac{48}{16} - \frac{13}{16}\right)$

$\qquad = \frac{25}{8} \div \frac{35}{16}$

$\qquad = \frac{\cancel{25}^5}{\cancel{8}_1} \cdot \frac{\cancel{16}^2}{\cancel{35}_7} = \frac{10}{7} = 1\frac{3}{7}$

35. **(a)** $3\frac{7}{10} - 3\frac{2}{5} = 3\frac{7}{10} - 3\frac{4}{10} = \frac{3}{10}$

The difference is $\frac{3}{10}$ sec.

(b) $3\frac{2}{5} + 3\frac{1}{2} + 3\frac{7}{10} + 3\frac{4}{5}$

$\qquad = 3\frac{4}{10} + 3\frac{5}{10} + 3\frac{7}{10} + 3\frac{8}{10}$

$\qquad = 12\frac{24}{10} = 14\frac{4}{10} = 14\frac{2}{5}$

$14\frac{2}{5} \div 4 = \frac{72}{5} \div \frac{4}{1} = \frac{\cancel{72}^{18}}{5} \cdot \frac{1}{\cancel{4}_1} = \frac{18}{5} = 3\frac{3}{5}$

The average is $3\frac{3}{5}$ sec.

37. **(a)** $11\frac{1}{4} + 10 + 7\frac{1}{2} + 9\frac{1}{2} + 4\frac{3}{4} + 8$

$\qquad = 11\frac{1}{4} + 10 + 7\frac{2}{4} + 9\frac{2}{4} + 4\frac{3}{4} + 8$

$\qquad = 49\frac{8}{4}$

$\qquad = 51$

The total weight loss is 51 lb.

(b) $51 \div 6 = \frac{51}{6} = 8\frac{3}{6} = 8\frac{1}{2}$

The average is $8\frac{1}{2}$ lb.

(c) $11\frac{1}{4} - 4\frac{3}{4} = 10\frac{5}{4} - 4\frac{3}{4} = 6\frac{2}{4} = 6\frac{1}{2}$

The difference is $6\frac{1}{2}$ lb.

39. $15\frac{5}{8} - 11\frac{3}{4} = 15\frac{5}{8} - 11\frac{6}{8} = 14\frac{13}{8} - 11\frac{6}{8} = 3\frac{7}{8}$

The stock dropped $\$3\frac{7}{8}$.

41. $\frac{1}{3} \cdot 80,250 = \frac{1}{3} \cdot \frac{80,250}{1} = \frac{80,250}{3} = 26,750$

George will receive $26,750.

43. $15\frac{1}{4} \div 4 = \frac{61}{4} \div \frac{4}{1} = \frac{61}{4} \cdot \frac{1}{4} = \frac{61}{16} = 3\frac{13}{16}$

Each piece is $3\frac{13}{16}$ ft.

45. $3 - \left(\frac{1}{4} + \frac{1}{3} + \frac{1}{6}\right) = 3 - \left(\frac{3}{12} + \frac{4}{12} + \frac{2}{12}\right)$

$\qquad = 3 - \frac{9}{12} = 2\frac{3}{12} = 2\frac{1}{4}$

$2\frac{1}{4}$ lb of cheese was eaten.

47. $65 \div 3\frac{1}{4} = \frac{65}{1} \div \frac{13}{4} = \frac{65}{1} \cdot \frac{4}{13} = \frac{260}{13} = 20$

20 loaves can be made.

49. $6\dfrac{1}{2}+\dfrac{3}{4}=6\dfrac{2}{4}+\dfrac{3}{4}=6\dfrac{5}{4}=7\dfrac{1}{4}$

The new rate is $7\dfrac{1}{4}$ points.

51. $3\cdot\left(2\dfrac{1}{2}+1\dfrac{1}{4}\right)=3\cdot\left(\dfrac{5}{2}+\dfrac{5}{4}\right)=3\cdot\left(\dfrac{10}{4}+\dfrac{5}{4}\right)$

$\qquad=\dfrac{3}{1}\cdot\dfrac{15}{4}=\dfrac{45}{4}=11\dfrac{1}{4}$

Stephanie will need $11\dfrac{1}{4}$ yd for the

dresses.

53. $6\dfrac{1}{2}-\left(\dfrac{1}{2}+\dfrac{1}{3}\right)\left(6\dfrac{1}{2}\right)=\dfrac{13}{2}-\left(\dfrac{3}{6}+\dfrac{2}{6}\right)\left(\dfrac{13}{2}\right)$

$\qquad=\dfrac{13}{2}-\dfrac{5}{6}\cdot\dfrac{13}{2}$

$\qquad=\dfrac{13}{2}-\dfrac{65}{12}=\dfrac{78}{12}-\dfrac{65}{12}$

$\qquad=\dfrac{13}{12}=1\dfrac{1}{12}$

Wilma has $1\dfrac{1}{12}$ lb left.

55. $30\cdot\left(22\dfrac{3}{4}-17\dfrac{2}{3}\right)=30\cdot\left(\dfrac{91}{4}-\dfrac{53}{3}\right)$

$\qquad=30\cdot\left(\dfrac{273}{12}-\dfrac{212}{12}\right)$

$\qquad=30\cdot\left(\dfrac{61}{12}\right)=\dfrac{1830}{12}$

$\qquad=152\dfrac{1}{2}$

Joan saves $152\dfrac{1}{2}$ gal.

57. $\left(2+9\dfrac{1}{2}+7+9\dfrac{1}{2}+2\right)-14\dfrac{2}{3}=30-14\dfrac{2}{3}$

$\qquad=29\dfrac{3}{3}-14\dfrac{2}{3}$

$\qquad=15\dfrac{1}{3}$

She needs $15\dfrac{1}{3}$ ft more.

59. $8\cdot\left(12\dfrac{1}{2}\right)=\dfrac{\overset{4}{\cancel{8}}}{1}\cdot\dfrac{25}{\underset{1}{\cancel{2}}}=100$

The perimeter is 100 in.

61. $35\dfrac{1}{3}+20\dfrac{1}{2}+20\dfrac{1}{2}=35\dfrac{2}{6}+20\dfrac{3}{6}+20\dfrac{3}{6}$

$\qquad=75\dfrac{8}{6}=76\dfrac{2}{6}=76\dfrac{1}{3}$

Matt needs $76\dfrac{1}{3}$ ft of gutter.

63. $2\cdot\left(35\dfrac{7}{8}\right)\left(14\dfrac{1}{4}\right)=\dfrac{\overset{1}{\cancel{2}}}{1}\cdot\dfrac{287}{\underset{4}{\cancel{8}}}\cdot\dfrac{57}{4}$

$\qquad=\dfrac{16{,}359}{16}$

$\qquad=1022\dfrac{7}{16}$

The area of the whole roof is $1022\dfrac{7}{16}$ ft^2.

65. (a) $\dfrac{1}{2}\left(5\dfrac{1}{2}\right)\left(13\dfrac{1}{5}\right)+(16)\left(13\dfrac{1}{5}\right)$

$\qquad=\dfrac{1}{\underset{1}{\cancel{2}}}\cdot\dfrac{11}{2}\cdot\dfrac{\overset{33}{\cancel{66}}}{5}+\dfrac{16}{1}\cdot\dfrac{66}{5}$

$\qquad=\dfrac{363}{10}+\dfrac{1056}{5}=\dfrac{363}{10}+\dfrac{2112}{10}$

$\qquad=\dfrac{2475}{10}=247\dfrac{1}{2}$

The area is $247\dfrac{1}{2}$ m^2.

(b) $5\dfrac{1}{2}+16+13\dfrac{1}{5}+16+14\dfrac{3}{10}$

$\qquad=5+16+13+16+14+\dfrac{1}{2}+\dfrac{1}{5}+\dfrac{3}{10}$

$\qquad=64+\dfrac{5}{10}+\dfrac{2}{10}+\dfrac{3}{10}=64\dfrac{10}{10}=65$

The perimeter is 65 m.

67. Door and front plus back:

$$2 \cdot \left[\left(6\frac{1}{2}\right)\left(4\frac{1}{3}\right) + \frac{1}{2}\left(6\frac{1}{2}\right)\left(1\frac{1}{2}\right) \right]$$

$$= 2 \cdot \left(\frac{13}{2} \cdot \frac{13}{3} + \frac{1}{2} \cdot \frac{13}{2} \cdot \frac{3}{2} \right)$$

$$= 2 \cdot \left(\frac{169}{6} + \frac{39}{8} \right) = 2 \cdot \left(\frac{676}{24} + \frac{117}{24} \right)$$

$$= \frac{\overset{1}{\cancel{2}}}{1} \cdot \frac{793}{\underset{12}{\cancel{24}}} = \frac{793}{12} \text{ m}^2$$

Sides:

$$2 \cdot (10)\left(4\frac{1}{3}\right) = 20 \cdot \frac{13}{3} = \frac{20}{1} \cdot \frac{13}{3} = \frac{260}{3} \text{ m}^2$$

$$\text{Total area} = \frac{793}{12} + \frac{260}{3} = \frac{793}{12} + \frac{1040}{12}$$

$$= \frac{1833}{12} = 152\frac{3}{4} \text{ m}^2$$

Chapter 3   Review Exercises

Section 3.1

1. 5 books + 3 books = 8 books

3. 25 mi − 13 mi = 12 mi

5. Fractions with the same denominators are considered like fractions.

7. $\dfrac{5}{6} + \dfrac{4}{6} = \dfrac{\overset{3}{\cancel{9}}}{\underset{2}{\cancel{6}}} = \dfrac{3}{2}$

9. $\dfrac{5}{12} + \dfrac{1}{12} = \dfrac{6}{12} = \dfrac{1}{2}$

11. $\dfrac{15}{7} - \dfrac{6}{7} = \dfrac{9}{7}$

13. $\dfrac{3}{8} \cdot \dfrac{3}{2} + \dfrac{3}{16} = \dfrac{9}{16} + \dfrac{3}{16} = \dfrac{12}{16} = \dfrac{3}{4}$

15. $\dfrac{21}{13} - \dfrac{5}{2} \div \dfrac{13}{4} = \dfrac{21}{13} - \dfrac{5}{\underset{1}{\cancel{2}}} \cdot \dfrac{\overset{2}{\cancel{4}}}{13} = \dfrac{21}{13} - \dfrac{10}{13} = \dfrac{11}{13}$

17. $\dfrac{11}{4} + \dfrac{13}{4} + \dfrac{11}{4} + \dfrac{13}{4} = \dfrac{48}{4} = 12$ in. or 1 ft

Section 3.2

19. (a) 7, 14, 21, 28

(b) 13, 26, 39, 52

(c) 22, 44, 66, 88

21. (a) 1, 2, 4, 5, 10, 20, 25, 50, 100

(b) 1, 5, 13, 65

(c) 1, 2, 5, 7, 10, 14, 35, 70

23. 30: 30, 60, 90, 120, 150, 180
25: 25, 50, 75, 100, 125, 150
LCM: 150

25. $105 = 3 \cdot 5 \cdot 7$
$28 = 2 \cdot 2 \cdot 7$
LCM: $2 \cdot 2 \cdot 3 \cdot 5 \cdot 7 = 420$

27. The LCM of 3 and 4 is 12. They will meet on the 12th day.

29. $\dfrac{9}{5} = \dfrac{9 \cdot 7}{5 \cdot 7} = \dfrac{63}{35}$

31. $\dfrac{17}{15} = \dfrac{17 \cdot 10}{15 \cdot 10} = \dfrac{170}{150}$

33. $\dfrac{5}{6} = \dfrac{5 \cdot 3}{6 \cdot 3} = \dfrac{15}{18}$

$\dfrac{7}{9} = \dfrac{7 \cdot 2}{9 \cdot 2} = \dfrac{14}{18}$

$\dfrac{15}{18} > \dfrac{14}{18}$, so $\dfrac{5}{6} > \dfrac{7}{9}$

35. $\dfrac{7}{10} = \dfrac{147}{210}$

$\dfrac{72}{105} = \dfrac{144}{210}$

$\dfrac{8}{15} = \dfrac{112}{210}$

$\dfrac{27}{35} = \dfrac{162}{210}$

$112 < 144 < 147 < 162$, so

$\dfrac{8}{15} < \dfrac{72}{105} < \dfrac{7}{10} < \dfrac{27}{35}$

Section 3.3

37. $\dfrac{9}{10} - \dfrac{61}{100} = \dfrac{90}{100} - \dfrac{61}{100} = \dfrac{29}{100}$

39. $\dfrac{3}{26} + \dfrac{5}{13} = \dfrac{3}{26} + \dfrac{10}{26} = \dfrac{13}{26} = \dfrac{1}{2}$

41. $4 - \dfrac{37}{20} = \dfrac{80}{20} - \dfrac{37}{20} = \dfrac{43}{20}$

43. $\dfrac{0}{17} + \dfrac{1}{34} = 0 + \dfrac{1}{34} = \dfrac{1}{34}$

45. $\dfrac{2}{15} + \dfrac{5}{8} - \dfrac{1}{3} = \dfrac{16}{120} + \dfrac{75}{120} - \dfrac{40}{120} = \dfrac{51}{120} = \dfrac{17}{40}$

47. $\left(\dfrac{2}{5} + \dfrac{1}{40}\right) \div \dfrac{15}{8} - \dfrac{4}{25}$

$= \left(\dfrac{16}{40} + \dfrac{1}{40}\right) \div \dfrac{15}{8} - \dfrac{4}{25}$

$= \dfrac{17}{\overset{}{\underset{5}{\cancel{40}}}} \cdot \dfrac{\overset{1}{\cancel{8}}}{15} - \dfrac{4}{25} = \dfrac{17}{75} - \dfrac{4}{25}$

$= \dfrac{17}{75} - \dfrac{12}{75} = \dfrac{5}{75} = \dfrac{1}{15}$

49. (a) $\dfrac{25}{16} + \dfrac{63}{16} + \dfrac{13}{4} = \dfrac{25}{16} + \dfrac{63}{16} + \dfrac{52}{16}$

$= \dfrac{140}{16} = \dfrac{35}{4}$ or $8\dfrac{3}{4}$ m

(b) $\dfrac{1}{2}\left(\dfrac{63}{16}\right)\left(\dfrac{5}{4}\right) = \dfrac{315}{128}$ or $2\dfrac{59}{128}$ m^2

Section 3.4

51.
$$
\begin{array}{rcl}
9\dfrac{8}{9} & = & 9\dfrac{56}{63} \\[2mm]
+\,1\dfrac{2}{7} & = & +\,1\dfrac{18}{63} \\[2mm]
\hline
 & & 10\dfrac{74}{63} = 11\dfrac{11}{63}
\end{array}
$$

53.
$$
\begin{array}{rclcl}
7\dfrac{5}{24} & = & 7\dfrac{5}{24} & = & 6\dfrac{29}{24} \\[2mm]
-\,4\dfrac{7}{12} & = & -\,4\dfrac{14}{24} & = & -\,4\dfrac{14}{24} \\[2mm]
\hline
 & & & & 2\dfrac{15}{24} = 2\dfrac{5}{8}
\end{array}
$$

55.
$$
\begin{array}{rcl}
5\dfrac{3}{8} & = & 5\dfrac{9}{24} \\[2mm]
-\,2\dfrac{1}{3} & = & -\,2\dfrac{8}{24} \\[2mm]
\hline
 & & 3\dfrac{1}{24}
\end{array}
$$

57.
$$
\begin{array}{rcl}
6\dfrac{4}{7} & = & 6\dfrac{8}{14} \\[2mm]
+\,5\dfrac{11}{14} & = & +\,5\dfrac{11}{14} \\[2mm]
\hline
 & & 11\dfrac{19}{14} = 12\dfrac{5}{14}
\end{array}
$$

59.
$$
\begin{array}{rcl}
6 & = & 5\dfrac{5}{5} \\[2mm]
-\,2\dfrac{3}{5} & = & -\,2\dfrac{3}{5} \\[2mm]
\hline
 & & 3\dfrac{2}{5}
\end{array}
$$

61.
$$42\frac{1}{8} = 42\frac{2}{16}$$
$$+\ 21\frac{13}{16} = +\ 21\frac{13}{16}$$
$$63\frac{15}{16}$$

63. Estimate: $2 + 4 + 2 = 8$

Exact: $2\frac{1}{4} + 4\frac{2}{9} + 1\frac{29}{36} = 2\frac{9}{36} + 4\frac{8}{36} + 1\frac{29}{36}$

$$= 7\frac{46}{36} = 8\frac{10}{36} = 8\frac{5}{18}$$

65. Estimate: $65 - 15 = 50$

Exact: $65\frac{1}{8} - 14\frac{9}{10} = 65\frac{5}{40} - 14\frac{36}{40}$

$$= 64\frac{45}{40} - 14\frac{36}{40} = 50\frac{9}{40}$$

67. $4\frac{1}{2} + 3\frac{2}{3} = 4\frac{3}{6} + 3\frac{4}{6} = 7\frac{7}{6} = 8\frac{1}{6}$

Corry drove a total of $8\frac{1}{6}$ hr.

Section 3.5

69. $1\frac{1}{5} + 4\frac{9}{10} \cdot 2\frac{2}{7} = \frac{6}{5} + \frac{\overset{7}{\cancel{49}}}{\underset{5}{\cancel{10}}} \cdot \frac{\overset{8}{\cancel{16}}}{\underset{1}{\cancel{7}}}$

$$= \frac{6}{5} + \frac{56}{5} = \frac{62}{5} = 12\frac{2}{5}$$

71. $\left(8\frac{1}{9} - 6\frac{2}{3}\right) \div 9\frac{3}{4} = \left(8\frac{1}{9} - 6\frac{6}{9}\right) \div 9\frac{3}{4}$

$$= \left(7\frac{10}{9} - 6\frac{6}{9}\right) \div 9\frac{3}{4}$$

$$= \left(1\frac{4}{9}\right) \div 9\frac{3}{4} = \frac{13}{9} \div \frac{39}{4}$$

$$= \frac{\overset{1}{\cancel{13}}}{9} \cdot \frac{4}{\underset{3}{\cancel{39}}} = \frac{4}{27}$$

73. $\left(1\frac{1}{5}\right)^2 \cdot \left(4\frac{1}{2} + 3\frac{5}{6}\right) = \left(\frac{6}{5}\right)^2 \cdot \left(4\frac{3}{6} + 3\frac{5}{6}\right)$

$$= \frac{36}{25} \cdot \left(7\frac{8}{6}\right)$$

$$= \frac{\overset{6}{\cancel{36}}}{25} \cdot \frac{\overset{2}{\cancel{50}}}{\underset{1}{\cancel{6}}} = 12$$

75. $\frac{9}{10}(160,000) = \frac{9}{\underset{1}{\cancel{10}}} \cdot \frac{\overset{16,000}{\cancel{160,000}}}{1} = 144,000$

The appraised value is \$144,000.

Chapter 3   Test

1. $\frac{4}{5} + \frac{3}{5} = \frac{7}{5}$

3. When subtracting like fractions, keep the same denominator and subtract the numerators. When multiplying fractions, multiply the denominators as well as the numerators.

5. $16 = 2 \cdot 2 \cdot 2 \cdot 2$
$24 = 2 \cdot 2 \cdot 2 \cdot 3$
$30 = 2 \cdot 3 \cdot 5$
LCM: $2 \cdot 2 \cdot 2 \cdot 2 \cdot 3 \cdot 5 = 240$

7. $\frac{11}{21} = \frac{11 \cdot 3}{21 \cdot 3} = \frac{33}{63}$

9. $33 < 35 < 36$, so $\frac{11}{21} < \frac{5}{9} < \frac{4}{7}$

11. $\frac{7}{3} - 2 = \frac{7}{3} - \frac{6}{3} = \frac{1}{3}$

13. $\frac{3}{5} + \frac{1}{15} = \frac{9}{15} + \frac{1}{15} = \frac{10}{15} = \frac{2}{3}$

15. $12 - 9\frac{10}{11} = 11\frac{11}{11} - 9\frac{10}{11} = 2\frac{1}{11}$

17. $15\frac{1}{6} - 12\frac{3}{8} - 1\frac{7}{24} = 15\frac{4}{24} - 12\frac{9}{24} - 1\frac{7}{24}$

$$= 14\frac{28}{24} - 12\frac{9}{24} - 1\frac{7}{24}$$

$$= 1\frac{12}{24} = 1\frac{1}{2}$$

19. $3\frac{1}{3} \div 2\frac{1}{2} + 5\frac{2}{3} = \frac{10}{3} \div \frac{5}{2} + \frac{17}{3}$

$$= \frac{\overset{2}{\cancel{10}}}{3} \cdot \frac{2}{\cancel{5}} + \frac{17}{3} = \frac{4}{3} + \frac{17}{3}$$

$$= \frac{21}{3} = 7$$

21. $\left(7\frac{1}{4} - 5\frac{1}{6}\right) \cdot 1\frac{3}{5} = \left(\frac{29}{4} - \frac{31}{6}\right) \cdot \frac{8}{5}$

$$= \left(\frac{87}{12} - \frac{62}{12}\right) \cdot \frac{8}{5}$$

$$= \frac{\overset{5}{\cancel{25}}}{\underset{3}{\cancel{12}}} \cdot \frac{\overset{2}{\cancel{8}}}{\underset{1}{\cancel{5}}} = \frac{10}{3} \text{ or } 3\frac{1}{3}$$

23. $4\frac{19}{40}(2000) = \frac{179}{\underset{1}{\cancel{40}}} \cdot \frac{\overset{50}{\cancel{2000}}}{1} = 8950$

The Ford Expedition can tow 8950 lb.

25. $14{,}000 - \left(\frac{3}{28} + \frac{1}{7}\right) \cdot 14{,}000$

$$= 14{,}000 - \left(\frac{3}{28} + \frac{4}{28}\right) \cdot 14{,}000$$

$$= 14{,}000 - \frac{7}{\underset{1}{\cancel{28}}} \cdot \frac{\overset{500}{\cancel{14{,}000}}}{1}$$

$$= 14{,}000 - 7 \cdot 500$$

$$= 14{,}000 - 3500 = 10{,}500$$

Justin has $10,500 for cabinets.

Chapters 1–3   Cumulative Review Exercises

1. Twenty-three million, four hundred thousand, eight hundred six

3. $\overset{\overset{6}{\cancel{7}}{\,}^{1}2}{\underline{-24}}$
48

5. $24\overline{)72}$ with quotient 3, -72, remainder 0

7. $4 \cdot 4 \cdot 5 \cdot 5 \cdot 5 \cdot 5 \cdot 8 \cdot 8 = 4^2 \cdot 5^4 \cdot 8^2$

9. 17, 19, 23, 29, 31

11. Numerator: 21; denominator: 17

13. $\frac{17}{22}$ had pepperoni, $22 - 17 = 5$, so $\frac{5}{22}$ did not have pepperoni.

15. **(a)** $2 + 3 + 9 + 0 = 14$, which is not divisible by 3, so 2390 is not divisible by 3.

(b) $1 + 2 + 4 + 5 = 12$, so 1245 is divisible by 3. The ones-place digit is 5, so 1245 is also divisible by 5.

(c) The ones-place digit is not 0 or 5, so 9321 is not divisible by 5.

17.

$$
\begin{array}{r}
5 \\
3\overline{)15} \\
3\overline{)45} \\
2\overline{)90} \\
2\overline{)180} \\
2\overline{)360}
\end{array}
$$

$360 = 2 \cdot 2 \cdot 2 \cdot 3 \cdot 3 \cdot 5 = 2^3 \cdot 3^2 \cdot 5$

19. $\dfrac{\overset{3}{\cancel{15}}}{\underset{8}{\cancel{16}}} \cdot \dfrac{\overset{1}{\cancel{2}}}{\underset{1}{\cancel{5}}} = \dfrac{3}{8}$

21. $\dfrac{13}{8} - \dfrac{7}{8} = \dfrac{6}{8} = \dfrac{3}{4}$

23. $4 - \dfrac{18}{5} = \dfrac{20}{5} - \dfrac{18}{5} = \dfrac{2}{5}$

25. $2\dfrac{3}{5} \div 1\dfrac{7}{10} = \dfrac{13}{5} \div \dfrac{17}{10} = \dfrac{13}{\underset{1}{\cancel{5}}} \cdot \dfrac{\overset{2}{\cancel{10}}}{17} = \dfrac{26}{17}$ or $1\dfrac{9}{17}$

27. $\dfrac{22}{7} \cdot 28 = \dfrac{22}{\underset{1}{\cancel{7}}} \cdot \dfrac{\overset{4}{\cancel{28}}}{1} = 88$

The distance around is approximately 88 cm.

29. $\dfrac{1}{2}\left(5\dfrac{1}{4}\right)(3) = \dfrac{1}{2} \cdot \dfrac{21}{4} \cdot \dfrac{3}{1} = \dfrac{63}{8}$ or $7\dfrac{7}{8}$ m^2

Chapter 4   Decimals

Chapter Opener Puzzle

4	1	2	5	3
2	3	4	1	5
1	a 2	5	3	c 4
3	5	1	4	2
d 5	4	b 3	2	1

Section 4.1   Decimal Notation and Rounding

Section 4.1   Practice Exercises

1. Answers will vary.

3. $10^2 = 100$

5. $10^4 = 10,000$

7. $\left(\dfrac{1}{10}\right)^2 = \dfrac{1}{100}$

9. $\left(\dfrac{1}{10}\right)^4 = \dfrac{1}{10,000}$

11. Tenths

13. Hundredths

15. Tens

17. Ten-thousandths

19. Thousandths

21. Ones

23. Nine-tenths

25. Twenty-three hundredths

27. Thirty-three thousandths

29. Four hundred seven ten-thousandths

31. Three and twenty-four hundredths

33. Five and nine-tenths

35. Fifty-two and three tenths

37. Six and two hundred nineteen thousandths

39. 8472.014

41. 700.07

43. 2,469,000.506

45. $3.7 = 3\dfrac{7}{10}$

47. $2.8 = 2\dfrac{8}{10} = 2\dfrac{4}{5}$

49. $0.25 = \dfrac{25}{100} = \dfrac{1}{4}$

51. $0.55 = \dfrac{55}{100} = \dfrac{11}{20}$

53. $20.812 = 20\dfrac{812}{1000} = 20\dfrac{203}{250}$

55. $15.0005 = 15\dfrac{5}{10,000} = 15\dfrac{1}{2000}$

57. $8.4 = \dfrac{84}{10} = \dfrac{42}{5}$

59. $3.14 = \dfrac{314}{100} = \dfrac{157}{50}$

61. $23.5 = \dfrac{235}{10} = \dfrac{47}{2}$

63. $11.91 = \dfrac{1191}{100}$

65. 34.2, 34.25, 34.29, 34.3

67. 0.042, 0.043, $\dfrac{4}{10}$, 0.42, 0.43

69. $6.312 < 6.321$

71. $11.21 > 11.2099$

73. $0.762 > 0.76$

75. $51.72 < 51.721$

77. a, b

79. 0.3444, 0.3493, 0.3558, 0.3585, 0.3664

81. These numbers are equivalent, but they represent different levels of accuracy.

83. 7.1

85. $49.943 \approx 49.9$

87. $33.416 \approx 33.42$

89. $9.0955 \approx 9.096$

91. $21.0239 \approx 21.0$

93. $6.9995 = 7.000$

95. $0.0079499 = 0.0079$

97. $0.00362005 \approx 0.0036$ mph

99. 971.0948
1000, 970, 971.1, 971.09, 971.095

101. 21.9754
0, 20, 22.0, 21.98, 21.975

103. 0.972

Section 4.2   Addition and Subtraction of Decimals

Section 4.2 Practice Exercises

1. Answers will vary.

3. b, c

5. $23.489 \approx 23.5$

7. $8.6025 \approx 8.603$

9. $2.82998 \approx 2.8300$

11.

Expression	Estimate
44.6	45
+ 18.6	+ 19
63.2	64

13.

Expression	Estimate
5.306	5
+ 3.645	+ 4
8.951	9

15.

Expression	Estimate
12.900	13
+ 3.091	+ 3
15.991	16

17.

78.9000
+ 0.9005
79.8005

19.

23.0000
+ 8.0148
31.0148

21.   34.0000
          23.0032
        + 5.6000
        ————————
          62.6032

23.   68.394
        + 32.020
        ————————
         100.414

25. 103.94
        + 24.50
        ————————
         128.44

27.   54.200
          23.993
         + 3.870
        ————————
          82.063

29. Expression   Estimate
         35.36   35
        − 21.12   − 21
        ————————   ————————
         14.24   14

31. Expression   Estimate
          7.24   7
        − 3.56   − 4
        ————————   ————————
          3.68   3

33. Expression   Estimate
         45.02   45
        − 32.70   − 33
        ————————   ————————
         12.32   12

35.   14.500
         − 8.823
        ————————
          5.677

37.   2.000
         − 0.123
        ————————
          1.877

39. 103.400
         − 45.050
        ————————
          58.350
         − 0.982
        ————————
          57.368

41.   55.9000
         − 34.2354
        ————————
          21.6646

43.   18.003
         − 3.238
        ————————
          14.765

45. 183.010
         − 23.452
        ————————
         159.558

47.   6.007
        + 12.740
        ————————
          18.747
         − 3.400
        ————————
          15.347

49.   23.370
         − 21.900
        ————————
           1.470
         + 5.111
        ————————
           6.581

51.   8.962
        + 51.000
        ————————
          59.962
         − 40.050
        ————————
          19.912

53.   5.3000
           5.0300
         + 5.0030
        ————————
          15.3330
         − 5.0003
        ————————
          10.3327

55.   5.8400
         + 5.0840
        ————————
          10.9240
         − 5.0084
        ————————
           5.9156

57.   10.000
         − 0.900
        ————————
           9.100
         − 0.090
        ————————
           9.010
         − 0.009
        ————————
           9.001

59. (a)   686.980
            − 365.256
           ————————
            321.724 days

(b)   224.70
 − 67.97
—————
 156.73 days

61. (a)

5.9	7.6	9.3
− 4.2	− 5.9	− 7.6
1.7	1.7	1.7

The water is rising 1.7 in./hr.

(b)   9.3
 + 1.7
————
 11.0

At 1 P.M. the level will be 11 in.

(c)   11.0
    1.7
 + 1.7
————
 14.4

At 3.00 P.M. the level will be 14.4 in.

63.   245.62
 −   52.48
—————
     193.14
 −   72.44
—————
     120.70
 −   108.34
—————
       12.36
 + 1084.90
—————
   1097.26
 −   23.87
—————
   1073.39
 +   200.00
—————
   1273.39

The ending balance was $1273.39.

65.   4.20
 − 2.85
————
    1.35

1.35 million cells per microliter.

67. 3 quarters and 1 dime:
$3 \times 1.75 = 5.25$
   5.25
 + 1.35
————
   6.60 mm
2 nickels and 2 pennies:
$2 \times 1.95 = 3.9$
$2 \times 1.55 = 3.1$
   3.9
 + 3.1
————
   7.0 mm
The pile containing 2 nickels and 2 pennies is higher.

69. x: 27.3
 − 18.4
————
    8.9 in.

y: 22.1
 − 6.7
————
   15.4 in.

$P =$ 27.3
    6.7
    8.9
   15.4
   18.4
 + 22.1
————
   98.8 in.

71. x: 4.875
 − 1.2
————
    3.675
 − 1.6
————
    2.075 ft

y: 3.62
 − 1.03
————
    2.59 ft

$P =$ 4.875
    3.620
    1.600
    2.590
    2.075
    2.590
    1.200
 + 3.620
—————
   22.170 ft

73.   7.25
    4.30
    9.75
    + 5.90
    ―――――
    27.20 mi

75. 6 + 0.5 + 0.5 = 7 mm

77. 103.17 − 97.27 = 5.90
    IBM increased by $5.90 per share.

79. 114.16 − 111.89 = 2.27
    Between February and March, FedEx
    increased the most, by $2.27 per share.

81. 114.16 − 106.07 = 8.09
    Between March and April, FedEx
    decreased the most, by $8.09 per share.

Section 4.3   Multiplication of Decimals

Section 4.3   Practice Exercises

1. (a) 248 or 250
   (b) 238
   (c) 239
   (d) 238 or 239

3. $10^3 = 1000$

5. $0.1^2 = 0.01$

7.   0.8
    × 0.5
    ―――
    0.40

9. 0.9
    × 4
    ―――
    3.6

11.   0.4
    × 2 0
    ―――
    8.0

13.   60
    × 0.003
    ――――
    0.180

15. 22.38
    × 0.8
    ――――
    17.904

17.   14
    × 0.002
    ――――
    0.028

19. 100

21. 30

23. 0.07

25. 0.2

27. Exact   Estimate
       8.3   8
     × 4.5   × 5
     ―――――   ――――
     4 15   40
    33 20
    ―――――
    37.35

29. Exact   Estimate
      0.58   0.6
     × 7.2   × 7
     ―――――   ―――
      116   4.2
    4 060
    ―――――
    4.176

31. Exact   Estimate
      5.92   6
     × 0.8   × 0.8
     ―――――   ―――――
     4.736   4.8

33. Exact   Estimate
     0.413   0.4
     ×   7   × 7
     ―――――   ―――
     2.891   2.8

35. Exact   Estimate
      35.9   40
     × 3.2   × 3
     ―――――   ―――――
      718   120
    107 70
    ―――――
    114.88

71

37. Exact
$$\begin{array}{r} 562 \\ \times\ 0.004 \\ \hline 2.248 \end{array}$$
Estimate
$$\begin{array}{r} 600 \\ \times\ 0.004 \\ \hline 2.400 \end{array}$$

39. Exact
$$\begin{array}{r} 0.0004 \\ \times\quad 3.6 \\ \hline 24 \\ 120 \\ \hline 0.00144 \end{array}$$
Estimate
$$\begin{array}{r} 0.0004 \\ \times\quad 4 \\ \hline 0.0016 \end{array}$$

41. The decimal point will move to the right 2 places.

43. **(a)** $5.1 \times 10 = 51$

 (b) $5.1 \times 100 = 510$

 (c) $5.1 \times 1000 = 5100$

 (d) $5.1 \times 10,000 = 51,000$

45. The decimal point will move to the left 1 place.

47. $34.9 \times 100 = 3490$

49. $96.59 \times 1000 = 96,590$

51. $93.3 \times 0.01 = 0.933$

53. $54.03 \times 0.001 = 0.05403$

55. $2.001 \times 10 = 20.01$

57. $0.5 \times 0.0001 = 0.00005$

59. 2.6 million $= 2.6 \times 1,000,000$
$$= 2,600,000$$

61. 400 thousand $= 400 \times 1000 = 400,000$

63. 20.549 billion $= 20.549 \times 1,000,000,000$
$$= \$20,549,000,000$$

65. **(a)**
$$\begin{array}{r} 6.3 \\ \times\ 32 \\ \hline 126 \\ 1890 \\ \hline 201.6 \text{ lb} \end{array}$$

 (b)
$$\begin{array}{r} 32 \\ \times\ 20 \\ \hline 640 \end{array}\text{ lb of } CO_2$$

67.
$$\begin{array}{ccc} 10.95 & 3.95 & 60 \\ \times\quad 20 & \times\quad 10 & \times\ 0.60 \\ \hline 219.00 & \overline{39.50} & \overline{36.00} \end{array}$$

$$\begin{array}{r} 219.00 \\ 39.50 \\ 36.00 \\ +\ 17.67 \\ \hline \$312.17 \end{array}$$

69.
$$\begin{array}{r} 50.20 \\ \times\quad 4 \\ \hline \$200.80 \end{array}\text{ for 4 Firestone tires.}$$

$$\begin{array}{r} 200.80 \\ -\ 197.99 \\ \hline \$2.81 \end{array}\text{ can be saved.}$$

71. $A = l \cdot w$
$A = (0.023)(0.05)$
$$\begin{array}{r} 0.023 \\ \times\quad 0.05 \\ \hline 0.00115 \end{array}$$
$= 0.00115 \text{ km}^2$

73. $A = l \cdot w$
$A = (15)(22.2)$
$$\begin{array}{r} 22.2 \\ \times\quad 15 \\ \hline 111\ 0 \\ 222\ 0 \\ \hline 333.0 \end{array}$$
$= 333 \text{ yd}^2$

75. $\$3.24 = 324¢$

77. $\$0.37 = 37¢$

79. $347¢ = \$3.47$

81. $2041¢ = \$20.41$

83. **(a)** $\$1.499 \approx \$1$

 (b) $\$1.499 \approx \$1.50$

72

85.
$$\begin{array}{r} 0.2 \\ \times\, 0.2 \\ \hline 0.04 \end{array}$$
$(0.2)^2 = 0.04,$ which is not equal to 0.4.

87.
$$\begin{array}{r} 0.4 \\ \times\, 0.4 \\ \hline 0.16 \end{array}$$
$(0.4)^2 = 0.16$

89.
$$\begin{array}{r} 1.3 \\ \times\, 1.3 \\ \hline 39 \\ 1\,30 \\ \hline 1.69 \end{array}$$
$(1.3)^2 = 1.69$

91. $(0.1)^3 = (0.1)(0.1)(0.1) = 0.001$

93.
$$\begin{array}{r} 0.2 \\ \times\, 0.2 \\ \hline 0.04 \\ \times\, 0.2 \\ \hline 0.008 \\ \times\, 0.2 \\ \hline 0.0016 \end{array}$$
$(0.2)^4 = 0.0016$

95. (a)
$$\begin{array}{r} 0.3 \\ \times\, 0.3 \\ \hline 0.09 \end{array}$$
$(0.3)^2 = 0.09$

(b) $\sqrt{0.09} = 0.3$

97. $\sqrt{0.01} = \sqrt{(0.1)^2} = 0.1$

99. $\sqrt{0.36} = \sqrt{(0.6)^2} = 0.6$

Section 4.4   Division of Decimals

Section 4.4   Practice Exercises

1. Answers will vary.

3. $5.28 \times 1000 = 5280$

5.
$$\begin{array}{r} 11.8 \\ \times\, 0.32 \\ \hline 236 \\ 3\,540 \\ \hline 3.776 \end{array}$$

7.
$$\begin{array}{r} 16.82 \\ -\, 14.80 \\ \hline 2.02 \end{array}$$

9.
$$9\,\overline{)\,8.1\,} \quad \begin{array}{r} 0.9 \\ \end{array}$$
$$\begin{array}{r} -8\,1 \\ \hline 0 \end{array}$$
$$\begin{array}{r} 0.9 \\ \times\ 9 \\ \hline 8.1 \end{array}$$

11.
$$6\,\overline{)\,1.08\,}$$
$$\begin{array}{r} 0.18 \\ -6 \\ \hline 48 \\ -48 \\ \hline 0 \end{array}$$
$$\begin{array}{r} 0.18 \\ \times\ 6 \\ \hline 1.08 \end{array}$$

13.
$$8\,\overline{)\,4.24\,}$$
$$\begin{array}{r} 0.53 \\ -40 \\ \hline 24 \\ -24 \\ \hline 0 \end{array}$$
$$\begin{array}{r} 0.53 \\ \times\ 8 \\ \hline 4.24 \end{array}$$

15.
$$5\,\overline{)\,105.5\,}$$
$$\begin{array}{r} 21.1 \\ -10 \\ \hline 05 \\ -5 \\ \hline 0\,5 \\ -5 \\ \hline 0 \end{array}$$
$$\begin{array}{r} 21.1 \\ \times\ 5 \\ \hline 105.5 \end{array}$$

73

17.
$$
\begin{array}{r}
1.96 \\
5\overline{)\,9.80} \\
\underline{-5} \\
4\,8 \\
\underline{-4\,5} \\
30 \\
\underline{-30} \\
0
\end{array}
$$

19.
$$
\begin{array}{r}
0.035 \\
8\overline{)0.280} \\
\underline{-24} \\
40 \\
\underline{-40} \\
0
\end{array}
$$

21.
$$
\begin{array}{r}
16.84 \\
5\overline{)\,84.20} \\
-5 \\
34 \\
-30 \\
4\,2 \\
-4\,0 \\
20 \\
-20 \\
0
\end{array}
$$

23.
$$
\begin{array}{r}
0.12 \\
50\overline{)\,6.00} \\
\underline{-50} \\
10 \\
\underline{-10} \\
0
\end{array}
$$

25.
$$
\begin{array}{r}
0.16 \\
25\overline{)\,4.00} \\
\underline{-25} \\
150 \\
\underline{-150} \\
0
\end{array}
$$

27.
$$
\begin{array}{r}
5.33... \\
3\overline{)\,16.00} \\
\underline{-15} \\
1\,0 \\
\underline{-9} \\
10 \\
\underline{-9} \\
1
\end{array}
$$
$5.33... = 5.\overline{3}$

29.
$$
\begin{array}{r}
3.166... \\
6\overline{)\,19.000} \\
\underline{-18} \\
1\,0 \\
\underline{-6} \\
40 \\
\underline{-36} \\
40 \\
\underline{-36} \\
4
\end{array}
$$
$3.166... = 3.1\overline{6}$

31.
$$
\begin{array}{r}
2.1515... \\
33\overline{)\,71.0000} \\
\underline{-66} \\
50 \\
\underline{-33} \\
170 \\
\underline{-165} \\
50 \\
\underline{-33} \\
170 \\
\underline{-165} \\
5
\end{array}
$$
$2.1515... = 2.\overline{15}$

33. $5.03 \div 0.01 = 503$

35. $0.992 \div 0.1 = 9.92$

37. $1.02\overline{)57.12}$

$$
\begin{array}{r}
56 \\
102\overline{)\,5712} \\
\underline{-510} \\
612 \\
\underline{-612} \\
0
\end{array}
$$

39. $0.8\overline{)2.38}$

$$
\begin{array}{r}
2.975 \\
8\overline{)\,23.800} \\
\underline{-16} \\
7\,8 \\
\underline{-7\,2} \\
60 \\
\underline{-56} \\
40 \\
\underline{-40} \\
0
\end{array}
$$

41. $0.3\overline{)62.5}$

$$
\begin{array}{r}
208.33... \\
3\overline{)\,625.00} \\
\underline{-6} \\
025 \\
\underline{-24} \\
1\,0 \\
\underline{-9} \\
10 \\
\underline{-9} \\
1
\end{array}
$$

$208.33... = 208.\overline{3}$

43. $0.13\overline{)6.305}$

$$
\begin{array}{r}
48.5 \\
13\overline{)\,630.5} \\
\underline{-52} \\
110 \\
\underline{-104} \\
6\,5 \\
\underline{-6\,5} \\
0
\end{array}
$$

45. $1.1 \div 0.001 = 1100$

47. $420.6 \div 0.01 = 42{,}060$

49. The decimal point will move to the left 2 places.

51. $3.923 \div 100 = 0.03923$

53. $98.02 \div 10 = 9.802$

55. $0.027 \div 100 = 0.00027$

57. $1.02 \div 1000 = 0.00102$

59. $2.\overline{4} = 2.4444...$

    (a) 2.4

    (b) 2.44

    (c) 2.444

61. $1.\overline{8} = 1.8888...$

    (a) 1.9

    (b) 1.89

    (c) 1.889

63. $3.\overline{62} = 3.6262...$

    (a) 3.6

    (b) 3.63

    (c) 3.626

65.

$$
\begin{array}{r}
0.257 \\
7\overline{)\,1.800} \\
\underline{-1\,4} \\
40 \\
\underline{-35} \\
50 \\
\underline{-49} \\
1
\end{array}
$$

$$
\begin{array}{r}
0.26 \\
\times \ \ 7 \\
\hline
1.82
\end{array}
$$

$0.257 \approx 0.26$

67. $3.7\overline{)54.9}$

$$
\begin{array}{r}
14.83 \\
37\overline{)\,549.00} \\
\underline{-37} \\
179 \\
\underline{-148} \\
31\,0 \\
\underline{-29\,6} \\
1\,40 \\
\underline{-1\,11} \\
29
\end{array}
$$

$$
\begin{array}{r}
14.8 \\
\times \ 3.7 \\
\hline
1036 \\
44\,40 \\
\hline
54.76
\end{array}
$$

$14.83 \approx 14.8$

69. $0.24\overline{)4.96}$

$$
\begin{array}{r}
20.6666 \\
24\overline{)\,496.0000} \\
\underline{-48} \\
16\,0 \\
\underline{-144} \\
1\,60 \\
\underline{-1\,44} \\
160 \\
\underline{-144} \\
160 \\
\underline{-144} \\
16
\end{array}
$$

$$
\begin{array}{r}
20.667 \\
\times \ \ 0.24 \\
\hline
82668 \\
4\,13340 \\
\hline
4.96008
\end{array}
$$

$20.6666 \approx 20.667$

Chapter 4   Decimals

71. $0.9\overline{)32.1}$

```
      35.666
  9) 321.000
    -27
     51
    -45
      6 0
     -5 4
       60
      -54
       60
      -54
        6
```

```
  35.67
×   0.9
32.103
```

$35.666 \approx 35.67$

73. $2.13\overline{)237.1}$

```
       111.31
  213) 23710.00
      -213
       241
      -213
        280
       -213
        670
       -639
        310
       -213
         97
```

```
   111.3
×   2.13
   3339
  11130
 222600
237.069
```

$111.31 \approx 111.3$

75. Unreasonable; $960

77. Unreasonable; $140,000

79. $560 − $50 = $510

```
      42.5
  12) 510.0
     -48
      30
     -24
       6 0
      -6 0
        0
```

The monthly payment is $42.50.

81. (a)

```
        12.5
  800)10,000.0
     -8 00
      2 000
     -1 600
       400 0
      -400 0
          0
```

13 bulbs would be needed (rounded up to the nearest whole unit).

(b)

```
  0.75
  ×13
  225
  750
$9.75
```

(c) The energy efficient fluorescent bulb would be more cost effective.

83.

```
          0.3420
  8399) 2873.0000
       -2519 7
        353 30
       -335 96
         17 340
        -16 798
          5420
```

Babe Ruth's batting average was 0.342.

85. $5.5\overline{)12}$

```
       2.18
  55) 120.00
     -110
      100
      -55
       45 0
      -44 0
        10
```

2.2 mph

87. 47.265

89.

```
     8.6
  × 12.4
    3 44
   17 20
   86 00
  106.64
```

b, d

76

91. $(2749.13)(418.2) = 1,149,686.166$

93. $(43.75)^2 = 1914.0625$

95. $21.5\overline{)2056.75} = 95.6627907$

97. Answers will vary.

99. (a) $\dfrac{3352}{12634} = 0.27$

(b) Yes, the claim is accurate. The decimal, 0.27 is close to 0.25, which is equal to $\frac{1}{4}$.

101. (a) $584,000,000 \div 365$
$= 1,600,000$ mi per day

(b) $1,600,000 \div 24 = 66,666.\overline{6}$ mph

Problem Recognition Exercises: Operations on Decimals

1. (a) $123.04 + 100 = 223.04$

(b) $123.04 \times 100 = 12,304$

(c) $123.04 - 100 = 23.04$

(d) $123.04 \div 100 = 1.2304$

(e) $123.04 + 0.01 = 123.05$

(f) $123.04 \times 0.01 = 1.2304$

(g) $123.04 \div 0.01 = 12,304$

(h) $123.04 - 0.01 = 123.03$

3. (a)
$$\begin{array}{r} 4.800 \\ +2.391 \\ \hline 7.191 \end{array}$$

(b)
$$\begin{array}{r} 2.391 \\ +4.800 \\ \hline 7.191 \end{array}$$

5. (a)
$$\begin{array}{r} 32.9 \\ \times 1.6 \\ \hline 1974 \\ 3290 \\ \hline 52.64 \end{array}$$

(b)
$$\begin{array}{r} 1.6 \\ \times 32.9 \\ \hline 144 \\ 320 \\ 4800 \\ \hline 52.64 \end{array}$$

7. (a)
$$\begin{array}{r} 21.6 \\ \times 4 \\ \hline 86.4 \end{array}$$

(b)
$$\begin{array}{r} 21.6 \\ \times 0.25 \\ \hline 1080 \\ 4320 \\ \hline 5.400 \end{array}$$

9. (a) $5.6\overline{)448}$
$$\begin{array}{r} 80 \\ 56\overline{)4480} \\ -448 \\ \hline 00 \\ -0 \\ \hline 0 \end{array}$$

(b)
$$\begin{array}{r} 5.6 \\ \times 80 \\ \hline 448.0 \end{array}$$

11.
$$\begin{array}{r} 0.125 \\ \times 8 \\ \hline 1.000 \end{array}$$

13. $0.07\overline{)280}$
$$\begin{array}{r} 4000 \\ 7\overline{)28000} \\ -28 \\ \hline 0000 \\ -000 \\ \hline 0 \end{array}$$

77

15. $490\overline{)98,000,000}$

$$\begin{array}{r} 200,000 \\ 490\overline{)98,000,000} \\ \underline{-98\ 0} \\ 000000 \\ \underline{-00000} \\ 0 \end{array}$$

17.
$$\begin{array}{r} 4500 \\ \times 300,000 \\ \hline 1,350,000,000 \end{array}$$

19.
$$\begin{array}{r} 83.4000 \\ -78.9999 \\ \hline 4.4001 \end{array}$$

Section 4.5   Fractions as Decimals

Section 4.5   Practice Exercises

1. Answers will vary.

3. $\dfrac{39}{100} = 0.39$

5. $\dfrac{71}{10,000} = 0.0071$

7. $0.0016 = \dfrac{16}{10,000} = \dfrac{1}{625}$

9. $0.125 = \dfrac{125}{1000} = \dfrac{1}{8}$

11. $\dfrac{2}{5} = \dfrac{2\cdot 2}{5\cdot 2} = \dfrac{4}{10} = 0.4$

13. $\dfrac{49}{50} = \dfrac{49\cdot 2}{50\cdot 2} = \dfrac{98}{100} = 0.98$

15. $\dfrac{7}{25} = \dfrac{7\cdot 4}{25\cdot 4} = \dfrac{28}{100} = 0.28$

17. $\dfrac{316}{500} = \dfrac{316\cdot 2}{500\cdot 2} = \dfrac{632}{1000} = 0.632$

19.
$$\begin{array}{r} 0.875 \\ 8\overline{)7.000} \\ \underline{-64} \\ 60 \\ \underline{-56} \\ 40 \\ \underline{-40} \\ 0 \end{array}$$
$\dfrac{7}{8} = 0.875$

21.
$$\begin{array}{r} 3.2 \\ 5\overline{)16.0} \\ \underline{-15} \\ 10 \\ \underline{-10} \\ 0 \end{array}$$
$\dfrac{16}{5} = 3.2$

23.
$$\begin{array}{r} 0.25 \\ 12\overline{)3.00} \\ \underline{-24} \\ 60 \\ \underline{-60} \\ 0 \end{array}$$
$5\dfrac{3}{12} = 5 + 0.25 = 5.25$

25.
$$\begin{array}{r} 0.2 \\ 5\overline{)1.0} \\ \underline{-10} \\ 0 \end{array}$$
$1\dfrac{1}{5} = 1 + 0.2 = 1.2$

27.
$$\begin{array}{r} 0.75 \\ 24\overline{)18.00} \\ \underline{-16\ 8} \\ 1\ 20 \\ \underline{-1\ 20} \\ 0 \end{array}$$
$\dfrac{18}{24} = 0.75$

29.

$$16 \overline{) \begin{array}{l} 3.3125 \\ 53.0000 \end{array}} \qquad \frac{53}{16} = 3.3125$$

$$\begin{array}{r} \underline{-48} \\ 5\,0 \\ \underline{-4\,8} \\ 20 \\ \underline{-16} \\ 40 \\ \underline{-32} \\ 80 \\ \underline{-80} \\ 0 \end{array}$$

31.

$$20 \overline{) \begin{array}{l} 0.45 \\ 9.00 \end{array}} \qquad 7\frac{9}{20} = 7 + 0.45 = 7.45$$

$$\begin{array}{r} \underline{-8\,0} \\ 1\,00 \\ \underline{-1\,00} \\ 0 \end{array}$$

33.

$$25 \overline{) \begin{array}{l} 0.88 \\ 22.00 \end{array}} \qquad \frac{22}{25} = 0.88$$

$$\begin{array}{r} \underline{-20\,0} \\ 200 \\ \underline{-200} \\ 0 \end{array}$$

35.

$$9 \overline{) \begin{array}{l} 0.88... \\ 8.00 \end{array}} \qquad 3\frac{8}{9} = 3.\overline{8}$$

$$\begin{array}{r} \underline{-72} \\ 80 \\ \underline{-72} \\ 8 \end{array}$$

37.

$$15 \overline{) \begin{array}{l} 0.466... \\ 7.000 \end{array}} \qquad \frac{7}{15} = 0.4\overline{6}$$

$$\begin{array}{r} \underline{-60} \\ 1\,00 \\ \underline{-90} \\ 100 \\ \underline{-90} \\ 10 \end{array}$$

39.

$$36 \overline{) \begin{array}{l} 0.5277... \\ 19.0000 \end{array}} \qquad \frac{19}{36} = 0.52\overline{7}$$

$$\begin{array}{r} \underline{-18\,0} \\ 1\,00 \\ \underline{-72} \\ 280 \\ \underline{-252} \\ 280 \\ \underline{-252} \\ 28 \end{array}$$

41.

$$11 \overline{) \begin{array}{l} 0.5454... \\ 6.0000 \end{array}} \qquad \frac{6}{11} = 0.\overline{54}$$

$$\begin{array}{r} \underline{-5\,5} \\ 50 \\ \underline{-44} \\ 60 \\ \underline{-55} \\ 50 \\ \underline{-44} \\ 6 \end{array}$$

43.

$$111 \overline{) \begin{array}{l} 0.126126... \\ 14.000000 \end{array}} \qquad \frac{14}{111} = 0.\overline{126}$$

$$\begin{array}{r} \underline{-111} \\ 2\,90 \\ \underline{-2\,22} \\ 680 \\ \underline{-666} \\ 140 \\ \underline{-111} \\ 290 \\ \underline{-222} \\ 680 \\ \underline{-666} \\ 14 \end{array}$$

45.

$$
\begin{array}{r}
1.13636\ldots \\
22\overline{)25.00000} \\
-22 \\
\hline
3\,0 \\
-2\,2 \\
\hline
8\,0 \\
-66 \\
\hline
140 \\
-132 \\
\hline
80 \\
-66 \\
\hline
140 \\
-132 \\
\hline
8
\end{array}
$$

$\dfrac{25}{22} = 1.1\overline{36}$

47.

$$
\begin{array}{r}
0.1428 \\
7\overline{)1.0000} \\
-7 \\
\hline
30 \\
-28 \\
\hline
20 \\
-14 \\
\hline
60 \\
-56 \\
\hline
4
\end{array}
$$

$\dfrac{1}{7} = 0.143$

49.

$$
\begin{array}{r}
0.076 \\
13\overline{)1.000} \\
-91 \\
\hline
90 \\
-78 \\
\hline
12
\end{array}
$$

$\dfrac{1}{13} = 0.08$

51.

$$
\begin{array}{r}
0.93 \\
16\overline{)15.00} \\
-14\,4 \\
\hline
60 \\
-48 \\
\hline
12
\end{array}
$$

$\dfrac{15}{16} \approx 0.9$

53.

$$
\begin{array}{r}
0.714 \\
7\overline{)5.000} \\
-4\,9 \\
\hline
10 \\
-7 \\
\hline
30 \\
-28 \\
\hline
2
\end{array}
$$

$\dfrac{5}{7} \approx 0.71$

55.

$$
\begin{array}{r}
1.19 \\
21\overline{)25.00} \\
-21 \\
\hline
40 \\
-2\,1 \\
\hline
1\,90 \\
-1\,89 \\
\hline
1
\end{array}
$$

$\dfrac{25}{21} \approx 1.2$

57. (a) $\dfrac{1}{9} = 0.\overline{1}$

(b) $\dfrac{2}{9} = 0.\overline{2}$

(c) $\dfrac{4}{9} = 0.\overline{4}$

(d) $\dfrac{5}{9} = 0.\overline{5}$

If we memorize that $\dfrac{1}{9} = 0.\overline{1}$, then

$\dfrac{2}{9} = 2 \cdot \dfrac{1}{9} = 2 \cdot 0.\overline{1} = 0.\overline{2}$, and so on.

59. (a) $0.45 = \dfrac{45}{100} = \dfrac{9}{20}$

(b) $1\dfrac{5}{8} = \dfrac{13}{8}$

$$
\begin{array}{r}
1.625 \\
8\overline{)13.000} \\
-8 \\
\hline
5\,0 \\
-4\,8 \\
\hline
20 \\
-16 \\
\hline
40 \\
-40 \\
\hline
0
\end{array}
$$

$= 1.625$

(c) $0.\overline{7} = \dfrac{7}{9}$

(d)

$$
\begin{array}{r}
0.4545\ldots \\
11\overline{)5.0000} \\
-4\,4 \\
\hline
60 \\
-55 \\
\hline
50 \\
-44 \\
\hline
60 \\
-55 \\
\hline
5
\end{array}
$$

$\dfrac{5}{11} = 0.\overline{45}$

61. (a) $0.\overline{3} = \dfrac{1}{3}$

(b) $2.125 = \dfrac{2125}{1000} = \dfrac{17}{8}$ or $2\dfrac{1}{8}$

(c)

$$
\begin{array}{r}
0.86363... \\
22\overline{\smash{\big)}\ 19.00000} \\
\underline{-17\ 6} \\
1\ 40 \\
\underline{-1\ 32} \\
80 \\
\underline{-66} \\
140 \\
\underline{-132} \\
80 \\
\underline{-66} \\
14
\end{array}
$$

$\dfrac{19}{22} = 0.8\overline{63}$

(d) $\dfrac{42}{25} = \dfrac{42 \cdot 4}{25 \cdot 4} = \dfrac{168}{100} = 1.68$

63. McGraw-Hill:

$69.25 = \dfrac{6925}{100} = \dfrac{277}{4}$ or $69\dfrac{1}{4}$

Walgreens: $44.95 = \dfrac{4495}{100} = \dfrac{899}{20}$ or $44\dfrac{19}{20}$

Home Depot: $38\dfrac{1}{2} = 38.50$

General Electric:

$37\dfrac{11}{25} = 37\dfrac{11 \cdot 4}{25 \cdot 4} = 37 + \dfrac{44}{100} = 37.44$

65. $0.2 = \dfrac{1}{5}$

67. $0.2 < 0.\overline{2}$

69. $\dfrac{1}{3} = 0.\overline{3} > 0.3$

71. $4\dfrac{1}{4} = 4.25 < 4.\overline{25}$

73. $0.\overline{5} = \dfrac{5}{9}$

75. $0.27 < 0.\overline{27} = \dfrac{3}{11}$

77.

79.

81. $0.\overline{9} = \dfrac{9}{9} = 1$

83. $6.\overline{9} = 6 + 0.\overline{9} = 6 + 1 = 7$

Section 4.6   Order of Operations and Applications of Decimals

Section 4.6   Practice Exercises

1. Answers will vary.

3.
$$
\begin{array}{r}
34.1 \\
\times\ \ 9.2 \\
\hline
6\ 82 \\
306\ 90 \\
\hline
313.72
\end{array}
$$

5. $\dfrac{34}{9} + \dfrac{5}{27} = \dfrac{34 \cdot 3}{9 \cdot 3} + \dfrac{5}{27} = \dfrac{102}{27} + \dfrac{5}{27} = \dfrac{107}{27}$

7. $\dfrac{55}{16} \div \dfrac{11}{4} = \dfrac{\overset{5}{\cancel{55}}}{\underset{4}{\cancel{16}}} \cdot \dfrac{\overset{1}{\cancel{4}}}{\underset{1}{\cancel{11}}} = \dfrac{5}{4}$

9.
$$
\begin{array}{r}
13.00 \\
-\ 6.04 \\
\hline
6.96
\end{array}
$$

11. $(3.7 - 1.2)^2$

$$
\begin{array}{r}
3.7 \\
-1.2 \\
\hline
2.5
\end{array}
$$

$= (2.5)^2$

$$
\begin{array}{r}
2.5 \\
\times 2.5 \\
\hline
1\,25 \\
5\,00 \\
\hline
6.25
\end{array}
$$

$= 6.25$

13. $16.25 - \left(18.2 - 15.7\right)^2$

$$
\begin{array}{r}
18.2 \\
-15.7 \\
\hline
2.5
\end{array}
$$

$= (2.5)^2$

$$
\begin{array}{r}
2.5 \\
\times 2.5 \\
\hline
125 \\
500 \\
\hline
6.25
\end{array}
$$

$= 16.25 - 6.25$

$$
\begin{array}{r}
16.25 \\
-6.25 \\
\hline
10.00
\end{array}
$$

$= 10$

15. $12.46 - 3.05 - 0.8^2$

$$
\begin{array}{r}
0.8 \\
\times 0.8 \\
\hline
0.64
\end{array}
$$

$= 12.46 - 3.05 - 0.64$

$$
\begin{array}{r}
12.46 \\
-3.05 \\
\hline
9.41
\end{array}
$$

$= 9.41 - 0.64$

$$
\begin{array}{r}
9.41 \\
-0.64 \\
\hline
8.77
\end{array}
$$

$= 8.77$

17. $63.75 - 9.5(4)$

$$
\begin{array}{r}
9.5 \\
\times\ 4 \\
\hline
38.0
\end{array}
$$

$= 63.75 - 38$

$$
\begin{array}{r}
63.75 \\
-38.00 \\
\hline
25.75
\end{array}
$$

$= 25.75$

19. $6.8 \div 2 \div 1.7$

$$
\begin{array}{r}
3.4 \\
2\overline{)\,6.8} \\
\underline{-6} \\
0\,8 \\
\underline{-8} \\
0
\end{array}
$$

$= 3.4 \div 1.7$

$$
1.7\overline{)3.4}
$$

$$
\begin{array}{r}
2 \\
17\overline{)\,34} \\
\underline{-34} \\
0
\end{array}
$$

$= 2$

21. $2.2 + [9.34 + (1.2)^2]$

$$
\begin{array}{r}
1.2 \\
\times 1.2 \\
\hline
24 \\
1\,20 \\
\hline
1.44
\end{array}
$$

$= 2.2 + (9.34 + 1.44)$
$= 2.2 + 9.34 + 1.44$

$$
\begin{array}{r}
2.20 \\
9.34 \\
+1.44 \\
\hline
12.98
\end{array}
$$

$= 12.98$

23. $89.8 \div 1\frac{1}{3} = 89.8 \div \frac{4}{3} = 89.8 \times \frac{3}{4}$

$\qquad\qquad = 89.8 \times 0.75$

$$
\begin{array}{r}
89.8 \\
\times\ 0.75 \\
\hline
4490 \\
62\,860 \\
\hline
67.350
\end{array}
$$

$= 67.35$

25. $20.04 \div \dfrac{4}{5} = 20.04 \div 0.8$

$$0.8 \overline{)20.04}$$

$$8 \overline{)200.40}$$
$$\underline{-16}$$
$$40$$
$$\underline{-40}$$
$$0\ 40$$
$$\underline{-40}$$
$$0$$

$$= 25.05$$

27. $14.4 \times \left(\dfrac{7}{4} - \dfrac{1}{8} \right) = 14.4 \times \left(\dfrac{14}{8} - \dfrac{1}{8} \right)$

$$= 14.4 \times \left(\dfrac{13}{8} \right)$$

$$= \dfrac{\overset{18}{\cancel{144}}}{10} \times \dfrac{13}{\underset{1}{\cancel{8}}}$$

$$= \dfrac{234}{10}$$
$$= 23.4$$

29. $2.3 \times \dfrac{5}{9} = \dfrac{23}{\underset{2}{\cancel{10}}} \times \dfrac{\overset{1}{\cancel{5}}}{9} = \dfrac{23}{18}$

$$18 \overline{)23.000}$$
$$\underline{-18}$$
$$5\ 0$$
$$\underline{-3\ 6}$$
$$1\ 40$$
$$\underline{-1\ 26}$$
$$140$$
$$\underline{-126}$$
$$14$$

$$\dfrac{23}{18} \approx 1.28$$

31. $6.5 \div \dfrac{3}{5} = 6\dfrac{1}{2} \div \dfrac{3}{5} = \dfrac{13}{2} \cdot \dfrac{5}{3} = \dfrac{65}{6}$

$$6 \overline{)65.000}$$
$$\underline{-6}$$
$$05\ 0$$
$$\underline{-4\ 8}$$
$$20$$
$$\underline{-18}$$
$$20$$
$$\underline{-18}$$
$$2$$

$$\dfrac{65}{6} \approx 10.83$$

33. $(42.81 - 30.01) \div \dfrac{9}{2}$

$$42.81$$
$$\underline{-\ 30.01}$$
$$12.80$$

$$= 12.8 \div \dfrac{9}{2} = \dfrac{128}{\underset{5}{\cancel{10}}} \cdot \dfrac{\overset{1}{\cancel{2}}}{9} = \dfrac{128}{45}$$

$$45 \overline{)128.000}$$
$$\underline{-90}$$
$$38\ 0$$
$$\underline{-36\ 0}$$
$$2\ 00$$
$$\underline{-1\ 80}$$
$$200$$
$$\underline{-180}$$
$$20$$

$$\dfrac{128}{45} \approx 2.84$$

35. $\dfrac{2}{9} \times 4.21 = \dfrac{2}{9} \times \dfrac{421}{\underset{50}{\cancel{100}}} = \dfrac{421}{450}$

$$450 \overline{)421.0000}$$
$$\underline{-405\ 0}$$
$$16\ 00$$
$$\underline{-13\ 50}$$
$$2\ 500$$
$$\underline{-2\ 250}$$
$$2500$$
$$\underline{-2250}$$
$$250$$

$$\dfrac{421}{450} = 0.93\overline{5}$$

37. $5.32 \div \dfrac{6}{5} = \dfrac{\overset{266}{\cancel{532}}}{\underset{20}{\cancel{100}}} \times \dfrac{\overset{1}{\cancel{5}}}{\underset{3}{\cancel{6}}} = \dfrac{266}{60}$

$$60\overline{\smash{)}\,266.000} \qquad \dfrac{266}{60} = 4.4\overline{3}$$

$$\begin{array}{r} 4.433... \\ 60\overline{\smash{)}\,266.000} \\ \underline{-240} \\ 260 \\ \underline{-240} \\ 200 \\ \underline{-180} \\ 200 \\ \underline{-180} \\ 20 \end{array}$$

39. (a)
$$\begin{array}{r} 21{,}816.6 \\ -\ 21{,}345.6 \\ \hline 471.0 \end{array}$$
They drove 471 mi.

(b) $471 \div 7\dfrac{1}{2} = 471 \div \dfrac{15}{2} = \dfrac{471}{1} \times \dfrac{2}{15} = \dfrac{942}{15}$

$$\begin{array}{r} 62.8 \\ 15\overline{\smash{)}\,942.0} \\ \underline{-90} \\ 42 \\ \underline{-30} \\ 12\,0 \\ \underline{-12\,0} \\ 0 \end{array}$$

The average speed is 62.8 mph.

41.
$$\begin{array}{r} 597 \\ -\ 450 \\ \hline 147 \end{array}\ \text{minutes over the included amount}$$

$$\begin{array}{r} 147 \\ \times\ 0.40 \\ \hline 58.80 \end{array} \qquad \begin{array}{r} 39.95 \\ +\ 58.80 \\ \hline 98.75 \end{array}$$
Jorge will be charged $98.75.

43. $\dfrac{1}{4} \times 60 = \dfrac{1}{\cancel{4}} \times \dfrac{\overset{15}{\cancel{60}}}{1} = 15$ g for breakfast

$$\begin{array}{r} 15.0 \\ +\ 20.7 \\ \hline 35.7 \end{array}\ \text{g so far today} \qquad \begin{array}{r} 60.0 \\ -\ 35.7 \\ \hline 24.3 \end{array}\ \text{g left}$$
She has 24.3 g left for dinner.

45.
$$\begin{array}{r} 4.79 \\ \times\ \ 3 \\ \hline 14.37 \end{array}\ \text{cost for 3 packages}$$
$$\begin{array}{r} 14.37 \\ +\ 0.86 \\ \hline 15.23 \end{array}\ \text{total cost}$$
$$\begin{array}{r} 20.00 \\ -\ 15.23 \\ \hline 4.77 \end{array}\ \text{change}$$
Caren should get $4.77 in change.

47.
$$\begin{array}{r} 92 \\ 84 \\ 77 \\ +\ 62 \\ \hline 315 \end{array} \qquad \begin{array}{r} 78.75 \\ 4\overline{\smash{)}\,315.00} \\ \underline{-28} \\ 35 \\ \underline{-32} \\ 3\,0 \\ \underline{-2\,8} \\ 20 \\ \underline{-20} \\ 0 \end{array}$$

Duncan's average is 78.75.

49.
$$\begin{array}{r} 6.6 \\ 18.1 \\ 18.8 \\ 16.8 \\ +\ 12.4 \\ \hline 72.7 \end{array} \qquad \begin{array}{r} 14.54 \\ 5\overline{\smash{)}\,72.70} \\ \underline{-5} \\ 22 \\ \underline{-20} \\ 27 \\ \underline{-25} \\ 20 \\ \underline{-20} \\ 0 \end{array}$$

The average snowfall per month is 14.54 in.

51. Answers will vary.

53. (a) $\text{BMI} = \dfrac{703W}{h^2} = \dfrac{703(220)}{(72)^2} = \dfrac{703(220)}{5184}$

$$= \dfrac{154{,}660}{5184} \approx 29.8$$

(b) Overweight

55. $0.\overline{3} \times 0.3 + 3.375 = \dfrac{3}{9} \times \dfrac{3}{10} + 3.375$

$= \dfrac{9}{90} + 3.375$

$= \dfrac{1}{10} + 3.375$

$= 0.1 + 3.375 = 3.475$

57. $(0.\overline{8} + 0.\overline{4}) \times 0.39 = \left(\dfrac{8}{9} + \dfrac{4}{9}\right) \times 0.39$

$= \left(\dfrac{12}{9}\right) \times 0.39$

$= \dfrac{4}{3} \times 0.39 = \dfrac{4 \times 0.39}{3}$

```
  0.39
×    4
─────
  1.56
```

$= \dfrac{1.56}{3}$

```
      0.52
3) 1.56
  −1 5
  ────
     06
    −6
    ──
      0
```

$= 0.52$

59. (a)
```
   132.05
 ×    50
 ───────
 6602.50
```

```
         237.5
278) 66025.0
    −556
    ─────
     1042
     −834
     ─────
      2085
     −1946
     ─────
       139 0
      −139 0
      ──────
           0
```

Deanna can buy 237 shares.

(b)
```
  27.80
 ×0.5
 ─────
 13.90
```
$13.90 will be leftover.

61. (a)
```
  145,000
 − 25,000
 ────────
  120,000
```
Marty will have to finance $120,000.

(b)
```
    12
 × 30
 ────
  360
```
There are 360 months in 30 years.

(c)
```
    798.36
 ×     360
 ─────────
 287,409.6
```
He will pay $287,409.60.

(d)
```
  287,409.60
 − 120,000.00
 ───────────
  167,409.60
```
He will pay $167,409.60 in interest.

63. $\dfrac{1}{3}(80,460.60) = 26,820.20$

```
  80,460.60
 − 26,820.20
 ──────────
  53,640.40
```

$53,640.40 \div 4 = 13,410.10$
Each person will get approximately $13,410.10.

Chapter 4   Review Exercises

Section 4.1

1. 32.16
The 3 is in the tens place, the 2 is in the ones place, the 1 is in the tenths place, and the 6 is in the hundredths place.

3. Five and seven-tenths

5. Fifty-one and eight thousandths

7. 33,015.047

9. $4.8 = 4\dfrac{8}{10} = 4\dfrac{4}{5}$

11. $1.3 = \dfrac{13}{10}$

13. $15.032 > 15.03$

15. $4.3875, 4.3953, 4.4839, 4.5000, 4.5142$

17. 34.890

19. a, b

Section 4.2

21.
$$
\begin{array}{r}
45.030 \\
+\ 4.713 \\
\hline
49.743
\end{array}
$$

23.
$$
\begin{array}{r}
34.89 \\
-\ 29.44 \\
\hline
5.45
\end{array}
$$

25.
$$
\begin{array}{r}
221.00 \\
-\ 23.04 \\
\hline
197.96
\end{array}
$$

27.
$$
\begin{array}{r}
17.300 \\
+\ 3.109 \\
\hline
20.409 \\
-\ 12.600 \\
\hline
7.809
\end{array}
$$

29. x:
$$
\begin{array}{r}
53.4 \\
-\ 48.9 \\
\hline
4.5 \text{ in.}
\end{array}
$$
y:
$$
\begin{array}{r}
47.10 \\
-\ 42.03 \\
\hline
5.07 \text{ in.}
\end{array}
$$

$P = 47.1 + 53.4 + 42.03 + 4.5 + 5.07 + 48.9$
$\quad = 201$ in.

31.
$$
\begin{array}{r}
1.40 \\
1.00 \\
0.09 \\
+\ 1.25 \\
\hline
3.74
\end{array}
$$
The total rainfall was 3.74 in.

Section 4.3

33.
$$
\begin{array}{r}
57.01 \\
\times\ \ 1.3 \\
\hline
17\,103 \\
57\,010 \\
\hline
74.113
\end{array}
$$

35.
$$
\begin{array}{r}
7.7 \\
\times\ \ 45 \\
\hline
38\,5 \\
308\,0 \\
\hline
346.5
\end{array}
$$

37. $1.0034 \times 100 = 100.34$

39. $104.22 \times 0.01 = 1.0422$

41. $432,000$

43.
$$
\begin{array}{r}
23 \\
\times\ 0.07 \\
\hline
1.61
\end{array}
$$
The call will cost $1.61.

45. **(a)**
$$
\begin{array}{r}
36.4 \\
\times\ \ 200 \\
\hline
7280.0
\end{array}
$$
7280 people

(b)
$$
\begin{array}{r}
90.0 \\
\times\ \ 200 \\
\hline
18,000.0
\end{array}
$$
18,000 people

Section 4.4

47. $1.5\overline{)64.2}$

$$
\begin{array}{r}
42.8 \\
15\overline{)642.0} \\
\underline{-60} \\
42 \\
\underline{-30} \\
120 \\
\underline{-120} \\
0
\end{array}
$$

49. $0.3\overline{)2.63}$

$$
\begin{array}{r}
8.766... \\
3\overline{)26.300} \\
-24 \\
\hline
2\ 3 \\
-2\ 1 \\
\hline
20 \\
-18 \\
\hline
20 \\
-18 \\
\hline
2
\end{array}
$$

$8.766... = 8.7\overline{6}$

51. $1.2\overline{)0.036}$

$$
\begin{array}{r}
0.03 \\
12\overline{)0.36} \\
-36 \\
\hline
0
\end{array}
$$

53. $90.234 \div 10 = 9.0234$

55. $2.6 \div 0.01 = 260$

57.
$$
\begin{array}{r}
11.622... \\
9\overline{)104.600} \\
-9 \\
\hline
14 \\
-9 \\
\hline
5\ 6 \\
-5\ 4 \\
\hline
20 \\
-18 \\
\hline
20 \\
-18 \\
\hline
2
\end{array}
$$

$11.622... \approx 11.62$

59. (a)
$$
\begin{array}{r}
0.499 \\
12\overline{)5.990} \\
-48 \\
\hline
119 \\
-108 \\
\hline
110 \\
-108 \\
\hline
2
\end{array}
$$

$0.50 per roll

(b)
$$
\begin{array}{r}
0.572 \\
4\overline{)2.290} \\
-20 \\
\hline
29 \\
-28 \\
\hline
10 \\
-8 \\
\hline
2
\end{array}
$$

$0.57 per roll

(c) The 12-pack is the better buy.

Section 4.5

61. $\dfrac{7}{20} = \dfrac{7 \cdot 5}{20 \cdot 5} = \dfrac{35}{100} = 0.35$

63. $2\dfrac{2}{5} = 2\dfrac{4}{10} = 2.4$

65. $\dfrac{24}{125} = \dfrac{24 \cdot 8}{125 \cdot 8} = \dfrac{192}{1000} = 0.192$

67.
$$
\begin{array}{r}
0.5833... \\
12\overline{)7.0000} \\
-6\ 0 \\
\hline
1\ 00 \\
-96 \\
\hline
40 \\
-36 \\
\hline
40 \\
-36 \\
\hline
4
\end{array}
$$

$\dfrac{7}{12} = 0.58\overline{3}$

69.
$$
\begin{array}{r}
0.31818... \\
22\overline{)7.00000} \\
-6\ 6 \\
\hline
40 \\
-22 \\
\hline
180 \\
-176 \\
\hline
40 \\
-22 \\
\hline
180 \\
-176 \\
\hline
4
\end{array}
$$

$4\dfrac{7}{22} = 4 + 0.3\overline{18} = 4.3\overline{18}$

71.
$$17\overline{\smash{\big)}5.000}$$
$$\underline{-34}$$
$$1\,60$$
$$\underline{-1\,53}$$
$$70$$
$$\underline{-68}$$
$$2$$

quotient: 0.294

$\dfrac{5}{17} \approx 0.29$

73.
$$3\overline{\smash{\big)}11.000}$$
$$\underline{-9}$$
$$2\,0$$
$$\underline{-1\,8}$$
$$20$$
$$\underline{-18}$$
$$20$$
$$\underline{-18}$$
$$2$$

quotient: 3.666

$\dfrac{11}{3} \approx 3.67$

75. $0.\overline{2} = \dfrac{2}{9}$

77. $3.\overline{3} = 3 + 0.\overline{3} = 3 + \dfrac{1}{3} = 3\dfrac{1}{3}$

79. Sun: $5.2 = 5\dfrac{2}{10} = 5\dfrac{1}{5}$

Sony: $55.53 = 55\dfrac{53}{100}$

Verizon: $41\dfrac{4}{25} = 41\dfrac{16}{100} = 41.16$

81. $2.25 = 2\dfrac{1}{4} = \dfrac{9}{4}$

83. 0.28

Section 4.6

85. $7.5 \div \dfrac{3}{2} = 7\dfrac{1}{2} \div \dfrac{3}{2} = \dfrac{\overset{5}{\cancel{15}}}{\cancel{2}_1} \cdot \dfrac{\overset{1}{\cancel{2}}}{\cancel{3}_1} = 5$

87. $3.14(5)^2 = 3.14(25)$

$$
\begin{array}{r}
3.14 \\
\times\ 25 \\
\hline
15\,70 \\
62\,80 \\
\hline
78.50
\end{array}
$$

$= 78.5$

89.
$$
\begin{array}{r}
189.95 \\
199.95 \\
+\ 219.95 \\
\hline
609.85 \\
-\ 519.95 \\
\hline
89.90
\end{array}
$$

$89.90 will be saved by buying the combo package.

Chapter 4   Test

1. (a) Tens place

(b) Hundredths place

3. $1.26 = 1\dfrac{26}{100} = 1\dfrac{13}{50} = \dfrac{63}{50}$

5. b is correct.

7.
$$
\begin{array}{r}
34.09 \\
-\ 12.80 \\
\hline
21.29
\end{array}
$$

9.
$$5\overline{\smash{\big)}25.40}$$
$$\underline{-25}$$
$$0\,40$$
$$\underline{-40}$$
$$0$$

quotient: 5.08

11.
$$
\begin{array}{r}
12.0300 \\
+\ 0.1943 \\
\hline
12.2243
\end{array}
$$

13.
$$\begin{array}{r} 42.7 \\ \times\ 10.3 \\ \hline 12\,81 \\ 427\,00 \\ \hline 439.81 \end{array}$$

15. $579.23 \times 100 = 57{,}923$

17. $2.931 \div 1000 = 0.002931$

19. (a) 50,500,000 votes

(b) 51,000,000 votes

(c) The difference is approximately 500,000 in favor of Al Gore.

21. Cost of purchase:

$$\begin{array}{r} 36.625 \\ \times\ 200 \\ \hline 7325.000 \end{array} \qquad \begin{array}{r} 7325.00 \\ +\ 4.25 \\ \hline 7329.25 \end{array}$$

Receipt from sale:

$$\begin{array}{r} 52.16 \\ \times\ 200 \\ \hline 10432.00 \end{array} \qquad \begin{array}{r} 10432.00 \\ -\ 8.00 \\ \hline 10424.00 \end{array}$$

Net profit:
$$\begin{array}{r} 10424.00 \\ -7329.25 \\ \hline 3094.75 \end{array}$$
He made \$3094.75.

23.
$$\begin{array}{r} 4.8 \\ 23\overline{)110.4} \\ \underline{-92} \\ 18\,4 \\ \underline{-18\,4} \\ 0 \end{array} \qquad \begin{array}{r} 5.2 \\ 26\overline{)135.2} \\ \underline{-130} \\ 5\,2 \\ \underline{-5\,2} \\ 0 \end{array}$$

$4.8 + 5.2 = 10.0$
He will use 10 gal of gas.

25.

27.
$$\frac{7}{3}\left(5.25 - \frac{3}{4}\right)^2 = \frac{7}{3}\left(\frac{21}{4} - \frac{3}{4}\right)^2 = \frac{7}{3}\left(\frac{18}{4}\right)^2$$
$$= \frac{7}{3}\left(\frac{9}{2}\right)^2 = \frac{7}{3} \times \frac{81}{4}$$
$$= \frac{7}{\overset{}{\cancel{3}}}\left(\frac{\overset{27}{\cancel{81}}}{4}\right) = \frac{189}{4} = 47.25$$

Chapters 1–4   Cumulative Review Exercises

1. $(17 + 12) - (8 - 3) \cdot 3 = (29) - (5) \cdot 3$
$$= 29 - 15 = 14$$

3.
$$\begin{array}{r} 3902 \\ 34 \\ +\ 904 \\ \hline 4840 \end{array}$$

5.
$$\begin{array}{r} 23{,}444 \\ \times\ 103 \\ \hline 70\,332 \\ 2\,344\,400 \\ \hline 2{,}414{,}732 \approx 2{,}415{,}000 \end{array}$$

7. To check a division problem, multiply the whole-number part of the quotient and the divisor. Then add the remainder to get the dividend.
$$20 \times 225 + 30 = 4530$$

9. $\dfrac{1}{5} \cdot \dfrac{6}{11} = \dfrac{6}{55}$

11. $\left(\dfrac{7}{10}\right)^2 = \left(\dfrac{7}{10}\right)\left(\dfrac{7}{10}\right) = \dfrac{49}{100}$

13. $\dfrac{8}{3} \div 4 = \dfrac{8}{3} \cdot \dfrac{1}{4} = \dfrac{8}{12} = \dfrac{2}{3}$

15. $\frac{2}{5}(15,000) = \frac{30,000}{5} = 6000$

$$\begin{array}{r} 15,000 \\ -\ 6\,000 \\ \hline 9\,000 \end{array}$$

There is \$9000 left.

17. $\frac{7}{10} + \frac{27}{100} = \frac{70}{100} + \frac{27}{100} = \frac{97}{100}$

19. $5 - \frac{2}{7} = \frac{35}{7} - \frac{2}{7} = \frac{33}{7}$

21. $A = l \cdot w = \frac{5}{8} \cdot \frac{3}{8} = \frac{15}{64}$ ft^2

$$P = 2l + 2w = 2\left(\frac{5}{8}\right) + 2\left(\frac{3}{8}\right) = \frac{10}{8} + \frac{6}{8}$$

$$= \frac{16}{8} = 2 \text{ ft}$$

23. $\begin{array}{r} 50.90 \\ +\ 123.23 \\ \hline 174.13 \end{array}$

25. $\begin{array}{r} 301.1 \\ \times\ 0.25 \\ \hline 15\ 055 \\ 60\ 220 \\ \hline 75.275 \end{array}$

27. $\frac{4}{3}(3.14)(9)^2 = \frac{4}{3}(3.14)(81)$

$$= \frac{4}{3}(81)(3.14) = (108)(3.14)$$

$$\begin{array}{r} 3.14 \\ \times\ 1\ 08 \\ \hline 25\ 12 \\ 314\ 00 \\ \hline 339.12 \end{array}$$

$$= 339.12$$

29. (a) $\begin{array}{r} 0.004 \\ \times\ 938.12 \\ \hline 8 \\ 40 \\ 3200 \\ 12000 \\ 3\ 60000 \\ \hline 3.75248 \end{array}$

(b) $\begin{array}{r} 938.12 \\ \times\ 0.004 \\ \hline 3.75248 \end{array}$

(c) Commutative property of multiplication

Chapter 5   Ratio and Proportion

Chapter Opener Puzzle

1. t r o i a − ratio

2. r s i l a i m − similar

3. e a r t − rate

4. o o r r i p p n t o − proportion

5. n i q t e a u o − equation

$$\frac{5}{6}$$

It is difficult to move a square because it has a **square root**.

Section 5.1   Ratios

Section 5.1 Practice Exercises

1. Answers will vary.

3. 5 : 6 and

5. 11 to 4 and $\dfrac{11}{4}$

7. 1 : 2 and 1 to 2

9. (a) $\dfrac{3}{2}$

   (b) $\dfrac{2}{3}$

   (c) Total = 3 + 2 = 5

    $\dfrac{3}{5}$

11. (a) $\dfrac{21}{52}$

   (b) Cars that are not silver = 52 − 21 = 31

    $\dfrac{21}{31}$

13. $\dfrac{4 \text{ yr}}{6 \text{ yr}} = \dfrac{2}{3}$

15. $\dfrac{5 \text{ mi}}{25 \text{ mi}} = \dfrac{1}{5}$

17. $\dfrac{8 \text{ m}}{2 \text{ m}} = \dfrac{4}{1}$

19. $\dfrac{33 \text{ cm}}{15 \text{ cm}} = \dfrac{11}{5}$

21. $\dfrac{\$60}{\$50} = \dfrac{6}{5}$

23. $\dfrac{18 \text{ in.}}{36 \text{ in.}} = \dfrac{1}{2}$

25. $\dfrac{3.6}{2.4} = \dfrac{3.6 \times 10}{2.4 \times 10} = \dfrac{36}{24} = \dfrac{3}{2}$

27. $\dfrac{8}{9\frac{1}{3}} = \dfrac{8}{\frac{28}{3}} = \dfrac{8}{1} \div \dfrac{28}{3} = \dfrac{8}{1} \cdot \dfrac{3}{28} = \dfrac{24}{28} = \dfrac{6}{7}$

29. $\dfrac{16\frac{4}{5}}{18\frac{9}{10}} = \dfrac{\frac{84}{5}}{\frac{189}{10}} = \dfrac{84}{5} \div \dfrac{189}{10} = \dfrac{\overset{4}{\cancel{84}}}{\underset{1}{\cancel{5}}} \cdot \dfrac{\overset{2}{\cancel{10}}}{\underset{9}{\cancel{189}}} = \dfrac{8}{9}$

31. $\dfrac{16.80}{2.40} = \dfrac{16.80 \times 100}{2.40 \times 100} = \dfrac{1680}{240} = \dfrac{7}{1}$

33. $\dfrac{\frac{1}{2}}{4} = \dfrac{1}{2} \div 4 = \dfrac{1}{2} \cdot \dfrac{1}{4} = \dfrac{1}{8}$

35. $\dfrac{10.25}{8.2} = \dfrac{10.25 \times 100}{8.2 \times 100} = \dfrac{1025}{820} = \dfrac{5}{4}$

37. Increase = $90 - 66 = 24$

$$\frac{24}{66} = \frac{4}{11}$$

39. (a) $1\frac{1}{3}$ ft $= 12 + 4 = 16$ in.

$$\frac{6}{16} = \frac{3}{8}$$

(b) 6 in. $= \frac{1}{2}$ ft

$$\frac{\frac{1}{2}}{1\frac{1}{3}} = \frac{\frac{1}{2}}{\frac{4}{3}} = \frac{1}{2} \div \frac{4}{3} = \frac{1}{2} \cdot \frac{3}{4} = \frac{3}{8}$$

41. $\dfrac{400,000}{4,400,000} = \dfrac{4}{44} = \dfrac{1}{11}$

43. Increase $= 4,400,000 - 400,000$
$$= 4,000,000$$

$$\frac{4,000,000}{400,000} = \frac{10}{1}$$

45. $\dfrac{60}{128} = \dfrac{15}{32}$

47. $\dfrac{60}{183} = \dfrac{20}{61}$

49. $\dfrac{1\frac{1}{2}}{2\frac{1}{4}} = \dfrac{\frac{3}{2}}{\frac{9}{4}} = \dfrac{3}{2} \div \dfrac{9}{4} = \dfrac{3}{2} \cdot \dfrac{4}{9} = \dfrac{12}{18} = \dfrac{2}{3}$

51. $\dfrac{0.3}{1.2} = \dfrac{0.3 \times 10}{1.2 \times 10} = \dfrac{3}{12} = \dfrac{1}{4}$

53. $8 + 5 = 13$ units

55. (a) $\dfrac{3}{2} = 1.5$

(b) $\dfrac{5}{3} = 1.\overline{6}$

(c) $\dfrac{8}{5} = 1.6$

(d) $\dfrac{13}{8} = 1.625$

Yes, they are approaching 1.618.

57. Answers will vary.

Section 5.2   Rates

Section 5.2   Practice Exercises

1. Answers will vary.

3. $3 : 5$ and $\dfrac{3}{5}$

5. $\dfrac{36¢}{27¢} = \dfrac{4}{3}$

7. $\dfrac{1.08 \text{ mi}}{2.04 \text{ mi}} = \dfrac{1.08 \times 100}{2.04 \times 100} = \dfrac{108}{204} = \dfrac{9}{17}$

9. $\dfrac{\$32}{5 \text{ ft}^2}$

11. $\dfrac{234 \text{ mi}}{4 \text{ hr}} = \dfrac{117 \text{ mi}}{2 \text{ hr}}$

13. $\dfrac{\$58}{8 \text{ hr}} = \dfrac{\$29}{4 \text{ hr}}$

15. $\dfrac{13 \text{ pages}}{26 \text{ sec}} = \dfrac{1 \text{ page}}{2 \text{ sec}}$

17. $\dfrac{130 \text{ calories}}{8 \text{ crackers}} = \dfrac{65 \text{ calories}}{4 \text{ crackers}}$

19. $\dfrac{\$30}{4 \text{ trays}} = \dfrac{\$15}{2 \text{ trays}}$

21. a, c, d

23. $\dfrac{452 \text{ mi}}{4 \text{ days}} = \dfrac{113 \text{ mi}}{1 \text{ day}}$ or 113 mi/day

25. $\dfrac{480 \text{ km}}{5 \text{ hr}} = 96 \text{ km/hr}$

27. $\dfrac{\$660}{12 \text{ payments}} = \$55 \text{ per payment}$

$$\begin{array}{r} 55 \\ 12\overline{)660} \\ \underline{-60} \\ 60 \\ \underline{-60} \\ 0 \end{array}$$

29. $\dfrac{\$1.50}{4 \text{ lb}} \approx \$0.38 \text{ lb}$

$$\begin{array}{r} 0.375 \\ 4\overline{)1.500} \\ \underline{-1\,2} \\ 30 \\ \underline{-28} \\ 20 \\ \underline{-20} \\ 0 \end{array}$$

31. $\dfrac{\$1,792,000}{7 \text{ people}} = \$256,000 \text{ per person}$

$$\begin{array}{r} 256,000 \\ 7\overline{)1,792,000} \\ \underline{-1\,4} \\ 39 \\ \underline{-35} \\ 42 \\ \underline{-42} \\ 0 \end{array}$$

33. $\dfrac{500 \text{ m}}{35 \text{ sec}} \approx 14.3 \text{ m/sec}$

$$\begin{array}{r} 14.28 \\ 35\overline{)500.00} \\ \underline{-35} \\ 150 \\ \underline{-140} \\ 100 \\ \underline{-70} \\ 300 \\ \underline{-280} \\ 20 \end{array}$$

35. $\dfrac{\$4.99}{100 \text{ oz}} = 0.0499 \approx \$0.050 \text{ per oz}$

37. $\dfrac{\$1.99}{2 \text{ liters}} = \$0.995 \text{ per liter}$

39. $\dfrac{\$210}{4 \text{ tires}} = \$52.50 \text{ per tire}$

41. $\dfrac{\$32.50}{6 \text{ bodysuits}} \approx \$5.417 \text{ per bodysuit}$

43. **(a)** $\dfrac{\$3.00}{40 \text{ oz}} = \$0.075/\text{oz}$

   (b) $\dfrac{\$2.10}{28 \text{ oz}} = \$0.075/\text{oz}$

   (c) Both sizes cost the same amount per ounce.

45. $\dfrac{\$1.19}{29 \text{ oz}} \approx \$0.041 \text{ per ounce}$

$\dfrac{\$0.77}{15 \text{ oz}} \approx \$0.051 \text{ per ounce}$

The larger can is the better buy.

47. Coca-Cola: $\dfrac{65}{20} = 3.25 \text{ g/fl oz}$

   Mello Yello: $\dfrac{47}{12} = 3.92 \text{ g/fl oz}$

   Ginger Ale: $\dfrac{24}{8} = 3 \text{ g/fl oz}$

   Mello Yello has the greatest amount per fluid oz.

49. Coca-Cola: $\dfrac{240}{20} = 12 \text{ cal/fl oz}$

   Mello Yello: $\dfrac{170}{12} = 14.2 \text{ cal/fl oz}$

   Ginger Ale: $\dfrac{90}{8} = 11.25 \text{ cal/fl oz}$

   Ginger Ale has the least number of calories per fluid oz.

51. $\dfrac{5,310,000}{18} = 295,000 \text{ vehicles per year}$

53. (a) $\dfrac{22,000,000}{10} = 2,200,000$ per year

(b) $\dfrac{10,200,000}{5} = 2,040,000$ per year

(c) Mexico

55. cheetah: $\dfrac{120}{4.1} \approx 29$ m/sec

antelope: $\dfrac{50}{2.1} \approx 24$ m/sec

The cheetah is faster.

57. (a) $\dfrac{328 \text{ wins}}{156 \text{ losses}} \approx 2.1$ wins/loss

(b) $\dfrac{250 \text{ wins}}{162 \text{ losses}} \approx 1.5$ wins/loss

(c) Shula

59. $\dfrac{\$3.61}{32 \text{ oz}} \approx \$0.113$ per oz

$\dfrac{\$2.19}{16 \text{ oz}} \approx \$0.137$ per oz

$\dfrac{\$1.19}{8 \text{ oz}} \approx \$0.149$ per oz

The best buy is the 32-oz jar.

61. (a) $\dfrac{\$4.99}{24 \times 12 \text{ oz}} = \dfrac{4.99}{288} \approx \$0.017$ per oz

(b) $\dfrac{\$5.00}{12 \times 8 \text{ oz}} = \dfrac{5.00}{96} \approx \$0.052$ per oz

The case of 24 12-oz cans for \$4.99 is the better buy.

Section 5.3   Proportions

Section 5.3   Practice Exercises

1. Pages 330 – 331

3. $\dfrac{3 \text{ ft}}{45 \text{ ft}} = \dfrac{1}{15}$

5. $\dfrac{6 \text{ apples}}{2 \text{ pies}} = \dfrac{3 \text{ apples}}{1 \text{ pie}}$

7. $\dfrac{264 \text{ mi}}{36 \text{ gal}} = \dfrac{22 \text{ mi}}{3 \text{ gal}}$

9. $\dfrac{4}{16} = \dfrac{5}{20}$

11. $\dfrac{25}{15} = \dfrac{10}{6}$

13. $\dfrac{2}{3} = \dfrac{4}{6}$

15. $\dfrac{30}{25} = \dfrac{12}{10}$

17. $\dfrac{\$6.25}{1 \text{ hr}} = \dfrac{\$187.50}{30 \text{ hr}}$

19. $\dfrac{1 \text{ in.}}{7 \text{ mi}} = \dfrac{5 \text{ in.}}{35 \text{ mi}}$

21. $\dfrac{5}{18} \; \blacklozenge \; \dfrac{4}{16}$

$(5)(16) \; \blacklozenge \; (18)(4)$

$80 \neq 72$

No

23. $\dfrac{16}{24} \; \blacklozenge \; \dfrac{2}{3}$

$(16)(3) \; \blacklozenge \; (24)(2)$

$48 = 48$

Yes

25.
$$\frac{2\frac{1}{2}}{3\frac{2}{3}} \blacklozenge \frac{15}{22}$$

$$\left(2\frac{1}{2}\right)(22) \blacklozenge \left(3\frac{2}{3}\right)(15)$$

$$\frac{5}{\cancel{2}_1} \cdot \frac{\cancel{22}^{11}}{1} \blacklozenge \frac{11}{\cancel{3}_1} \cdot \frac{\cancel{15}^5}{1}$$

$$55 = 55$$

Yes

27.
$$\frac{2}{3.2} \blacklozenge \frac{10}{16}$$
$$(2)(16) \blacklozenge (3.2)(10)$$
$$32 = 32$$
Yes

29.
$$\frac{48}{18} \blacklozenge \frac{24}{9}$$
$$(48)(9) \blacklozenge (18)(24)$$
$$432 = 432$$
Yes

$$\frac{2\frac{3}{8}}{1\frac{1}{2}} \blacklozenge \frac{9\frac{1}{2}}{6}$$

$$\left(2\frac{3}{8}\right)(6) \blacklozenge \left(1\frac{1}{2}\right)\left(9\frac{1}{2}\right)$$

$$\frac{19}{\cancel{8}_4} \cdot \frac{\cancel{6}^3}{1} \blacklozenge \frac{3}{2} \cdot \frac{19}{2}$$

$$\frac{57}{4} = \frac{57}{4}$$

31.
Yes

33.
$$\frac{6.3}{9} \blacklozenge \frac{12.6}{16}$$
$$(6.3)(16) \blacklozenge (9)(12.6)$$
$$100.8 \neq 113.4$$
No

35. Divide by 2

37. Divide by 5

39. Divide by 8

41. Divide by 0.6

43. $\frac{x}{40} = \frac{1}{8}; x = 5$

$$\frac{5}{40} \blacklozenge \frac{1}{8}$$
$$(5)(8) \blacklozenge 40$$
$$40 = 40$$
Yes

45. $\frac{12.4}{31} = \frac{8.2}{y}; y = 20$

$$\frac{12.4}{31} \blacklozenge \frac{8.2}{20}$$
$$(12.4)(20) \blacklozenge (31)(8.2)$$
$$248 \neq 254.2$$
No

47. $\frac{12}{16} = \frac{3}{x}$ Check: $\frac{12}{16} \blacklozenge \frac{3}{4}$
$$12x = (16)(3) \qquad (12)(4) \blacklozenge (16)(3)$$
$$12x = 48 \qquad\qquad 48 = 48 ✓$$
$$\frac{12x}{12} = \frac{48}{12}$$
$$x = 4$$

49. $\frac{9}{21} = \frac{x}{7}$ Check: $\frac{9}{21} \blacklozenge \frac{3}{7}$
$$(9)(7) = 21x \qquad (9)(7) \blacklozenge (21)(3)$$
$$63 = 21x \qquad\qquad 63 = 63 ✓$$
$$\frac{63}{21} = \frac{21x}{21}$$
$$3 = x$$

51. $\frac{p}{12} = \frac{25}{4}$ Check: $\frac{75}{12} \blacklozenge \frac{25}{4}$
$$4p = (12)(25) \qquad (75)(4) \blacklozenge (12)(25)$$
$$4p = 300 \qquad\qquad 300 = 300 ✓$$
$$\frac{4p}{4} = \frac{300}{4}$$
$$p = 75$$

53. $\frac{6}{n} = \frac{4}{8}$ Check: $\frac{6}{12} \blacklozenge \frac{4}{8}$
$$(6)(8) = 4n \qquad (6)(8) \blacklozenge (12)(4)$$
$$48 = 4n \qquad\qquad 48 = 48 ✓$$
$$\frac{48}{4} = \frac{4n}{4}$$
$$12 = n$$

55. $\dfrac{2}{3} = \dfrac{t}{18}$ Check: $\dfrac{2}{3} \blacklozenge \dfrac{12}{18}$

$(2)(18) = 3t$ $(2)(18) \blacklozenge (3)(12)$

$36 = 3t$ $36 = 36 \checkmark$

$\dfrac{36}{3} = \dfrac{3t}{3}$

$12 = t$

57. $\dfrac{25}{100} = \dfrac{9}{y}$

$25y = (100)(9)$

$25y = 900$

$\dfrac{25y}{25} = \dfrac{900}{25}$

$y = 36$

Check: $\dfrac{25}{100} \blacklozenge \dfrac{9}{36}$

$(25)(36) \blacklozenge (100)(9)$

$900 = 900 \checkmark$

59. $\dfrac{17}{12} = \dfrac{4\frac{1}{4}}{x}$

$17x = \left(4\dfrac{1}{4}\right)(12)$

$17x = \dfrac{17}{\cancel{4}} \cdot \dfrac{\cancel{12}^{\,3}}{1}$

$17x = 51$

$\dfrac{17x}{17} = \dfrac{51}{17}$

$x = 3$

Check: $\dfrac{17}{12} \blacklozenge \dfrac{4\frac{1}{4}}{3}$

$(17)(3) \blacklozenge (12)\left(4\dfrac{1}{4}\right)$

$51 \blacklozenge \cancel{12}^{\,3} \cdot \dfrac{17}{\cancel{4}}_{1}$

$51 = 51 \checkmark$

61. $\dfrac{m}{12} = \dfrac{5}{8}$

$8m = (12)(5)$

$8m = 60$

$\dfrac{8m}{8} = \dfrac{60}{8}$

$m = \dfrac{15}{2}$ or $7\dfrac{1}{2}$ or 7.5

Check: $\dfrac{7.5}{12} \blacklozenge \dfrac{5}{8}$

$(7.5)(8) \blacklozenge (12)(5)$

$60 = 60 \checkmark$

63. $\dfrac{3.125}{5} = \dfrac{18.75}{k}$

$(3.125)k = (5)(18.75)$

$3.125k = 93.75$

$\dfrac{3.125k}{3.125} = \dfrac{93.75}{3.125}$

$k = 30$

Check: $\dfrac{3.125}{5} \blacklozenge \dfrac{18.75}{30}$

$(3.125)(30) \blacklozenge (5)(18.75)$

$93.75 = 93.75 \checkmark$

65. $\dfrac{0.5}{h} = \dfrac{1.8}{9}$

$(0.5)(9) = 1.8h$

$4.5 = 1.8h$

$\dfrac{4.5}{1.8} = \dfrac{1.8h}{1.8}$

$2.5 = h$

Check: $\dfrac{0.5}{2.5} \blacklozenge \dfrac{1.8}{9}$

$(0.5)(9) \blacklozenge (2.5)(1.8)$

$4.5 = 4.5 \checkmark$

67.

$$\frac{\frac{3}{8}}{6.75} = \frac{x}{72}$$

$$6.75x = \left(\frac{3}{8}\right)(72)$$

$$6.75x = \frac{3}{\cancel{8}_1} \cdot \frac{\cancel{72}^9}{1}$$

$$6.75x = 27$$

$$\frac{6.75x}{6.75} = \frac{27}{6.75}$$

$$x = 4$$

Check: $\dfrac{\frac{3}{8}}{6.75} \;\blacklozenge\; \dfrac{4}{72}$

$$\left(\frac{3}{8}\right)(72) \;\blacklozenge\; (4)(6.75)$$

$$\frac{3}{\cancel{8}_1} \cdot \cancel{72}^9 \;\blacklozenge\; 27$$

$$27 = 27 \;\checkmark$$

69.

$$\frac{4}{\frac{1}{10}} = \frac{\frac{1}{2}}{z}$$

$$4z = \left(\frac{1}{10}\right)\left(\frac{1}{2}\right)$$

$$4z = \frac{1}{20}$$

$$\frac{1}{4} \cdot 4z = \frac{1}{4} \cdot \frac{1}{20}$$

$$z = \frac{1}{80}$$

Check: $\dfrac{4}{\frac{1}{10}} \;\blacklozenge\; \dfrac{\frac{1}{2}}{\frac{1}{80}}$

$$(4)\left(\frac{1}{80}\right) \;\blacklozenge\; \left(\frac{1}{10}\right)\left(\frac{1}{2}\right)$$

$$\frac{1}{20} = \frac{1}{20} \;\checkmark$$

Section 5.4   Applications of Proportions and Similar Figures

Section 5.4   Practice Exercises

1. Answers will vary.

3.

$$\frac{3}{13} \;\blacklozenge\; \frac{15}{65}$$

$$(3)(65) \;\blacklozenge\; (13)(15)$$

$$195 = 195$$

$$\frac{3}{13} = \frac{15}{65}$$

5.

$$\frac{12}{7} \;\blacklozenge\; \frac{35}{19}$$

$$(12)(19) \;\blacklozenge\; (7)(35)$$

$$228 \neq 245$$

$$\frac{12}{7} \neq \frac{35}{19}$$

7.

$$\frac{4}{3} = \frac{n}{5}$$

$$(4)(5) = 3n$$

$$20 = 3n$$

$$\frac{20}{3} = \frac{3n}{3}$$

$$n = \frac{20}{3} \text{ or } 6\frac{2}{3} \text{ or } 6.\overline{6}$$

9.
$$\frac{3\frac{1}{2}}{k} = \frac{2\frac{1}{3}}{4}$$

$$\left(3\frac{1}{2}\right)(4) = \left(2\frac{1}{3}\right)k$$

$$\frac{7}{2} \cdot \frac{4}{1} = \frac{7}{3}k$$

$$\frac{28}{2} = \frac{7}{3}k$$

$$\frac{3}{\cancel{7}_{1}} \cdot \frac{\cancel{28}^{4}}{2} = \frac{3}{7} \cdot \frac{7}{3}k$$

$$\frac{12}{2} = k$$

$$6 = k$$

11.
$$\frac{3}{2.1} = \frac{7}{y}$$

$$3y = (2.1)(7)$$

$$3y = 14.7$$

$$\frac{3y}{3} = \frac{14.7}{3}$$

$$y = 4.9$$

13. Let x represent the number of miles.
$$\frac{244 \text{ mi}}{4 \text{ gal}} = \frac{x \text{ mi}}{10 \text{ gal}}$$

$$(244)(10) = 4x$$

$$2440 = 4x$$

$$\frac{2440}{4} = \frac{4x}{4}$$

$$610 = x$$

Pam can drive 610 mi on 10 gal of gas.

15. Let x represent the amount of crushed rock needed.
$$\frac{x \text{ kg}}{24 \text{ kg}} = \frac{3.25}{1}$$

$$x = (24)(3.25)$$

$$x = 78$$

78 kg of crushed rock will be required.

17. Let x represent the distance from Sacramento to Modesto.
$$\frac{8 \text{ cm}}{91 \text{ mi}} = \frac{7 \text{ cm}}{x \text{ mi}}$$

$$8x = (91)(7)$$

$$8x = 637$$

$$\frac{8x}{8} = \frac{637}{8}$$

$$x = 79.625$$

The actual distance is about 80 mi.

19. Let x represent the number of male students.
$$\frac{6200 \text{ female}}{x \text{ male}} = \frac{31}{19}$$

$$(6200)(19) = 31x$$

$$117{,}800 = 31x$$

$$\frac{117{,}800}{31} = \frac{31x}{31}$$

$$3800 = x$$

There are 3800 male students.

21. Let x represent the number of heads.
$$\frac{x \text{ heads}}{630 \text{ flips}} = \frac{1}{2}$$

$$2x = (630)(1)$$

$$2x = 630$$

$$\frac{2x}{2} = \frac{630}{2}$$

$$x = 315$$

Heads would come up about 315 times.

23. Let x represent the number of earned runs.
$$\frac{x \text{ runs}}{9 \text{ innings}} = \frac{42 \text{ runs}}{126 \text{ innings}}$$

$$126x = (9)(42)$$

$$126x = 378$$

$$\frac{126x}{126} = \frac{378}{126}$$

$$x = 3$$

There would be approximately 3 earned runs for a 9-inning game.

25. Let x represent the number of Euros.

$$\frac{37.0 \text{ Euros}}{\$50} = \frac{x \text{ Euros}}{\$900}$$

$$(37.0)(900) = 50x$$

$$33{,}300 = 50x$$

$$\frac{33{,}300}{50} = \frac{50x}{50}$$

$$666 = x$$

Pierre can buy 666 Euros.

27. Let x represent the number of falls.

$$\frac{x \text{ falls}}{60 \text{ visits}} = \frac{3 \text{ falls}}{4 \text{ visits}}$$

$$4x = (60)(3)$$

$$4x = 180$$

$$\frac{4x}{4} = \frac{180}{4}$$

$$x = 45$$

45 visits would be a result of falls.

29. Let x represent the number of fish.

$$\frac{21 \text{ fish}}{100 \text{ fish}} = \frac{75 \text{ fish}}{x \text{ fish}}$$

$$21x = (100)(75)$$

$$21x = 7500$$

$$\frac{21x}{21} = \frac{7500}{21}$$

$$x \approx 357$$

There are approximately 357 bass in the lake.

31. Let x represent the number of bison.

$$\frac{200 \text{ bison}}{x \text{ bison}} = \frac{6 \text{ bison}}{120 \text{ bison}}$$

$$(200)(120) = 6x$$

$$24{,}000 = 6x$$

$$\frac{24{,}000}{6} = \frac{6x}{6}$$

$$4000 = x$$

There are approximately 4000 bison in the park.

33.
$$\frac{16}{32} = \frac{12}{x}$$

$$16x = (32)(12)$$

$$16x = 384$$

$$\frac{16x}{16} = \frac{384}{16}$$

$$x = 24 \text{ cm}$$

$$\frac{16}{32} = \frac{18}{y}$$

$$16y = (32)(18)$$

$$16y = 576$$

$$\frac{16y}{16} = \frac{576}{16}$$

$$y = 36 \text{ cm}$$

35.
$$\frac{x}{1.5} = \frac{6}{9}$$

$$9x = (1.5)(6)$$

$$9x = 9$$

$$\frac{9x}{9} = \frac{9}{9}$$

$$x = 1 \text{ yd}$$

$$\frac{6}{9} = \frac{7}{y}$$

$$6y = (9)(7)$$

$$6y = 63$$

$$\frac{6y}{6} = \frac{63}{6}$$

$$y = 10.5 \text{ yd}$$

37.
$$\frac{h}{3} = \frac{10}{2\frac{1}{2}}$$

$$\left(2\frac{1}{2}\right)h = (3)(10)$$

$$\frac{5}{2}h = 30$$

$$\frac{2}{5} \cdot \frac{5}{2}h = \frac{2}{5} \cdot \frac{30}{1}$$

$$h = \frac{60}{5}$$

$$h = 12$$

The flagpole is 12 ft high.

39. Let x represent the height of the platform.

$$\frac{1.6}{1} = \frac{x}{1.5}$$

$$(1.6)(1.5) = 1x$$

$$2.4 = x$$

The platform is 2.4 m tall.

41.
$$\frac{4}{10} = \frac{7}{x}$$

$$4x = (10)(7)$$

$$4x = 70$$

$$\frac{4x}{4} = \frac{70}{4}$$

$$x = 17.5 \text{ in.}$$

43.

$$\frac{1.6}{2} = \frac{4.8}{x}$$
$$1.6x = (2)(4.8)$$
$$1.6x = 9.6$$
$$\frac{1.6x}{1.6} = \frac{9.6}{1.6}$$
$$x = 6 \text{ ft}$$

$$\frac{2}{1.6} = \frac{10}{y}$$
$$2y = (1.6)(10)$$
$$2y = 16$$
$$\frac{2y}{2} = \frac{16}{2}$$
$$y = 8 \text{ ft}$$

45.

$$\frac{8 \text{ in.}}{28 \text{ ft}} = \frac{6 \text{ in.}}{x \text{ ft}}$$
$$8x = (28)(6)$$
$$8x = 168$$
$$\frac{8x}{8} = \frac{168}{8}$$
$$x = 21 \text{ ft}$$

$$\frac{8 \text{ in.}}{28 \text{ ft}} = \frac{6 \text{ in.}}{y \text{ ft}}$$
$$8y = (28)(6)$$
$$8y = 168$$
$$\frac{8y}{8} = \frac{168}{8}$$
$$y = 21 \text{ ft}$$

$$\frac{8 \text{ in.}}{28 \text{ ft}} = \frac{15.2 \text{ in.}}{z \text{ ft}}$$
$$8z = (28)(15.2)$$
$$8z = 425.6$$
$$\frac{8z}{8} = \frac{425.6}{8}$$
$$z = 53.2 \text{ ft}$$

47. Let x represent the number of crimes.

$$\frac{4743 \text{ crimes}}{100,000 \text{ people}} = \frac{x \text{ crimes}}{3,500,000 \text{ people}}$$
$$(4743)(3,500,000) = 100,000x$$
$$16,600,500,000 = 100,000x$$
$$\frac{16,600,500,000}{100,000} = \frac{100,000x}{100,000}$$
$$166,005 = x$$

There were approximately 166,005 crimes committed.

49. Let x represent the number of women with breast cancer.

$$\frac{110 \text{ cases}}{100,000 \text{ women}} = \frac{x \text{ cases}}{14,000,000 \text{ women}}$$
$$(110)(14,000,000) = 100,000x$$
$$1,540,000,000 = 100,000x$$
$$\frac{1,540,000,000}{100,000} = \frac{100,000x}{100,000}$$
$$15,400 = x$$

Approximately 15,400 women would be expected to have breast cancer.

Chapter 5   Review Exercises

Section 5.1

1. 5 to 4 and $\dfrac{5}{4}$

3. 8 : 7 and 8 to 7

5. (a) $\dfrac{4}{5}$

  (b) $\dfrac{5}{4}$

  (c) Total bottles = 4 + 5 = 9
  $$\frac{5}{9}$$

7. $\dfrac{52 \text{ cards}}{13 \text{ cards}} = \dfrac{4}{1}$

9. $\dfrac{80 \text{ ft}}{200 \text{ ft}} = \dfrac{2}{5}$

11. $\dfrac{1\frac{1}{2} \text{ hr}}{\frac{1}{3} \text{ hr}} = \dfrac{\frac{3}{2}}{\frac{1}{3}} = \dfrac{3}{2} \div \dfrac{1}{3} = \dfrac{3}{2} \cdot \dfrac{3}{1} = \dfrac{9}{2}$

13. $\dfrac{\$2.56}{\$1.92} = \dfrac{2.56 \times 100}{1.92 \times 100} = \dfrac{256}{192} = \dfrac{4}{3}$

15. (a) $1200 + 320 = 1520$
    This year's enrollment is 1520 students.

  (b) $\dfrac{320 \text{ students}}{1520 \text{ students}} = \dfrac{4}{19}$

17. $\dfrac{12}{60} = \dfrac{1}{5}$

Section 5.2

19. $\dfrac{20 \text{ hot dogs}}{45 \text{ min}} = \dfrac{4 \text{ hot dogs}}{9 \text{ min}}$

21. $\dfrac{130{,}000 \text{ tons}}{1800 \text{ ft}} = \dfrac{650 \text{ tons}}{9 \text{ ft}}$

23. All unit rates have a denominator of 1, and reduced rates may not.

25. $\dfrac{14°}{3.5 \text{ hr}} = 4° \text{ per hour}$

27. $\dfrac{66 \text{ min}}{6 \text{ lawns}} = 11 \text{ min/lawn}$

29. $\dfrac{\$10.00}{3 \text{ towels}} = \$3.333 \text{ per towel}$

31. (a) $\dfrac{\$2.49}{32 \text{ oz}} = \$0.078/\text{oz}$

    (b) $\dfrac{\$3.59}{48 \text{ oz}} = \$0.075/\text{oz}$

    (c) The 48-oz jar is the best buy.

33. $\dfrac{\$6.99}{24 \text{ rolls}} \approx \$0.29 \text{ per roll}$

$$
\begin{array}{r}
0.291 \\
24\overline{)\,6.990} \\
-48 \\
\hline
219 \\
-216 \\
\hline
30 \\
-24 \\
\hline
6
\end{array}
$$

$\$0.29 - 17¢ = \$0.29 - \$0.17 = \$0.12$
The difference is about $0.12 or 12¢ per roll.

35. (a) $250{,}000 - 130{,}000 = 120{,}000$
    There was an increase of 120,000 hybrid vehicles.

    (b) $\dfrac{120{,}000}{12} = 10{,}000$
    There will be 10,000 additional hybrid vehicles per month.

Section 5.3

37. $\dfrac{16}{14} = \dfrac{12}{10\frac{1}{2}}$

39. $\dfrac{5}{3} = \dfrac{10}{6}$

41. $\dfrac{\$11}{1 \text{ hr}} = \dfrac{\$88}{8 \text{ hr}}$

43. $\dfrac{64}{81} \, \blacklozenge \, \dfrac{8}{9}$
$(64)(9) \, \blacklozenge \, (81)(8)$
$576 \neq 648$
No

45. $\dfrac{5.2}{3} \, \blacklozenge \, \dfrac{15.6}{9}$
$(5.2)(9) \, \blacklozenge \, (3)(15.6)$
$46.8 = 46.8$
Yes

47. $\dfrac{2\frac{1}{8}}{4\frac{3}{4}} \, \blacklozenge \, \dfrac{3\frac{2}{5}}{7\frac{3}{5}}$

$\left(2\frac{1}{8}\right)\left(7\frac{3}{5}\right) \, \blacklozenge \, \left(4\frac{3}{4}\right)\left(3\frac{2}{5}\right)$

$\dfrac{17}{\cancel{8}_{4}} \cdot \dfrac{\overset{19}{\cancel{38}}}{5} \, \blacklozenge \, \dfrac{19}{4} \cdot \dfrac{17}{5}$

$\dfrac{323}{20} = \dfrac{323}{20}$

Yes

49. $\dfrac{4.25}{8} \, \blacklozenge \, \dfrac{5.25}{10}$
$(4.25)(10) \, \blacklozenge \, (8)(5.25)$
$42.5 \neq 42$
No

51. $\dfrac{100}{16} = \dfrac{25}{x}$
$100x = (16)(25)$
$100x = 400$
$\dfrac{100x}{100} = \dfrac{400}{100}$
$x = 4$

53.
$$\frac{1\frac{6}{7}}{b} = \frac{13}{21}$$

$$\left(1\frac{6}{7}\right)(21) = 13b$$

$$\frac{13}{\cancel{7}_{1}} \cdot \frac{\cancel{21}^{3}}{1} = 13b$$

$$39 = 13b$$

$$\frac{39}{13} = \frac{13b}{13}$$

$$3 = b$$

55.
$$\frac{2.5}{6.8} = \frac{5}{h}$$

$$2.5h = (6.8)(5)$$

$$2.5h = 34$$

$$\frac{2.5h}{2.5} = \frac{34}{2.5}$$

$$h = 13.6$$

Section 5.4

57. Let x represent the number of human years.

$$\frac{1 \text{ dog year}}{7 \text{ human years}} = \frac{12 \text{ dog years}}{x \text{ human years}}$$

$$1x = (7)(12)$$

$$x = 84$$

The human equivalent is 84 years.

59. Let x represent the population of Alabama.

$$\frac{59{,}800 \text{ births}}{x \text{ people}} = \frac{13}{1000}$$

$$(59{,}800)(1000) = 13x$$

$$59{,}800{,}000 = 13x$$

$$\frac{59{,}800{,}000}{13} = \frac{13x}{13}$$

$$4{,}600{,}000 = x$$

Alabama had approximately 4,600,000 people.

61.
$$\frac{x}{13.5} = \frac{40}{54} \qquad \frac{y}{46} = \frac{54}{40}$$

$$54x = (13.5)(40) \qquad 40y = (46)(54)$$

$$54x = 540 \qquad 40y = 2484$$

$$\frac{54x}{54} = \frac{540}{54} \qquad \frac{40y}{40} = \frac{2484}{40}$$

$$x = 10 \text{ in.} \qquad y = 62.1 \text{ in.}$$

63.
$$\frac{x}{4} = \frac{2}{5} \qquad \frac{y}{4.5} = \frac{2}{5}$$

$$5x = (4)(2) \qquad 5y = (4.5)(2)$$

$$5x = 8 \qquad 5y = 9$$

$$\frac{5x}{5} = \frac{8}{5} \qquad \frac{5y}{5} = \frac{9}{5}$$

$$x = 1.6 \text{ yd} \qquad y = 1.8 \text{ yd}$$

Chapter 5   Test

1. 25 to 521, 25 : 521, $\dfrac{25}{521}$

3. $\dfrac{26}{8} = \dfrac{13}{4}$

5. $\dfrac{65 \text{ cm}}{104 \text{ cm}} = \dfrac{5}{8}$

7. (a) $30 \text{ sec} = \dfrac{1}{2} \text{ min}$

$$\frac{\frac{1}{2}}{1\frac{1}{2}} = \frac{\frac{1}{2}}{\frac{3}{2}} = \frac{1}{2} \div \frac{3}{2} = \frac{1}{\cancel{2}} \cdot \frac{\cancel{2}^{1}}{3} = \frac{1}{3}$$

(b) $1\dfrac{1}{2} \text{ min} = 90 \text{ sec}$

$$\frac{30}{90} = \frac{1}{3}$$

9. $\dfrac{20 \text{ lb}}{6 \text{ weeks}} = \dfrac{10 \text{ lb}}{3 \text{ weeks}}$

11. $\dfrac{2145 \text{ g}}{100 \text{ cm}^3} = 21.45 \text{ g/cm}^3$

13. $\dfrac{\$6.72}{30 \text{ oz}} = \$0.22 \text{ per oz}$

$$
\begin{array}{r}
0.224 \\
30)\overline{6.720} \\
-6\,0 \\
\hline
72 \\
-60 \\
\hline
120 \\
-120 \\
\hline
0
\end{array}
$$

15. $\dfrac{\$3.00}{2 \times 30 \text{ tablets}} = \dfrac{3}{60} = \$0.05/\text{tablet}$

$\dfrac{\$1.92}{24 \text{ capsules}} = \$0.08/\text{capsule}$

The generic pain reliever is the better buy.

17. $\dfrac{42}{15} = \dfrac{28}{10}$

19. $\dfrac{\$15}{1 \text{ hr}} = \dfrac{\$75}{5 \text{ hr}}$

21. $\dfrac{25}{p} = \dfrac{45}{63}$

$(25)(63) = 45p$

$1575 = 45p$

$\dfrac{1575}{45} = \dfrac{45p}{45}$

$35 = p$

23. $\dfrac{n}{9} = \dfrac{3\frac{1}{3}}{6}$

$6n = (9)\left(3\dfrac{1}{3}\right)$

$6n = \dfrac{\overset{3}{\cancel{9}}}{1} \cdot \dfrac{10}{\underset{1}{\cancel{3}}}$

$6n = 30$

$\dfrac{6n}{6} = \dfrac{30}{6}$

$n = 5$

25. Let x represent the time to download the file.

$\dfrac{1.6 \text{ MB}}{2.5 \text{ min}} = \dfrac{4.8 \text{ MB}}{x \text{ min}}$

$1.6x = (2.5)(4.8)$

$1.6x = 12$

$\dfrac{1.6x}{1.6} = \dfrac{12}{1.6}$

$x = 7.5$

It will take 7.5 min.

27. Let x represent the number of goldfish.

$\dfrac{8}{x} = \dfrac{3}{10}$

$(8)(10) = 3x$

$80 = 3x$

$\dfrac{80}{3} = \dfrac{3x}{3}$

$27 \approx x$

There are approximately 27 fish in her pond.

29. $\dfrac{x}{10} = \dfrac{24}{15}$

$15x = (10)(24)$

$15x = 240$

$\dfrac{15x}{15} = \dfrac{240}{15}$

$x = 16 \text{ cm}$

Chapters 1–5   Cumulative Review Exercises

1. Five hundred three thousand, forty-two

3. $226 \times 100{,}000 = 22{,}600{,}000$

5.

$$16\overline{)355.0000} \;\; 22.1875$$

$$\underline{-32}$$
$$35$$
$$\underline{-32}$$
$$30$$
$$\underline{-16}$$
$$140$$
$$\underline{-128}$$
$$120$$
$$\underline{-112}$$
$$80$$
$$\underline{-80}$$
$$0$$

7.

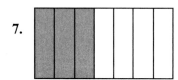

9. $\dfrac{13}{2}\cdot\dfrac{3}{7}=\dfrac{39}{14}$

11. $\dfrac{1}{4}(6)=\dfrac{6}{4}=\dfrac{3}{2}=1\dfrac{1}{2}$

$6-1\dfrac{1}{2}=4\dfrac{1}{2}$

Bruce has $4\dfrac{1}{2}$ in. of sandwich left.

13. $\dfrac{8}{9}+3=\dfrac{8}{9}+\dfrac{27}{9}=\dfrac{35}{9}$

15. Longer walls: $8\dfrac{1}{4}-2\dfrac{1}{2}=\dfrac{33}{4}-\dfrac{5}{2}$

$$=\dfrac{33}{4}-\dfrac{10}{4}=\dfrac{23}{4}\text{ ft}$$

Wall with door: $4\dfrac{1}{3}-2\dfrac{3}{4}=\dfrac{13}{3}-\dfrac{11}{4}$

$$=\dfrac{52}{12}-\dfrac{33}{12}=\dfrac{19}{12}\text{ ft}$$

$$\dfrac{23}{4}+\dfrac{23}{4}+\dfrac{19}{12}=\dfrac{69}{12}+\dfrac{69}{12}+\dfrac{19}{12}$$

$$=\dfrac{157}{12}\text{ or }13\dfrac{1}{12}$$

Emil needs $13\dfrac{1}{12}$ ft of wallpaper border.

17. $6\dfrac{5}{9}=\dfrac{6\times9+5}{9}=\dfrac{59}{9}$

There are 59 ninths.

19.
$$23.880$$
$$\underline{+\,11.300}$$
$$35.180$$
$$\underline{-\;\;7.123}$$
$$28.057$$

21. $43.923\times100=4392.3$

23.
$$29.20$$
$$10.75$$
$$30.50$$
$$34.20$$
$$\underline{+\,26.25}$$
$$130.90\text{ cm}$$

25. $\dfrac{1950}{150}=\dfrac{13}{1}$

27. $\dfrac{1,200,000}{9600\text{ mi}^2}=125\text{ people/mi}^2$

29. $\dfrac{13}{11.7}=\dfrac{5}{x}$

$$13x=(11.7)(5)$$
$$13x=58.5$$
$$\dfrac{13x}{13}=\dfrac{58.5}{13}$$
$$x=4.5$$

Chapter 6   Percents

Chapter Opener Puzzle

```
 1        2
 b        m
 3                  4
 m  a  r  k  u  p   h
 s        l         u
 e        t         n
       5
       d  i  v  i  d  e
          p         r
          l         e
          y         d
```

Section 6.1   Percents and Their Fraction and Decimal Forms

Section 6.1   Practice Exercises

1. Answers will vary.

3. $\dfrac{48}{100} = 48\%$

5. $\dfrac{50}{100} = 50\%$

7. $\dfrac{25}{100} = 25\%$

9. $\dfrac{\$2}{\$100} = 2\%$

11. $\dfrac{70}{100} = 70\%$

13. Replace the symbol % by $\times \dfrac{1}{100}$ (or ÷ 100).
Then reduce the fraction to lowest terms.

15. $3\% = 3 \times \dfrac{1}{100} = \dfrac{3}{100}$

17. $84\% = 84 \times \dfrac{1}{100} = \dfrac{84}{100} = \dfrac{21}{25}$

19. $25\% = 25 \times \dfrac{1}{100} = \dfrac{25}{100} = \dfrac{1}{4}$

21. $3.4\% = 3.4 \times \dfrac{1}{100} = \dfrac{3.4}{100} = \dfrac{34}{1000} = \dfrac{17}{500}$

23. $115\% = 115 \times \dfrac{1}{100} = \dfrac{115}{100} = \dfrac{23}{20}$ or $1\dfrac{3}{20}$

25. $175\% = 175 \times \dfrac{1}{100} = \dfrac{175}{100} = \dfrac{7}{4}$ or $1\dfrac{3}{4}$

27. $0.5\% = 0.5 \times \dfrac{1}{100} = \dfrac{0.5}{100} = \dfrac{5}{1000} = \dfrac{1}{200}$

29. $0.25\% = 0.25 \times \dfrac{1}{100} = \dfrac{0.25}{100} = \dfrac{25}{10,000}$
$= \dfrac{1}{400}$

31. $66\dfrac{2}{3}\% = \dfrac{200}{3} \times \dfrac{1}{100} = \dfrac{200}{300} = \dfrac{2}{3}$

33. $24\dfrac{1}{2}\% = \dfrac{49}{2} \times \dfrac{1}{100} = \dfrac{49}{200}$

35. Replace the % symbol by × 0.01 (or ÷ 100).

37. $72\% = 72 \times 0.01 = 0.72$

39. $66\% = 66 \times 0.01 = 0.66$

41. $12.9\% = 12.9 \times 0.01 = 0.129$

43. $41.05\% = 41.05 \times 0.01 = 0.4105$

45. $201\% = 201 \times 0.01 = 2.01$

47. $0.75\% = 0.75 \times 0.01 = 0.0075$

49. $16\frac{1}{4}\% = 16.25 \times 0.01 = 0.1625$

51. $62\frac{1}{5}\% = 62\frac{2}{10} \times 0.01 = 62.2 \times 0.01 = 0.622$

53. $\frac{1}{4} = \frac{25}{100} = 25\%$

55. $\frac{1}{1} = 100\%$

57. $\frac{3}{2} = \frac{150}{100} = 150\%$

59. $10\% = 10 \times \frac{1}{100} = \frac{10}{100} = \frac{1}{10}$
d

61. $75\% = 75 \times \frac{1}{100} = \frac{75}{100} = \frac{3}{4}$
b

63. $150\% = 150 \times \frac{1}{100} = \frac{150}{100} = \frac{3}{2}$
a

65. $33\frac{1}{3}\% = 33.\overline{3} \times 0.01 = 0.\overline{3}$
d

67. $50\% = 50 \times 0.01 = 0.50$
b

69. $80\% = 80 \times 0.01 = 0.80$
c

71. $22.5\% = 22.5 \times 0.01 = 0.225$

$22.5\% = 22.5 \times \frac{1}{100} = \frac{225}{10} \times \frac{1}{100}$
$= \frac{225}{1000} = \frac{9}{40}$

73. $4.3\% = 4.3 \times 0.01 = 0.043$

$4.3\% = 4.3 \times \frac{1}{100} = \frac{43}{10} \times \frac{1}{100} = \frac{43}{1000}$

75. $0.6\% = 0.6 \times 0.01 = 0.006$

$0.6\% = 0.6 \times \frac{1}{100} = \frac{6}{10} \times \frac{1}{100} = \frac{6}{1000}$
$= \frac{3}{500}$

77. $35\% = 35 \times 0.01 = 0.35$

$35\% = 35 \times \frac{1}{100} = \frac{35}{100} = \frac{7}{20}$

79. $40\% = 40 \times 0.01 = 0.4$

$40\% = 40 \times \frac{1}{100} = \frac{40}{100} = \frac{2}{5}$

$42\% = 42 \times 0.01 = 0.42$

$42\% = 42 \times \frac{1}{100} = \frac{42}{100} = \frac{21}{50}$

$59\% = 59 \times 0.01 = 0.59$

$59\% = 59 \times \frac{1}{100} = \frac{59}{100}$

$73\% = 73 \times 0.01 = 0.73$

$73\% = 73 \times \frac{1}{100} = \frac{73}{100}$

Section 6.2   Fractions and Decimals and Their Percent Forms

Section 6.2   Practice Exercises

1. Answers will vary.

3. $130\% = 130 \times \dfrac{1}{100} = \dfrac{130}{100} = \dfrac{13}{10}$ or $1\dfrac{3}{10}$

5. $0.5\% = 0.5 \times \dfrac{1}{100} = \dfrac{5}{10} \times \dfrac{1}{100} = \dfrac{5}{1000} = \dfrac{1}{200}$

7. $6\dfrac{1}{3}\% = 6.\overline{3} \times 0.01 = 0.06\overline{3}$

9. $0.3\% = 0.3 \times 0.01 = 0.003$

11. $1.62 \times 100\% = 162\%$

13. $0.26 \times 100\% = 26\%$

15. $\dfrac{5}{4} \times 100\% = \dfrac{5}{4} \times \dfrac{100}{1}\% = \dfrac{500}{4}\% = 125\%$

17. $\dfrac{77}{100} \times 100\% = \dfrac{77}{\cancel{100}} \times \dfrac{\overset{1}{\cancel{100}}}{1}\% = 77\%$

19. $0.27 = 0.27 \times 100\% = 27\%$

21. $0.19 = 0.19 \times 100\% = 19\%$

23. $1.75 = 1.75 \times 100\% = 175\%$

25. $0.124 = 0.124 \times 100\% = 12.4\%$

27. $0.006 = 0.006 \times 100\% = 0.6\%$

29. $1.014 = 1.014 \times 100\% = 101.4\%$

31. $\dfrac{71}{100} = \dfrac{71}{100} \times 100\% = \dfrac{71}{\cancel{100}} \times \dfrac{\overset{1}{\cancel{100}}}{1}\%$

   $= 71\%$

33. $\dfrac{19}{20} = \dfrac{19}{20} \times 100\% = \dfrac{19}{\cancel{20}} \times \dfrac{\overset{5}{\cancel{100}}}{1}\% = 95\%$

35. $\dfrac{7}{8} = \dfrac{7}{8} \times 100\% = \dfrac{7}{8} \times \dfrac{100}{1}\% = \dfrac{700}{8}\%$

   $= 87.5$ or $87\dfrac{1}{2}\%$

37. $\dfrac{13}{16} = \dfrac{13}{16} \times 100\% = \dfrac{13}{16} \times \dfrac{100}{1}\% = \dfrac{1300}{16}\%$

   $= 81.25\%$ or $81\dfrac{1}{4}\%$

39. $\dfrac{5}{6} = \dfrac{5}{6} \times 100\% = \dfrac{5}{6} \times \dfrac{100}{1}\% = \dfrac{500}{6}\%$

   $= 83.\overline{3}\%$ or $83\dfrac{1}{3}\%$

41. $\dfrac{4}{9} = \dfrac{4}{9} \times 100\% = \dfrac{4}{9} \times \dfrac{100}{1}\% = \dfrac{400}{9}\%$

   $= 44.\overline{4}\%$ or $44\dfrac{4}{9}\%$

43. $\dfrac{1}{4} = \dfrac{1}{4} \times 100\% = \dfrac{1}{4} \times \dfrac{100}{1}\% = \dfrac{100}{4}\% = 25\%$

45. $\dfrac{1}{10} = \dfrac{1}{10} \times 100\% = \dfrac{1}{10} \times \dfrac{100}{1}\% = \dfrac{100}{10}\%$

   $= 10\%$

47. $\dfrac{2}{3} = \dfrac{2}{3} \times 100\% = \dfrac{2}{3} \times \dfrac{100}{1}\% = \dfrac{200}{3}\%$

   $= 66.\overline{6}\%$ or $66\dfrac{2}{3}\%$

49. $1\dfrac{3}{4} = \dfrac{7}{4} \times 100\% = \dfrac{7}{\underset{1}{\cancel{4}}} \times \dfrac{\overset{25}{\cancel{100}}}{1}\% = 175\%$

51. $\dfrac{27}{20} = \dfrac{27}{20} \times 100\% = \dfrac{27}{\underset{1}{\cancel{20}}} \times \dfrac{\overset{5}{\cancel{100}}}{1}\% = 135\%$

53. $\dfrac{11}{9} = \dfrac{11}{9} \times 100\% = \dfrac{11}{9} \times \dfrac{100}{1}\% = \dfrac{1100}{9}\%$

   $= 122.\overline{2}\%$ or $122\dfrac{2}{9}\%$

55. $1\frac{2}{3} = \frac{5}{3} = \frac{5}{3} \times 100\% = \frac{5}{3} \times \frac{100}{1}\% = \frac{500}{3}\%$

$\qquad = 166.\overline{6}\% \text{ or } 166\frac{2}{3}\%$

57. $\frac{3}{7} = \frac{3}{7} \times 100\%$

$\qquad = \frac{3}{7} \times \frac{100}{1}\%$

$\qquad = \frac{300}{7}\%$

$\qquad \approx 42.9\%$

$$\begin{array}{r} 42.85 \\ 7\overline{)300.00} \\ -28 \\ \overline{20} \\ -14 \\ \overline{6\,0} \\ -5\,6 \\ \overline{40} \\ -35 \\ \overline{5} \end{array}$$

59. $\frac{1}{13} = \frac{1}{13} \times 100\%$

$\qquad = \frac{1}{13} \times \frac{100}{1}\%$

$\qquad = \frac{100}{13}\%$.

$\qquad \approx 7.7\%$

$$\begin{array}{r} 7.69 \\ 13\overline{)100.00} \\ -91 \\ \overline{9\,0} \\ -7\,8 \\ \overline{1\,20} \\ -1\,17 \\ \overline{3} \end{array}$$

61. $\frac{5}{11} = \frac{5}{11} \times 100\%$

$\qquad = \frac{5}{11} \times \frac{100}{1}\%$

$\qquad = \frac{500}{11} \times$

$\qquad \approx 45.5\%$

$$\begin{array}{r} 45.45 \\ 11\overline{)500.00} \\ -44 \\ \overline{60} \\ -55 \\ \overline{5\,0} \\ -4\,4 \\ \overline{60} \\ -55 \\ \overline{5} \end{array}$$

63. $\frac{13}{15} = \frac{13}{15} \times 100\%$

$\qquad = \frac{13}{15} \times \frac{100}{1}\%$

$\qquad = \frac{1300}{15}\%$

$\qquad \approx 86.7\%$

$$\begin{array}{r} 86.66 \\ 15\overline{)1300.00} \\ -120 \\ \overline{100} \\ -90 \\ \overline{10\,0} \\ -9\,0 \\ \overline{1\,00} \\ -90 \\ \overline{10} \end{array}$$

65. The fraction $\frac{1}{2} = 0.5$ and

$\qquad \frac{1}{2}\% = 0.5\% = 0.005.$

67. $25\% = 0.25$ and $0.25\% = 0.0025.$

69. a, c

71. a, c

73. (a) $\frac{1}{4} = \frac{25}{100} = 0.25$

$\qquad \frac{1}{4} \times 100\% = \frac{100}{4}\% = 25\%$

(b) $0.92 = \frac{92}{100} = \frac{23}{25}$

$\qquad 0.92 \times 100\% = 92\%$

(c) $15\% = 15 \times \frac{1}{100} = \frac{15}{100} = \frac{3}{20}$

$\qquad 15\% = 15 \times 0.01 = 0.15$

(d) $1.6 = \frac{16}{10} = \frac{8}{5} \text{ or } 1\frac{3}{5}$

$\qquad 1.6 \times 100\% = 160\%$

(e) $\frac{1}{100} = 0.01$

$\qquad \frac{1}{100} = 1\%$

(f) $0.5\% = 0.5 \times \frac{1}{100} = \frac{5}{10} \times \frac{1}{100} = \frac{5}{1000}$

$\qquad = \frac{1}{200}$

$\qquad 0.5\% = 0.5 \times 0.01 = 0.005$

75. (a) $14\% = 14 \times \dfrac{1}{100} = \dfrac{14}{100} = \dfrac{7}{50}$

$14\% = 14 \times 0.01 = 0.14$

(b) $0.87 = \dfrac{87}{100}$

$0.87 = 0.87 \times 100\% = 87\%$

(c) $1 = \dfrac{1}{1}$ or 1

$1 = 1 \times 100\% = 100\%$

(d) $\dfrac{1}{3} = 0.\overline{3}$

$\dfrac{1}{3} = \dfrac{1}{3} \times 100\% = \dfrac{100}{3}\%$

$= 33.\overline{3}\%$ or $33\dfrac{1}{3}\%$

(e) $0.2\% = 0.2 \times \dfrac{1}{100} = \dfrac{2}{10} \times \dfrac{1}{100} = \dfrac{2}{1000}$

$= \dfrac{1}{500}$

$0.2\% = 0.2 \times 0.01 = 0.002$

(f) $\dfrac{19}{20} = \dfrac{95}{100} = 0.95$

$\dfrac{19}{20} = \dfrac{19}{20} \times 100\% = \dfrac{19}{\cancel{20}} \times \dfrac{\cancel{100}^{\,5}}{1}\%$

$= 95\%$

77. $1.4 = 1.4 \times 100\% = 140\%$

$1.4 > 100\%$

79. $0.052 = 0.052 \times 100\% = 5.2\%$

$0.052 < 50\%$

Section 6.3   Percent Proportions and Applications

Section 6.3 Practice Exercises

1. Answers will vary.

3. $0.55 = 0.55 \times 100\% = 55\%$

5. $0.0006 \times 100\% = 0.06\%$

7. $\dfrac{5}{2} = \dfrac{5}{2} \times 100\% = \dfrac{5}{2} \times \dfrac{100}{1}\% = \dfrac{500}{2}\%$

$= 250\%$

9. $62\dfrac{1}{2}\% = \dfrac{125}{2} \times \dfrac{1}{100} = \dfrac{125}{200} = \dfrac{5}{8}$

11. $77\% = 77 \times \dfrac{1}{100} = \dfrac{77}{100}$

13. $0.3\% = 0.3 \times 0.01 = 0.003$

15. Yes, because $\dfrac{7}{100} = 7\%$.

17. No, because no denominator is 100.

19. Yes, because $\dfrac{\frac{3}{4}}{100} = \dfrac{3}{4}\%$.

21.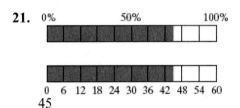

45

23.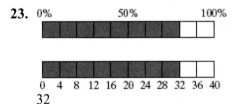

32

25. Amount: 12
base: 20
$p = 60$

27. Amount: 99
base: 200
$p = 49.5$

29. Amount: 50
base: 40
$p = 125$

31. Amount: 12
base: 120
$p = 10$
$$\frac{10}{100} = \frac{12}{120}$$

33. Amount: 72
base: 90
$p = 80$
$$\frac{80}{100} = \frac{72}{90}$$

35. Amount: 21,684
base: 20,850
$p = 104$
$$\frac{104}{100} = \frac{21,684}{20,850}$$

37. Amount: x
base: 200
$p = 54$
$$\frac{54}{100} = \frac{x}{200}$$
$$100x = (54)(200)$$
$$100x = 10,800$$
$$\frac{100x}{100} = \frac{10,800}{100}$$
$$x = 108 \text{ employees}$$

39. Amount: x
base: 40
$$p = \frac{1}{2}$$
$$\frac{\frac{1}{2}}{100} = \frac{x}{40}$$
$$100x = \left(\frac{1}{2}\right)(40)$$
$$100x = 20$$
$$\frac{100x}{100} = \frac{20}{100}$$
$$x = 0.2$$

41. Amount: x
base: 500
$p = 112$
$$\frac{112}{100} = \frac{x}{500}$$
$$100x = (112)(500)$$
$$100x = 56,000$$
$$\frac{100x}{100} = \frac{56,000}{100}$$
$$x = 560$$

43. Let x represent the amount Pedro pays in taxes.
base: 72,000
$p = 28$
$$\frac{28}{100} = \frac{x}{72,000}$$
$$100x = (28)(72,000)$$
$$100x = 2,016,000$$
$$\frac{100x}{100} = \frac{2,016,000}{100}$$
$$x = 20,160$$
Pedro pays \$20,160 in taxes.

45. Let x represent the number of votes cast for Mr. Ventura.
base: 2,060,000
$p = 37$
$$\frac{37}{100} = \frac{x}{2,060,000}$$
$$100x = (37)(2,060,000)$$
$$100x = 76,220,000$$
$$\frac{100x}{100} = \frac{76,220,000}{100}$$
$$x = 762,200$$
Jesse Ventura received approximately 762,200 votes.

47. Amount: 18
base: x
$p = 50$
$$\frac{50}{100} = \frac{18}{x}$$
$$50x = (100)(18)$$
$$50x = 1800$$
$$\frac{50x}{50} = \frac{1800}{50}$$
$$x = 36$$

49. Amount: 69
base: x
$p = 30$
$$\frac{30}{100} = \frac{69}{x}$$
$30x = (100)(69)$
$30x = 6900$
$$\frac{30x}{30} = \frac{6900}{30}$$
$x = 230$ lb

51. Amount: 9
base: x
$$p = \frac{2}{3}$$
$$\frac{\frac{2}{3}}{100} = \frac{9}{x}$$
$$\frac{2}{3}x = (100)(9)$$
$$\frac{2}{3}x = 900$$
$$\frac{3}{2} \cdot \frac{2}{3}x = \frac{3}{2} \cdot \frac{900}{1}$$
$$x = \frac{2700}{2}$$
$$x = 1350$$

53. Let x represent Albert's monthly income.
amount: 120
$p = 7.5$
$$\frac{7.5}{100} = \frac{120}{x}$$
$7.5x = (100)(120)$
$7.5x = 12,000$
$$\frac{7.5x}{7.5} = \frac{12,000}{7.5}$$
$x = 1600$
Albert makes $1600 per month.

55. Let x represent the total number of e-mails.
amount: 14
$p = 40$
$$\frac{40}{100} = \frac{14}{x}$$
$40x = (100)(14)$
$40x = 1400$
$$\frac{40x}{40} = \frac{1400}{40}$$
$x = 35$
Amiee has a total of 35 e-mails.

57. Amount: 42
base: 120
p unknown
$$\frac{p}{100} = \frac{42}{120}$$
$120p = (100)(42)$
$120p = 4200$
$$\frac{120p}{120} = \frac{4200}{120}$$
$p = 35$
35% of $120 is $42.

59. Amount: 84
base: 70
p unknown
$$\frac{p}{100} = \frac{84}{70}$$
$70p = (100)(84)$
$70p = 8400$
$$\frac{70p}{70} = \frac{8400}{70}$$
$p = 120$
120% of 70 is 84.

61. Amount: 280
base: 320
p unknown
$$\frac{p}{100} = \frac{280}{320}$$
$320p = (100)(280)$
$320p = 28,000$
$$\frac{320p}{320} = \frac{28,000}{320}$$
$p = 87.5$
87.5% of 320 mi is 280 mi.

63. Let p represent the percent of the questions answered correctly.
amount: 29
base: 40
$$\frac{p}{100} = \frac{29}{40}$$
$$40p = (100)(29)$$
$$40p = 2900$$
$$\frac{40p}{40} = \frac{2900}{40}$$
$$p = 72.5$$
She answered 72.5% correctly.

65. Amount: 120
base: 600
p unknown
$$\frac{p}{100} = \frac{120}{600}$$
$$600p = (100)(120)$$
$$600p = 12,000$$
$$\frac{600p}{600} = \frac{12,000}{600}$$
$$p = 20$$
20% of the officers are female.

67. Amount: 160
base: 600
p unknown
$$\frac{p}{100} = \frac{160}{600}$$
$$600p = (100)(160)$$
$$600p = 16,000$$
$$\frac{600p}{600} = \frac{16,000}{600}$$
$$p = 26.\overline{6}$$
Approximately 26.7% of the officers were promoted.

69. Let x represent the amount of rain that fell in August.
base: 56
$p = 125$
$$\frac{125}{100} = \frac{x}{56}$$
$$100x = (125)(56)$$
$$100x = 7000$$
$$\frac{100x}{100} = \frac{7000}{100}$$
$$x = 70$$
70 mm of rain fell in August.

71. Let x represent the number of freshmen admitted.
amount: 209
$p = 11$
$$\frac{11}{100} = \frac{209}{x}$$
$$11x = (100)(209)$$
$$11x = 20,900$$
$$\frac{11x}{11} = \frac{20,900}{11}$$
$$x = 1900$$
Approximately 1900 freshmen were admitted.

73. Let p represent the percent of shots made.
amount: 72
base: 173
$$\frac{p}{100} = \frac{72}{173}$$
$$173p = (100)(72)$$
$$173p = 7200$$
$$\frac{173p}{173} = \frac{7200}{173}$$
$$p \approx 41.6$$
Nowitzki had approximately 41.6% completion of three-point shots.

75. (a) Let x represent the number of five-person households that own a dog.
base: 200
$p = 53$
$$\frac{53}{100} = \frac{x}{200}$$
$100x = (53)(200)$
$100x = 10,600$
$$\frac{100x}{100} = \frac{10,600}{100}$$
$x = 106$
106 five-person households own a dog.

(b) Let x represent the number of three-person households that own a dog.
base: 50
$p = 46$
$$\frac{46}{100} = \frac{x}{50}$$
$100x = (46)(50)$
$100x = 2300$
$$\frac{100x}{100} = \frac{2300}{100}$$
$x = 23$
23 three-person households own a dog.

77. Let x represent the number of Chevys.
base: 215
$p = 34$
$$\frac{34}{100} = \frac{x}{215}$$
$100x = (34)(215)$
$100x = 7310$
$$\frac{100x}{100} = \frac{7310}{100}$$
$x = 73.10$
73 were Chevys.

79. Let x represent the total vehicles sold.
amount: 27
$p = 15$
$$\frac{15}{100} = \frac{27}{x}$$
$15x = (100)(27)$
$15x = 2700$
$$\frac{15x}{15} = \frac{2700}{15}$$
$x = 180$
There were 180 total vehicles.

81. Let x represent the amount spent on clothes. Let y represent the amount spent on dinner.
$$\frac{44}{100} = \frac{x}{600}$$
$100x = (44)(600)$
$100x = 26,400$
$$\frac{100x}{100} = \frac{26,400}{100}$$
$x = 264$

$$\begin{array}{r} 600 \\ -264 \\ \hline 336 \end{array} \text{ left after clothes}$$

$$\frac{20}{100} = \frac{y}{336}$$
$100y = (20)(336)$
$100y = 6720$
$$\frac{100y}{100} = \frac{6720}{100}$$
$y = 67.20$

$$\begin{array}{r} 264 \\ +67.20 \\ \hline 331.20 \end{array}$$
Carson spent \$331.20.

83. Step 1: \$57.65 ≈ \$58
Step 2: 10% of 58 is 5.8.
Step 3: 2 × 5.8 = 11.6
A 20% tip is \$11.60.

85. Step 1: \$42 is already a whole dollar amount.
Step 2: 10% of 42 is 4.2
Step 4: $\frac{1}{2}(4.2) = 2.1$
$2.1 + 4.2 = 6.3$
A 15% tip is \$6.30.

Section 6.4   Percent Equations and Applications

Section 6.4   Practice Exercises

1. Answers will vary.

3. Divide both sides of the equation by 26 to get $x = 2.5$.

5. $3x = 27$
    $$\frac{3x}{3} = \frac{27}{3}$$
    $$x = 9$$

7. $0.15x = 45$
    $$\frac{0.15x}{0.15} = \frac{45}{0.15}$$
    $$x = 300$$

9. $1.02x = 841.5$
    $$\frac{1.02x}{1.02} = \frac{841.5}{1.02}$$
    $$x = 825$$

11. Let x represent the unknown amount.
    $x = (35\%)(700)$
    $x = (0.35)(700)$
    $x = 245$

13. Let x represent the unknown amount.
    $x = (0.55\%)(900)$
    $x = (0.0055)(900)$
    $x = 4.95$

15. Let x represent the unknown amount.
    $x = (133\%)(600)$
    $x = (1.33)(600)$
    $x = 798$

17. 50% equals one-half of the number. So divide the number by 2.

19. $2 \times 14 = 28$

21. $\frac{1}{2} \times 40 = 20$

23. Let x represent the amount of active ingredient.
    $x = (6\%)(64)$
    $x = (0.06)(64)$
    $x = 3.84$
    There are 3.84 oz of sodium hypochlorite in household bleach.

25. Let x represent the number of completed passes.
    $x = (60\%)(8358)$
    $x = (0.6)(8358) \approx 5015$
    Marino completed approximately 5015 passes.

27. Let x represent the base.
    $18 = (0.40)x$
    $$\frac{18}{0.4} = \frac{0.4x}{0.4}$$
    $$45 = x$$

29. Let x represent the base.
    $(0.92)x = 41.4$
    $$\frac{0.92x}{0.92} = \frac{41.4}{0.92}$$
    $$x = 45$$

31. Let x represent the base.
    $3.09 = (1.03)x$
    $$\frac{3.09}{1.03} = \frac{1.03x}{1.03}$$
    $$3 = x$$

33. Let x represent the number tested.
    $47 = (0.04)x$
    $$\frac{47}{0.04} = \frac{0.04x}{0.04}$$
    $$1175 = x$$
    There were 1175 subjects tested.

35. Let x represent the total population.
    $61.6 = (0.22)x$
    $$\frac{61.6}{0.22} = \frac{0.22x}{0.22}$$
    $$280 = x$$
    At that time, the population was about 280 million.

37. $0.13 = 0.13 \times 100\% = 13\%$

39. $1.08 = 1.08 \times 100\% = 108\%$

41. $0.005 = 0.005 \times 100\% = 0.5\%$

43. $0.17 = 0.17 \times 100\% = 17\%$

45. Let x represent the percent.
$$x \cdot 480 = 120$$
$$\frac{480x}{480} = \frac{120}{480}$$
$$x = 0.25$$
$$x = 0.25 \times 100\%$$
$$x = 25\%$$

47. Let x represent the percent.
$$666 = x \cdot 740$$
$$\frac{666}{740} = \frac{740x}{740}$$
$$0.9 = x$$
$$x = 0.9 \times 100\%$$
$$x = 90\%$$

49. Let x represent the percent.
$$x \cdot 300 = 400$$
$$\frac{300x}{300} = \frac{400}{300}$$
$$x = 1.333$$
$$x = 1.333 \times 100\%$$
$$x = 133.3\%$$

51. Let x represent the percent.
$$x \cdot 120 = 84$$
$$\frac{120x}{120} = \frac{84}{120}$$
$$x = 0.7$$
$$x = 0.7 \times 100\%$$
$$x = 70\%$$
70% of the hot dogs were sold.

53. (a) $4 + 2 + 14 + 10 + 16 + 18 + 10 + 6$
$= 80$
There are 80 total employees.

(b) Let x represent the percent.
$$x \cdot 80 = 10$$
$$\frac{80x}{80} = \frac{10}{80}$$
$$x = 0.125$$
$$x = 0.125 \times 100\%$$
$$x = 12.5\%$$
12.5% missed 3 days of work.

(c) $2 + 14 + 10 + 16 + 18 = 60$
Let x represent the percent.
$$x \cdot 80 = 60$$
$$\frac{80x}{80} = \frac{60}{80}$$
$$x = 0.75$$
$$x = 0.75 \times 100\%$$
$$x = 75\%$$
75% missed 1 to 5 days of work.

55. Let x represent the total number of hospital stays.
$$6.3 = (0.18)x$$
$$\frac{6.3}{0.18} = \frac{0.18x}{0.18}$$
$$35 = x$$
There were 35 million total hospital stays that year.

57. Let x represent the percent.
$$x \cdot 87 = 11$$
$$\frac{87x}{87} = \frac{11}{87}$$
$$x \approx 0.126$$
$0.126 \times 100\% = 12.6\%$
Approximately 12.6% of Florida's panthers live in Everglades National Park.

59. Let x represent the number saving for their children's education.
$$x = (0.52)(800)$$
$$x = 416$$
416 parents would be expected to have started saving for their children's education.

61. Let x represent the total cost of the TV.
$$1440 = (0.60)x$$
$$\frac{1440}{0.60} = \frac{0.60x}{0.60}$$
$$2400 = x$$
The total cost of the TV is $2400.

63. Let x represent the minutes of commercials.
$$x = (0.26)(60)$$
$$x = 15.6$$
15.6 min of commercials would be expected.

65. Let x represent the number that made over $10 per hour.
$x = (0.635)(10,000,000)$
$x = 6,350,000$
6,350,000 people ages 25–34 made over $10/hr.

67. Let x represent the total workers ages 16–24.
$4,000,000 = (0.25)x$
$$\frac{4,000,000}{0.25} = \frac{0.25x}{0.25}$$
$16,000,000 = x$
There are a total of 16,000,000 workers in the 16–24 age group.

69. (a) $220 - 20 = 200$ beats per minute

(b) $(0.60)(200) = 120$
$(0.85)(200) = 170$
Between 120 and 170 beats per minute.

Problem Recognition Exercises: Percents

1. $\dfrac{41}{100} = 41\%$

3. $\dfrac{1}{3} = \dfrac{33\frac{1}{3}}{100} = 33\frac{1}{3}\%$

5. Greater than, since $104\% > 100\%$.

7. Greater than, since $11\% > 10\%$ and 10% of 90 is 9.

9. Let x represent the base.
$6 = (0.002)x$
$$\frac{6}{0.002} = \frac{0.002x}{0.002}$$
$3000 = x$

11. Let x represent the number.
$x = (0.12)(40)$
$x = 4.8$

13. Let x represent the base.
$$\frac{150}{100} = \frac{105}{x}$$
$150x = (100)(105)$
$150x = 10,500$
$$\frac{150x}{150} = \frac{10,500}{150}$$
$x = 70$

15. Let x represent the amount.
$x = (0.07)(90)$
$x = 6.3$

17. Let x represent the percent.
$x \cdot 60 = 180$
$$\frac{60x}{60} = \frac{180}{60}$$
$x = 3$
$x = 3 \times 100\%$
$x = 300\%$

19. Let x represent the base.
$75 = (0.001)x$
$$\frac{75}{0.001} = \frac{0.001x}{0.001}$$
$75,000 = x$

21. Let x represent the amount.
$x = (0.50)(50)$
$x = 25$

23. Let x represent the base.
$50 = (0.50)x$
$$\frac{50}{0.50} = \frac{0.50x}{0.50}$$
$100 = x$

25. Let x represent the percent.
$$x \cdot 250 = 2$$
$$\frac{250x}{250} = \frac{2}{250}$$
$$x = 0.008$$
$$x = 0.008 \times 100\%$$
$$x = 0.8\%$$

27. Let x represent the base.
$$26 = (0.10)x$$
$$\frac{26}{0.10} = \frac{0.10x}{0.10}$$
$$260 = x$$

29. Let x represent the amount.
$$x = (0.10)(26)$$
$$x = 2.6$$

31. Let x represent the percent.
$$x \cdot 248 = 186$$
$$\frac{248x}{248} = \frac{186}{248}$$
$$x = 0.75$$
$$x = 0.75 \times 100\%$$
$$x = 75\%$$

33. Let x represent the percent.
$$x \cdot 186 = 248$$
$$\frac{186x}{186} = \frac{248}{186}$$
$$x = 1.3\overline{3}$$
$$x = 1.3\overline{3} \times 100\%$$
$$x = 133.\overline{3}\% \text{ or } 133\tfrac{1}{3}\%$$

35. $0.10(82) = 8.2$

37. $0.20(82) = 16.4$

39. $2.00(82) = 164$

Section 6.5   Applications Involving Sales Tax, Commission, Discount, and Markup

Section 6.5   Practice Exercises

1. 395; 400; 404; answers will vary.

3. 12

5. 28

7. Let x represent the number.
$$52 = (0.002)x$$
$$\frac{52}{0.002} = \frac{0.002x}{0.002}$$
$$26,000 = x$$

9. Let x represent the percent.
$$\frac{x}{100} = \frac{6}{25}$$
$$25x = (100)(6)$$
$$25x = 600$$
$$\frac{25x}{25} = \frac{600}{25}$$
$$x = 24$$

11. Let x represent the amount.
$$x = (0.016)(550)$$
$$x = 8.8$$

13. $(0.05)(\$20.00) = \$1.00$
$\$20.00 + \$1.00 = \$21.00$

15. $\dfrac{\$0.50}{\$12.50} = 0.04 \text{ or } 4\%$
$\$12.50 + \$0.50 = \$13.00$

17. Let x represent the cost.
$$\$2.75 = (0.025)x$$
$$\frac{\$2.75}{0.025} = \frac{0.025x}{0.025}$$
$$\$110.00 = x$$
$$\$110.00 + \$2.75 = \$112.75$$

19. $\$58.30 - \$55.00 = \$3.30$
$$\frac{\$3.30}{\$55.00} = 0.06 = 6\%$$

21. Let x represent the amount of tax.
$$x = (0.05)(68.25)$$
$$x = 3.41$$
$$68.25 + 3.41 = 71.66$$
The total bill is $71.66.

23. Let x represent the tax rate.
$$16.80 = x \cdot \$240.00$$
$$\frac{16.80}{240.00} = \frac{240.00x}{240.00}$$
$$0.07 = x$$
The tax rate is 7%.

25. Let x represent the price of the fruit basket.
$$2.67 = (0.06)x$$
$$\frac{2.67}{0.06} = \frac{0.06x}{0.06}$$
$$44.5 = x$$
The price is $44.50.

27. $(0.05)(\$20,000.00) = \$1000.00$

29. $\dfrac{\$10,000.00}{\$125,000.00} = 0.08$ or 8%

31. Let x represent the total sales.
$$\$540.00 = (0.10)x$$
$$\frac{\$540.00}{0.10} = \frac{0.10x}{0.10}$$
$$\$5400.00 = x$$

33. Let x represent the amount of commission.
$$x = (0.07)(\$48,000)$$
$$x = \$3360$$
Zach made $3360 in commission.

35. Let x represent the commission rate.
$$300 = x \cdot 2000$$
$$\frac{300}{2000} = \frac{2000x}{2000}$$
$$0.15 = x$$
Rodney's commission rate is 15%.

37. Amount of commission
$$= \$67,000 - \$25,000$$
$$= \$42,000$$
Let x represent total sales.
$$42,000 = 0.03x$$
$$\frac{42,000}{0.03} = \frac{0.03x}{0.03}$$
$$1,400,000 = x$$
Her sales were $1,400,000.

39. $\$86,000 - \$60,000 = \$26,000.$
Commission on $60,000:
$(0.06)(60,000) = 3600$
Commission on $26,000:
$(0.085)(26,000) = 2210$
$3600 + 2210 = 5810$
Jeff's commissions totaled $5810.00.

41. Discount $= (0.15)(\$175.00) = \$26.25$
Sale price $= \$175.00 - \$26.25 = \$148.75$

43. Discount $= \$900.00 - \$630.00 = \$270.00$
$(\text{rate})(\$900.00) = \$270.00$
$$\text{rate} = \frac{\$270.00}{\$900.00}$$
$$= 0.30 \text{ or } 30\%$$

45. Original price $= \$77.00 + \$33.00 = \$110.00$
$(\text{rate})(\$110.00) = \$33.00$
$$\text{rate} = \frac{\$33.00}{\$110.00}$$
$$= 0.30 \text{ or } 30\%$$

47. $(0.40)(\text{original price}) = \$23.36$
$$\text{original price} = \frac{\$23.36}{0.40}$$
$$= \$58.40$$
Sale price $= \$58.40 - \$23.36 = \$35.04$

49. (a) Let x represent the amount of discount.
$x = (0.10)(\$550) = \$55$
The discount is $55.

(b) Sale price = $550 – $55 = $495
The discounted yearly membership will cost $495.

51. Discount = $229 – $183.20 = $45.80
Let x represent the discount rate.
$(x)(\$229) = \$45.80$
$$x = \frac{\$45.80}{\$229} = 0.2$$
The discount rate is 20%.

53. The set of dishes is not free. After the first discount, the price was 50% or one-half of $112, which is $56. Then the second discount is 50% or one-half of $56, which is $28.

55. Discount = $235 – $188 = $47
Let x represent the discount rate.
$(x)(\$235) = \$47$
$$x = \frac{\$47}{\$235} = 0.2$$
The discount is $47.00, and the discount rate is 20%.

57. Markup = $(0.05)(\$92.00) = \$4.60$
Retail price = $92.00 + $4.60 = $96.60

59. Markup = $118.80 – $110.00 = $8.80
(rate)($110.00)=$8.80
$$\text{rate} = \frac{\$8.80}{\$110.00}$$
$$= 0.08 \text{ or } 8\%$$

61. Original price = $422.50 – $97.50
= $325.00
(rate)($325.00) = $97.50
$$\text{rate} = \frac{\$97.50}{\$325.00}$$
$= 0.3$ or 30%

63. (0.20)(original price) = $9.00
$$\text{original price} = \frac{\$9.00}{0.20}$$
$= \$45.00$
Retail price = $45.00 + $9.00 = $54.00

65. (a) Let x represent the amount of markup.
$x = (0.18)(\$150.00) = \$27.00$
The markup is $27.00.

(b) Retail price = $150.00 + $27.00
= $177.00
The retail price is $177.00.

(c) Tax = $(0.07)(\$177.00) = \$12.39$
The total price is
$177.00 + $12.39 = $189.39.

67. Markup amount = $375 – $300 = $75
Let x represent the markup rate.
$\$75 = (x)(\$300)$
$$\frac{\$75}{\$300} = x$$
$0.25 = x$
The markup rate is 25%.

69. Original price = $123.20 – $43.20 = $80.00
Let x represent the markup rate.
$\$43.20 = (x)(\$80.00)$
$$\frac{\$43.20}{\$80.00} = x$$
$0.54 = x$
The markup rate, is 54%.

Section 6.6   Percent Increase and Decrease

Section 6.6   Practice Exercises

1. All items should be checked.

3. **(a)** Tax = (0.05)($65) = $3.25
$65 + $3.25 = $68.25
The total price will be $68.25.

(b) Let x represent the discount amount.
$x = (0.20)(\$65) = \$13$
Sale price = $65 − $13 = $52
Tax = (0.05)($52) = $2.60
$52 + $2.60 = $54.60
The total price with the 20% discount
would be $54.60.

(c) $68.25 − $54.60 = $13.65
Chris will save $13.65.

5. $425 − $200 = $225
Commission = (0.14)($225) = $31.50
Katie's commission is $31.50.

7. Multiply the decimal by 100% by moving
the decimal point 2 places to the right and
attaching the % sign.

9. 0.05 = 5%

11. 0.12 = 12%

13. **(a)** Increase
(b) 59 − 48 = 11

15. **(a)** Decrease
(b) 145 − 135 = 10

17. **(a)** Decrease
(b) 654 − 645 = 9

19. **(a)** Increase
(b) 79 − 67 = 12

21. $60 − $30
$$\frac{\$30}{\$30} \times 100\% = 1 \times 100\% = 100\%$$
c

23. Increase = 14 − 8 = 6
$$\text{Percent increase} = \frac{6}{8} \times 100\%$$
$$= 0.75 \times 100\%$$
$$= 75\%$$

25. Increase = 42,000 − 21,000 = 21,000
$$\text{Percent increase} = \frac{21,000}{21,000} \times 100\%$$
$$= 1 \times 100\%$$
$$= 100\%$$

27. Increase = 5500 − 5000 = 500
$$\text{Percent increase} = \frac{500}{5000} \times 100\%$$
$$= 0.10 \times 100\%$$
$$= 10\%$$

29. Increase = 170 − 165 = 5
$$\text{Percent increase} = \frac{5}{165} \times 100\%$$
$$\approx 0.03 \times 100\%$$
$$\approx 3\%$$

31. $62 − $31 = $31
$$\frac{\$31}{\$62} \times 100\% = 0.5 \times 100\% = 50\%$$
a

33. Decrease = $12.60 − $11.97 = $0.63
$$\text{Percent decrease} = \frac{\$0.63}{\$12.60} \times 100\%$$
$$= 0.05 \times 100\%$$
$$= 5\%$$

35. Decrease = 5 − 1.6 = 3.4
$$\text{Percent decrease} = \frac{3.4}{5} \times 100\%$$
$$= 0.68 \times 100\%$$
$$= 68\%$$

37. Decrease $= 79 - 59 = 20$

$$\text{Percent decrease} = \frac{20}{79} \times 100\%$$
$$= 0.253 \times 100\%$$
$$= 25.3\%$$

39. Decrease $= 12 - 10.2 = 1.8$

$$\text{Percent decrease} = \frac{1.8}{12} \times 100\%$$
$$= 0.15 \times 100\%$$
$$= 15\%$$

41. Change $= 110.8 - 100.3 = 10.5$

$$\text{Percent increase} = \frac{10.5}{100.3} \times 100\%$$
$$\approx 0.105 \times 100\%$$
$$= 10.5\%$$

43. Change $= 8.15 - 8.11 = 0.04$

$$\text{Percent decrease} = \frac{0.04}{8.15} \times 100\%$$
$$\approx 0.005 \times 100\%$$
$$= 0.5\%$$

45. Change $= 7.8 - 5.6 = 2.2$ million

$$\text{Percent increase} = \frac{2.2}{5.6} \times 100\%$$
$$\approx 0.393 \times 100\%$$
$$= 39.3\%$$

47. Change $= \$8.8 - \$5.7 = \$3.1$ trillion

$$\text{Percent increase} = \frac{\$3.1}{\$5.7} \times 100\%$$
$$\approx 0.544 \times 100\%$$
$$= 54.4\%$$

Section 6.7   Simple and Compound Interest

Section 6.7   Practice Exercises

1. Answers will vary.

3. Change $= 1099 - 987 = 112$

$$\text{Percent decrease} = \frac{112}{1099} \times 100\%$$
$$\approx 0.10 \times 100\%$$
$$= 10\%$$

5. Change $= 404 - 364 = 40$

$$\text{Percent increase} = \frac{40}{364} \times 100\%$$
$$\approx 0.11 \times 100\%$$
$$= 11\%$$

7. $I = Prt = (\$6000)(0.05)(3) = \$300(3)$
$= \$900$
$\$6000 + \$900 = \$6900$

9. $I = Prt = (\$5050)(0.06)(4) = \$303(4)$
$= \$1212$
$\$5050 + \$1212 = \$6262$

11. $I = Prt = (\$12,000)(0.04)\left(4\frac{1}{2}\right)$

$= \$480(4.5) = \$2160$
$\$12,000 + \$2160 = \$14,160$

13. $I = Prt = (\$10,500)(0.045)(4)$
$= \$472.50(4) = \$1890$
$\$10,500 + \$1890 = \$12,390$

15. **(a)** $I = Prt = (\$2500)(0.035)(4)$
$= \$87.50(4) = \$350$

 (b) $\$2500 + \$350 = \$2850$

17. **(a)** $I = Prt = (\$400)(0.08)(1.5) = \$32(1.5)$
$= \$48$

 (b) $\$400 + \$48 = \$448$

19. $I = Prt = (\$10,300)(0.04)(5) = \$412(5)$
$= \$2060$
$\$10,300 + \$2060 = \$12,360$

21. $I = Prt = (\$4500)(0.10)(2.5) = \$450(2.5)$
$= \$1125$
$\$4500 + \$1125 = \$5625$

23. There are $2(3) = 6$ total compounding periods.

25. There are $12(2) = 24$ total compounding periods.

27. (a) $I = Prt = \$500(0.04)(3) = \$20(3) = \$60$
$\$500 + \$60 = \$560$

(b)

Year	Interest	Total
1	($500)(0.04) = $20	$500 + $20 = $520
2	($520)(0.04) = $20.80	$520 + $20.80 = $540.80
3	($540.80)(0.04) = $21.63	$540.80 + $21.63 = $562.43

29. (a) $I = Prt = (\$8000)(0.04)(3) = \$960$
$\$8000 + \$960 = \$8960$

(b)

Year	$I = Prt$	Total in account
1	$I = ($8000)(0.04)(1) = 320	$8000 + $320 = $8320
2	$I = ($8320)(0.04)(1) = 332.80	$8320 + $332.80 = $8652.80
3	$I = ($8652.80)(0.04)(1) = 346.11	$8652.80 + $346.11 = $8998.91

(c) $\$8998.91 - \$8960 = \$38.91$

31. A = total amount in the account;
P = principal;
r = annual interest rate;
n = number of compounding periods per
     year;
t = time in years

33. $A = \$5000\left(1 + \dfrac{0.045}{1}\right)^{1\cdot5} \approx \$6230.91$

35. $A = \$6000\left(1 + \dfrac{0.05}{2}\right)^{2\cdot2} \approx \$6622.88$

37. $A = \$10,000\left(1 + \dfrac{0.06}{4}\right)^{4\cdot1.5} \approx \$10,934.43$

39. $A = \$14,000\left(1 + \dfrac{0.045}{12}\right)^{12\cdot3} \approx \$16,019.47$

Chapter 6   Review Exercises

Section 6.1

1. $\dfrac{75}{100} = 75\%$

3. $\dfrac{5}{4} = \dfrac{125}{100} = 125\%$

5. b, c

7. $30\% = \dfrac{30}{100} = \dfrac{3}{10}$
   f

9. $33\% = \dfrac{33}{100}$
   a

11. $66\dfrac{2}{3}\% = 66.\overline{6}\% = \dfrac{66.\overline{6}}{100} = \dfrac{2}{3}$
   c

13. $7.5\% = 7.5 \times 0.01 = 0.075$
   e

15. $50\% = 50 \times 0.01 = 0.5$
   f

17. $0.25\% = 0.25 \times 0.01 = 0.0025$
   d

19. $42\% = 42 \times \dfrac{1}{100} = \dfrac{42}{100} = \dfrac{21}{50}$

$42\% = 42 \times 0.01 = 0.42$

21. $6.15\% = 6.15 \times 0.01 = 0.0615$

23. $9.15\% = 9.15 \times \dfrac{1}{100} = \dfrac{915}{100} \times \dfrac{1}{100}$

$= \dfrac{183}{20} \times \dfrac{1}{100} = \dfrac{183}{2000}$

Section 6.2

25. $\dfrac{17}{100} = \dfrac{17}{100} \times 100\% = 17\%$

27. $\dfrac{4}{5} = \dfrac{4}{5} \times 100\% = 80\%$

29. $0.12 = 0.12 \times 100\% = 12\%$

31. $0.005 = 0.005 \times 100\% = 0.5\%$

33. $\dfrac{14}{16} = 0.875 \times 100\% = 87.5\%$

35. $\dfrac{3}{5} = 0.6 \times 100\% = 60\%$

37. $\dfrac{9}{20} = \dfrac{45}{100} = 0.45$

$\dfrac{9}{20} = \dfrac{9}{20} \times 100\% = 45\%$

39. $6\% = 6 \times \dfrac{1}{100} = \dfrac{6}{100} = \dfrac{3}{50}$

$6\% = 6 \times 0.01 = 0.06$

41. $\dfrac{9}{1000} = 0.009$

$\dfrac{9}{1000} = \dfrac{9}{1000} \times 100\% = \dfrac{9}{10}\% \text{ or } 0.9\%$

Section 6.3

43. Amount: 67.50
base: 150
$p = 45$

45. Amount: 30.24
base: 144
$p = 21$

47. $\dfrac{6}{8} = \dfrac{75}{100}$

49. $\dfrac{840}{420} = \dfrac{200}{100}$

51. Let x represent the amount.

$\dfrac{x}{50} = \dfrac{12}{100}$

$100x = (50)(12)$

$\dfrac{100x}{100} = \dfrac{600}{100}$

$x = 6$

53. Let x represent the percent.

$\dfrac{x}{100} = \dfrac{11}{88}$

$88x = (100)(11)$

$\dfrac{88x}{88} = \dfrac{1100}{88}$

$x = 12.5$

12.5%

55. Let x represent the base.

$\dfrac{33\frac{1}{3}}{100} = \dfrac{13}{x}$

$\dfrac{100}{3}x = (100)(13)$

$\dfrac{100}{3}x = 1300$

$\dfrac{3}{100} \cdot \dfrac{100}{3}x = \dfrac{3}{100} \cdot 1300$

$x = 39$

57. Let x represent the number of no-shows.

$\dfrac{x}{260} = \dfrac{4.2}{100}$

$100x = (260)(4.2)$

$\dfrac{100x}{100} = \dfrac{1092}{100}$

$x = 10.92$

Approximately 11 people would be no-shows.

59. Let x represent the percent spent on rent.
$$\frac{x}{100} = \frac{720}{1800}$$
$$1800x = (100)(720)$$
$$\frac{1800x}{1800} = \frac{72,000}{1800}$$
$$x = 40$$
Victoria spends 40% on rent.

Section 6.4

61. $0.18 \cdot 900 = x$
$162 = x$

63. $18.90 = x \cdot 63$
$$\frac{18.90}{63} = \frac{63x}{63}$$
$$0.3 = x$$
$$x = 30\%$$

65. $30 = 0.25 \cdot x$
$$\frac{30}{0.25} = \frac{0.25x}{0.25}$$
$$120 = x$$

67. Let x represent the original price.
$$\frac{0.80 \cdot x}{0.80} = \frac{54.40}{0.80}$$
$$x = 68$$
The original price is $68.00.

69. Let x represent the number of fat calories.
$x = 0.30 \cdot 2400$
$x = 720$
Elaine can consume 720 fat calories.

Section 6.5

71. Tax $= (0.06)(1279) = 76.74$
The sales tax is $76.74.

73. **(a)** $7.29 - $6.75 = $0.54
The tax is $0.54.

(b) Let x represent the tax rate.
$$\frac{x \cdot 6.75}{6.75} = \frac{0.54}{6.75}$$
$$x = 0.08$$
The tax rate is 8%.

75. Let x represent the rate.
$$\frac{11}{104} = \frac{x \cdot 104}{104}$$
$$0.106 \approx x$$
The commission rate was approximately 10.6%.

77. $8 \text{ hr} \times $8 = 64
$420 - $200 = $220
$0.05 \cdot $220 = $11
$64 + $11 = $75
Sela will earn $75 that day.

79. Discount $= (0.30)($28.95) \approx $8.69
Sale price $= $28.95 - $8.69 = $20.26
The discount is $8.69. The sale price is $20.26.

81. Markup $= $208 - $160 = $48
Let x represent the markup rate.
$$\frac{48}{160} = \frac{x \cdot 160}{160}$$
$$0.3 = x$$
The markup rate is 30%.

Section 6.6

83. **(a)** Increase
(b) $107.5 - 86 = 21.5$
$$\text{Percent increase} = \frac{21.5}{86} \times 100\%$$
$$= 0.25 \times 100\%$$
$$= 25\%$$

85. Increase $= 410 - 263 = 147$
$$\text{Percent increase} = \frac{147}{263} \times 100\%$$
$$\approx 0.559 \times 100\%$$
$$= 55.9\%$$

87. Increase = 224,000 − 128,000 = 96,000

$$\text{Percent increase} = \frac{96,000}{128,000} \times 100\%$$
$$= 0.75 \times 100\%$$
$$= 75\%$$

89. $I = Prt = (\$10,200)(0.03)(4) = \$306(4)$
$$= \$1224$$
$$\$10,200 + \$1224 = \$11,424$$

Section 6.7

91. $I = Prt = (\$2500)(0.05)(1.5) = \$125(1.5)$
$$= \$187.50$$
$$\$2500 + \$187.50 = \$2687.50$$
Jean-Luc will have to pay $2687.50.

93.

Year	Interest	Total
1	($6000)(0.04) = $240	$6000 + $240 = $6240
2	($6240)(0.04) = $249.60	$6240 + $249.60 = $6489.60
3	($6489.60)(0.04) ≈ $259.58	$6489.60 + $259.58 = $6749.18

95. $A = \$850\left(1 + \dfrac{0.08}{4}\right)^{4 \cdot 2} \approx \$995.91$

97. $A = \$11,000\left(1 + \dfrac{0.075}{1}\right)^{1 \cdot 6} \approx \$16,976.32$

Chapter 6   Test

1. $\dfrac{22}{100} = 22\%$

3. (a) $5.4\% = 5.4 \times 0.01 = 0.054$
$$5.4\% = 5.4 \times \frac{1}{100} = \frac{54}{10} \times \frac{1}{100}$$
$$= \frac{27}{5} \times \frac{1}{100} = \frac{27}{500}$$

(b) $0.15\% = 0.15 \times 0.01 = 0.0015$
$$0.15\% = 0.15 \times \frac{1}{100} = \frac{15}{100} \times \frac{1}{100}$$
$$= \frac{3}{20} \times \frac{1}{100} = \frac{3}{2000}$$

(c) $170\% = 170 \times 0.01 = 1.70$
$$170\% = 170 \times \frac{1}{100} = \frac{170}{100} = \frac{17}{10}$$

5. $1.5\% = 1.5 \times 0.01 = 0.015$
$$1.5\% = 1.5 \times \frac{1}{100} = \frac{15}{10} \times \frac{1}{100} = \frac{3}{2} \times \frac{1}{100} = \frac{3}{200}$$

7. Multiply the fraction by 100%.

9. $\dfrac{1}{250} = \dfrac{1}{250} \times 100\% = \dfrac{100}{250}\% = 0.4\%$

11. $\dfrac{5}{7} = \dfrac{5}{7} \times 100\% = \dfrac{500}{7}\% \approx 71.4\%$

13. $0.32 = 0.32 \times 100\% = 32\%$

15. $1.3 = 1.3 \times 100\% = 130\%$

17. $(0.24)(150) = 36$

19. $\dfrac{21}{0.06} = \dfrac{(0.06)x}{0.06}$
$$350 = x$$

21. $\dfrac{x \cdot 220}{220} = \dfrac{198}{220}$
$$x = 0.9$$
$$x = 90\%$$

23. (a) $740 - 10 = 730$ mg

(b) Let x represent the percent.
$$\frac{730}{740} = \frac{x \cdot 740}{740}$$
$$0.986 \approx x$$
98.6% is from the dressing.

25. $(0.21)(2000) = 420$ m^3

27. Commission $= (0.06)(\$3500) = \$210$
$\$400 + \$210 = \$610$
Charles will earn $610.

29. Increase $= \$0.44 - \$0.32 = \$0.12$
Percent increase $= \dfrac{\$0.12}{\$0.32} \times 100\%$
$= 0.375 \times 100\%$
$= 37.5\%$

31. $A = \$25,000\left(1 + \dfrac{0.045}{4}\right)^{4 \cdot 5} \approx \$31,268.76$

Chapters 1–6   Cumulative Review Exercises

1. Millions place

3. $\begin{array}{r} 34,882 \\ \times\ \ \ \ 100 \\ \hline 3,488,200 \end{array}$

5. $\begin{array}{r} 234 \\ 44 \\ 6 \\ +\ 2901 \\ \hline 3185 \end{array}$

7. (a) Improper
(b) Improper
(c) Proper
(d) Proper

9. $\dfrac{42}{25} \div \dfrac{7}{100} = \dfrac{\overset{6}{\cancel{42}}}{\underset{1}{\cancel{25}}} \cdot \dfrac{\overset{4}{\cancel{100}}}{\underset{1}{\cancel{7}}} = 24$

11. $16 \times \dfrac{1}{24} = \dfrac{16}{24} = \dfrac{2}{3}$

13. Perimeter $= \dfrac{8}{3} + \dfrac{10}{3} + 3 = \dfrac{18}{3} + 3 = 6 + 3$
$= 9$ km

15. $\begin{array}{rcl} \$14\dfrac{7}{10} & = & 14\dfrac{7}{10} \\[2mm] +\ \$1\dfrac{2}{5} & = & +1\dfrac{4}{10} \\ \hline & & 15\dfrac{11}{10} = 15 + 1\dfrac{1}{10} = \$16\dfrac{1}{10} \end{array}$

17. (a) 18, 36, 54, 72
(b) 1, 2, 3, 6, 9, 18

(c) $18 = 2 \cdot 3 \cdot 3 = 2 \cdot 3^2$

19. $\dfrac{3}{8} = 0.375$

21. $\dfrac{7}{9} = 0.\overline{7}$

23. $90\% - 24.7\% = 65.3\%$

25. $85 \times 0.001 = 0.085$

27. $85 \div 10 = 8.5$

29. $\dfrac{3}{4} = \dfrac{15}{p}$
$3p = (4)(15)$
$\dfrac{3p}{3} = \dfrac{60}{3}$
$p = 20$

31. $\dfrac{4\frac{1}{3}}{p} = \dfrac{12}{18}$
$12p = \left(4\dfrac{1}{3}\right)(18)$
$12p = \dfrac{13}{\underset{1}{\cancel{3}}} \cdot \dfrac{\overset{6}{\cancel{18}}}{1}$
$\dfrac{12p}{12} = \dfrac{78}{12}$
$p = 6\dfrac{1}{2}$

33. Let x represent the time to read 5 chapters.

$$\frac{\frac{1}{2}}{1} = \frac{x}{5}$$

$$1 \cdot x = \left(\frac{1}{2}\right)(5)$$

$$x = \frac{5}{2}$$

$$x = 2\frac{1}{2}$$

It will take $2\frac{1}{2}$ hr.

35. Let x represent the time to download 4.6 MB.

$$\frac{1.6 \text{ MB}}{2.5 \text{ min}} = \frac{4.6 \text{ MB}}{x \text{ min}}$$

$$1.6x = (2.5)(4.6)$$

$$\frac{1.6x}{1.6} = \frac{11.5}{1.6}$$

$$x = 7.1875$$

It will take about 7.2 min.

37. Increase $= 17.4 - 3.7 = 13.7$

$$\text{Percent increase} = \frac{13.7}{3.7} \times 100\%$$

$$\approx 3.70 \times 100\%$$

$$= 370\%$$

The increase will be about 370%.

39. $I = Prt = (\$13{,}000)(0.032)(5) = \$416(5)$

$$= \$2080$$

$\$13{,}000 + \$2080 = \$15{,}080$

Kevin will have $15,080.

Chapter 7   Measurement

Chapter Opener Puzzle

Section 7.1   Converting U.S. Customary Units of Length

Section 7.1   Practice Exercises

1. Answers will vary.

3. 5280 ft = 1 mi

5. 1 yd = 3 ft

7. $1 \text{ ft} = \frac{1}{3} \text{ yd}$

9. 2 yd = 2 × 1 yd = 2 × 3 ft = 6 ft

11. 6 ft = 6 × 1 ft = 6 × 12 in. = 72 in.

13. 2 mi = 2 × 1 mi = 2 × 5280 ft = 10,560 ft

15. $24 \text{ ft} = 24 \times 1 \text{ ft} = 24 \times \frac{1}{3} \text{ yd} = \frac{24}{3} \text{ yd}$
$\qquad = 8 \text{ yd}$

17. $9 \text{ in.} = 9 \times 1 \text{ in.} = 9 \times \frac{1}{12} \text{ ft} = \frac{9}{12} \text{ ft} = \frac{3}{4} \text{ ft}$

19. $1760 \text{ ft} = 1760 \times 1 \text{ ft} = 1760 \times \frac{1}{5280} \text{ mi}$
$\qquad = \frac{1760}{5280} \text{ mi} = \frac{1}{3} \text{ mi}$

21. b

23. a

25. $9 \text{ ft} = \frac{9 \text{ ft}}{1} \cdot \frac{1 \text{ yd}}{3 \text{ ft}} = \frac{9}{3} \text{ yd} = 3 \text{ yd}$

27. $3.5 \text{ ft} = \frac{3.5 \text{ ft}}{1} \cdot \frac{12 \text{ in.}}{1 \text{ ft}} = (3.5)(12) \text{ in.}$
$\qquad = 42 \text{ in.}$

29. $11,880 \text{ ft} = \frac{11,880 \text{ ft}}{1} \cdot \frac{1 \text{ mi}}{5280 \text{ ft}}$
$\qquad = \frac{11,880}{5280} \text{ mi} = 2\frac{1}{4} \text{ mi}$

31. $6 \text{ yd} = \frac{6 \text{ yd}}{1} \cdot \frac{3 \text{ ft}}{1 \text{ yd}} = (6)(3) \text{ ft} = 18 \text{ ft}$

33. $14 \text{ ft} = \frac{14 \text{ ft}}{1} \cdot \frac{1 \text{ yd}}{3 \text{ ft}} = \frac{14}{3} \text{ yd} = 4\frac{2}{3} \text{ yd}$

35. $320 \text{ mi} = \frac{320 \text{ mi}}{1} \cdot \frac{1760 \text{ yd}}{1 \text{ mi}}$
$\qquad = (320)(1760) \text{ yd} = 563,200 \text{ yd}$

37. $171 \text{ in.} = \dfrac{171 \text{ in.}}{1} \cdot \dfrac{1 \text{ ft}}{12 \text{ in.}} \cdot \dfrac{1 \text{ yd}}{3 \text{ ft}}$

$= \dfrac{171}{(12)(3)} \text{ yd} = \dfrac{171}{36} \text{ yd} = 4\dfrac{3}{4} \text{ yd}$

39. $2 \text{ yd} = \dfrac{2 \text{ yd}}{1} \cdot \dfrac{3 \text{ ft}}{1 \text{ yd}} \cdot \dfrac{12 \text{ in.}}{1 \text{ ft}} = (2)(3)(12) \text{ in.}$

$= 72 \text{ in.}$

41. $0.8 \text{ mi} = \dfrac{0.8 \text{ mi}}{1} \cdot \dfrac{5280 \text{ ft}}{1 \text{ mi}} \cdot \dfrac{12 \text{ in.}}{1 \text{ ft}}$

$= (0.8)(5280)(12) \text{ in.} = 50{,}688 \text{ in.}$

43. $12{,}672 \text{ in.} = \dfrac{12{,}672 \text{ in.}}{1} \cdot \dfrac{1 \text{ ft}}{12 \text{ in.}} \cdot \dfrac{1 \text{ mi}}{5280 \text{ ft}}$

$= \dfrac{12{,}672}{(12)(5280)} = \dfrac{1}{5} \text{ mi}$

45. $1.6 \text{ mi} = \dfrac{1.6 \text{ mi}}{1} \cdot \dfrac{5280 \text{ ft}}{1 \text{ mi}} \cdot \dfrac{12 \text{ in.}}{1 \text{ ft}}$

$= (1.6)(5280)(12) \text{ in.} = 101{,}376 \text{ in.}$

47. (a) $10 \text{ ft} = 10 \times 1 \text{ ft} = 10 \times 12 \text{ in.} = 120 \text{ in.}$
$10 \text{ ft } 8 \text{ in.} = 120 \text{ in.} + 8 \text{ in.} = 128 \text{ in.}$

(b) $8 \text{ in.} = 8 \times 1 \text{ in.} = 8 \times \dfrac{1}{12} \text{ ft} = \dfrac{2}{3} \text{ ft}$

$10 \text{ ft } 8 \text{ in.} = 10 \text{ ft} + \dfrac{2}{3} \text{ ft} = 10\dfrac{2}{3} \text{ ft}$

49. (a) $6 \text{ in.} = 6 \times 1 \text{ in.} = 6 \times \dfrac{1}{12} \text{ ft} = 0.5 \text{ ft}$

$3'6'' = 3 \text{ ft} + 0.5 \text{ ft} = 3.5 \text{ ft}$

(b) $3 \text{ ft} = 3 \times 1 \text{ ft} = 3 \times 12 \text{ in.} = 36 \text{ in.}$
$3'6'' = 36 \text{ in.} + 6 \text{ in.} = 42 \text{ in.}$

51. $6'2'' + 4'6'' = \quad 6 \text{ ft } 2 \text{ in.} \qquad 10'8''$
$\underline{+\ 4 \text{ ft } 6 \text{ in.}}$
$10 \text{ ft } 8 \text{ in. or}$

53. $\quad 5 \text{ ft } \ 2 \text{ in.}$
$\underline{+\ 6 \text{ ft } 10 \text{ in.}}$
$11 \text{ ft } 12 \text{ in.} = 11 \text{ ft} + 1 \text{ ft} = 12 \text{ ft}$

55. $4'9'' + 3'9'' =$
$\quad 4 \text{ ft } \ 9 \text{ in.}$
$\underline{+\ 3 \text{ ft } \ 9 \text{ in.}}$
$7 \text{ ft } 18 \text{ in.} = 7 \text{ ft} + 1 \text{ ft} + 6 \text{ in.}$
$= 8 \text{ ft } 6 \text{ in. or } 8'6''$

57. $\quad 3 \text{ ft } 2 \text{ in.} = \quad 2 \text{ ft } 14 \text{ in.}$
$\underline{-\ (1 \text{ ft } 5 \text{ in.})} = \underline{-\ (1 \text{ ft } \ 5 \text{ in.})}$
$\qquad\qquad\qquad\qquad 1 \text{ ft } \ 9 \text{ in.}$

59. $2(4 \text{ ft } 5 \text{ in.}) = 8 \text{ ft } 10 \text{ in.}$

61. $6(4 \text{ ft } 8 \text{ in.}) = 24 \text{ ft } 48 \text{ in.}$
$= 24 \text{ ft} + 4 \text{ ft} = 28 \text{ ft}$

63. $(6'4'') \div 2 = 3'2''$

65. $\dfrac{18 \text{ ft } 3 \text{ in.}}{3} = 6 \text{ ft } 1 \text{ in.}$

67. $9 \text{ in.} = 9 \times 1 \text{ in.} = 9 \times \dfrac{1}{2} \text{ ft} = \dfrac{3}{4} \text{ ft}$

$2 \text{ ft} + \dfrac{3}{4} \text{ ft} + 2 \text{ ft} + \dfrac{3}{4} \text{ ft} = 4 + \dfrac{6}{4} \text{ ft} = 5\dfrac{1}{2} \text{ ft}$

69. $10 \text{ ft } 6 \text{ in.} = 10.5 \text{ ft}$
$10.5 \text{ ft} + 3 \text{ ft} + 10.5 \text{ ft} + 3 \text{ ft} = 27 \text{ ft}$
$\dfrac{27 \text{ ft}}{1.5 \text{ ft}} = 18$
18 pieces of border are needed.

71. $8(6 \text{ ft } 4 \text{ in.}) = 48 \text{ ft } 32 \text{ in.}$
$= 48 \text{ ft} + 2 \text{ ft} + 8 \text{ in.}$
$= 50 \text{ ft } 8 \text{ in.}$
50 ft 8 in. is needed.

73. $4'6'' + 2'8'' =$
$\quad 4 \text{ ft } \ 6 \text{ in.}$
$\underline{+\ 2 \text{ ft } \ 8 \text{ in.}}$
$6 \text{ ft } 14 \text{ in.} = 6 \text{ ft} + 1 \text{ ft} + 2 \text{ in.} = 7 \text{ ft } 2 \text{ in.}$
The plumber used $7'2''$ of pipe.

75. $4 \text{ yd} = 4 \times 1 \text{ yd} = 4 \times 3 \text{ ft} = 12 \text{ ft}$
$4 \text{ yd} - 5 \text{ ft} = 12 \text{ ft} - 5 \text{ ft} = 7 \text{ ft}$
7 ft is left over.

77. $\dfrac{6 \text{ ft } 9 \text{ in.}}{3} = 2 \text{ ft } 3 \text{ in.}$

Each piece is 2 ft 3 in. long.

79. $5(6') + 4(3'3'') + 2(18'')$
$= 5(6 \text{ ft}) + 4(3.25 \text{ ft}) + 2(1.5 \text{ ft})$
$= 30 \text{ ft} + 13 \text{ ft} + 3 \text{ ft}$
$= 46 \text{ ft}$
The total length is 46'.

81. $32 \text{ in.} - 2(4 \text{ in.}) = 32 \text{ in.} - 8 \text{ in.}$
$= 24 \text{ in.} = 2 \text{ ft}$
$14(2)(2 \text{ ft}) = 56 \text{ ft}$
$5 \text{ yd} = 5 \times 3 \text{ ft} = 15 \text{ ft}$
$\dfrac{56 \text{ ft}}{15 \text{ ft}} = 3\dfrac{11}{15} \text{ rolls}$
She should buy 4 rolls.

83. $54 \text{ ft}^2 = \dfrac{54 \text{ ft}^2}{1} \cdot \dfrac{1 \text{ yd}}{3 \text{ ft}} \cdot \dfrac{1 \text{ yd}}{3 \text{ ft}} = \dfrac{54}{9} \text{ yd}^2$
$= 6 \text{ yd}^2$

85. $432 \text{ in.}^2 = \dfrac{432 \text{ in.}^2}{1} \cdot \dfrac{1 \text{ ft}}{12 \text{ in.}} \cdot \dfrac{1 \text{ ft}}{12 \text{ in.}}$
$= \dfrac{432}{144} \text{ ft}^2 = 3 \text{ ft}^2$

87. $5 \text{ ft}^2 = \dfrac{5 \text{ ft}^2}{1} \cdot \dfrac{12 \text{ in.}}{1 \text{ ft}} \cdot \dfrac{12 \text{ in.}}{1 \text{ ft}} = 720 \text{ in.}^2$

89. $3 \text{ yd}^2 = \dfrac{3 \text{ yd}^2}{1} \cdot \dfrac{3 \text{ ft}}{1 \text{ yd}} \cdot \dfrac{3 \text{ ft}}{1 \text{ yd}} = 27 \text{ ft}^2$

Section 7.2   Converting U.S. Customary Units of Time, Weight, and Capacity

Section 7.2   Practice Exercises

1. Answers will vary.

3. $12 \text{ ft} = \dfrac{12 \text{ ft}}{1} \cdot \dfrac{1 \text{ yd}}{3 \text{ ft}} = 4 \text{ yd}$
$12 \text{ ft} = \dfrac{12 \text{ ft}}{1} \cdot \dfrac{12 \text{ in.}}{1 \text{ ft}} = 144 \text{ in.}$

5. $6 \text{ yd} = \dfrac{6 \text{ yd}}{1} \cdot \dfrac{3 \text{ ft}}{1 \text{ yd}} = 18 \text{ ft}$
$= \dfrac{18 \text{ ft}}{1} \cdot \dfrac{12 \text{ in.}}{1 \text{ ft}} = 216 \text{ in.}$

7. $50 \text{ yd} = \dfrac{50 \text{ yd}}{1} \cdot \dfrac{3 \text{ ft}}{1 \text{ yd}} = 150 \text{ ft}$
$= \dfrac{150 \text{ ft}}{1} \cdot \dfrac{12 \text{ in.}}{1 \text{ ft}} = 1800 \text{ in.}$

9. $1 \text{ qt} = 2 \text{ pt}$

11. $1 \text{ lb} = 16 \text{ oz}$

13. $1 \text{ yr} = 365 \text{ days}$

15. $1 \text{ gal} = 4 \text{ qt}$

17. $2 \text{ yr} = \dfrac{2 \text{ yr}}{1} \cdot \dfrac{365 \text{ days}}{1 \text{ yr}} = 730 \text{ days}$

19. $90 \text{ min} = \dfrac{90 \text{ min}}{1} \cdot \dfrac{1 \text{ hr}}{60 \text{ min}} = \dfrac{90}{60} \text{ hr}$
$= \dfrac{3}{2} \text{ hr or } 1\dfrac{1}{2} \text{ hr}$

21. $180 \text{ sec} = \dfrac{180 \text{ sec}}{1} \cdot \dfrac{1 \text{ min}}{60 \text{ sec}} = 3 \text{ min}$

23. $72 \text{ hr} = \dfrac{72 \text{ hr}}{1} \cdot \dfrac{1 \text{ day}}{24 \text{ hr}} = \dfrac{72}{24} \text{ days} = 3 \text{ days}$

25. $3600 \text{ sec} = \dfrac{3600 \text{ sec}}{1} \cdot \dfrac{1 \text{ min}}{60 \text{ sec}} \cdot \dfrac{1 \text{ hr}}{60 \text{ min}}$
$= \dfrac{3600}{3600} \text{ hr} = 1 \text{ hr}$

27. $9 \text{ wk} = \dfrac{9 \text{ wk}}{1} \cdot \dfrac{7 \text{ day}}{1 \text{ wk}} \cdot \dfrac{24 \text{ hr}}{1 \text{ day}} = 1512 \text{ hr}$

29. 1:20:30 = 1 hr 20 min 30 sec

1 hr = 60 min

30 sec = 0.5 min

1 hr 20 min 30 sec

　　= 60 min + 20 min + 0.5 min

　　= 80.5 min

31. 2:55:15 = 2 hr 55 min 15 sec

$2 \text{ hr} = \dfrac{2 \text{ hr}}{1} \cdot \dfrac{60 \text{ min}}{1 \text{ hr}} = 120 \text{ min}$

$15 \sec = \dfrac{15 \sec}{1} \cdot \dfrac{1 \min}{60 \sec} = \dfrac{1}{4} \min = 0.25 \min$

2 hr 55 min 15 sec

　　= 120 min + 55 min + 0.25 min

　　= 175.25 min

33.
$$
\begin{array}{r}
1 \text{ hr} \quad 10 \text{ min} \\
45 \text{ min} \\
1 \text{ hr} \quad 20 \text{ min} \\
30 \text{ min} \\
50 \text{ min} \\
+\ 1 \text{ hr} \quad\quad \\
\hline
3 \text{ hr} \ 155 \text{ min}
\end{array}
$$
3 hr 155 min = 3 hr + 2 hr + 35 min

　　　　　　 = 5 hr 35 min

Gil ran for 5 hr 35 min.

35.
$$
\begin{array}{r}
15 \text{ min } 30 \text{ sec} \\
50 \text{ min } 20 \text{ sec} \\
+\ 28 \text{ min } 10 \text{ sec} \\
\hline
93 \text{ min } 60 \text{ sec}
\end{array}
$$
93 min 60 sec = 93 min + 1 min

　　　　　　　 = 94 min or 1 hr 34 min

The total time is 1 hr 34 min.

37. $32 \text{ oz} = \dfrac{32 \text{ oz}}{1} \cdot \dfrac{1 \text{ lb}}{16 \text{ oz}} = 2 \text{ lb}$

39. $2 \text{ tons} = \dfrac{2 \text{ tons}}{1} \cdot \dfrac{2000 \text{ lb}}{1 \text{ ton}} = 4000 \text{ lb}$

41. $4 \text{ lb} = \dfrac{4 \text{ lb}}{1} \cdot \dfrac{16 \text{ oz}}{1 \text{ lb}} = 64 \text{ oz}$

43. $3000 \text{ lb} = \dfrac{3000 \text{ lb}}{1} \cdot \dfrac{1 \text{ ton}}{2000 \text{ lb}}$

$\quad\quad\quad\quad = \dfrac{3000}{2000} \text{ tons}$

$\quad\quad\quad\quad = 1\dfrac{1}{2} \text{ tons or } 1.5 \text{ tons}$

45.
$$
\begin{array}{r}
6 \text{ lb } 10 \text{ oz} \\
+\ 3 \text{ lb } 14 \text{ oz} \\
\hline
9 \text{ lb } 24 \text{ oz}
\end{array}
$$
9 lb 24 oz = 9 lb + 1 lb + 8 oz

　　　　　 = 10 lb 8 oz

47.
$$
\begin{array}{r}
30 \text{ lb } 10 \text{ oz} \\
-\ 22 \text{ lb } \ \ 8 \text{ oz} \\
\hline
8 \text{ lb } \ \ 2 \text{ oz}
\end{array}
$$

49.
$$
\begin{array}{rcl}
10 \text{ lb} & = & 9 \text{ lb } 16 \text{ oz} \\
-\ (3 \text{ lb } 8 \text{ oz}) & & -\ (3 \text{ lb } \ \ 8 \text{ oz}) \\
\hline
& & 6 \text{ lb } \ \ 8 \text{ oz}
\end{array}
$$

51. 50(6 lb 4 oz) = 300 lb 200 oz

　　　　　　　　 = 300 lb + 12 lb + 8 oz

　　　　　　　　 = 312 lb 8 oz

The total weight is 312 lb 8 oz.

53. $2\dfrac{1}{2} \text{ tons} = \dfrac{2.5 \text{ tons}}{1} \cdot \dfrac{2000 \text{ lb}}{1 \text{ ton}} = 5000 \text{ lb}$

$\dfrac{5000 \text{ lb}}{2500 \text{ lb}} = 2$

The truck will have to make 2 trips.

55. $16 \text{ fl oz} = \dfrac{16 \text{ fl oz}}{1} \cdot \dfrac{1 \text{ c}}{8 \text{ fl oz}} = 2 \text{ c}$

57. $6 \text{ gal} = \dfrac{6 \text{ gal}}{1} \cdot \dfrac{4 \text{ qt}}{1 \text{ gal}} = 24 \text{ qt}$

59. $1 \text{ gal} = \dfrac{1 \text{ gal}}{1} \cdot \dfrac{4 \text{ qt}}{1 \text{ gal}} \cdot \dfrac{2 \text{ pt}}{1 \text{ qt}} \cdot \dfrac{2 \text{ c}}{1 \text{ pt}} = 16 \text{ c}$

61. $2 \text{ qt} = \dfrac{2 \text{ qt}}{1} \cdot \dfrac{1 \text{ gal}}{4 \text{ qt}} = \dfrac{1}{2} \text{ gal}$

63. $1 \text{ pt} = \dfrac{1 \text{ pt}}{1} \cdot \dfrac{2 \text{ c}}{1 \text{ pt}} \cdot \dfrac{8 \text{ fl oz}}{1 \text{ c}} = 16 \text{ fl oz}$

65. $2 \text{ T} = \dfrac{2 \text{ T}}{1} \cdot \dfrac{3 \text{ tsp}}{1 \text{ T}} = 6 \text{ tsp}$

67. $3 \text{ c} = \dfrac{3 \text{ c}}{1} \cdot \dfrac{8 \text{ fl oz}}{1 \text{ c}} = 24 \text{ fl oz}$

Yes, 3 c is 24 fl oz, so the 48-fl-oz jar will suffice.

69. $1 \text{ qt} = \dfrac{1 \text{ qt}}{1} \cdot \dfrac{2 \text{ pt}}{1 \text{ qt}} \cdot \dfrac{2 \text{ c}}{1 \text{ pt}} \cdot \dfrac{8 \text{ fl oz}}{1 \text{ c}} = 32 \text{ fl oz}$

$\dfrac{\$2.69}{24 \text{ fl oz}} \approx \$0.112 \text{ per fl oz}$

$\dfrac{\$3.29}{32 \text{ fl oz}} \approx \$0.103 \text{ per fl oz}$

The unit price for the 24-fl oz jar is about $0.112 per ounce, and the unit price for the 1-qt jar is about $0.103 per ounce; therefore the 1-qt jar is the better buy.

71. $32 \text{ fl oz} = \dfrac{32 \text{ fl oz}}{1} \cdot \dfrac{1 \text{ c}}{8 \text{ fl oz}} = 4 \text{ c}$

$= \dfrac{4 \text{ c}}{1} \cdot \dfrac{1 \text{ pt}}{2 \text{ c}} = 2 \text{ pt}$

$= 1 \text{ qt}$

$= \dfrac{1 \text{ qt}}{1} \cdot \dfrac{1 \text{ gal}}{4 \text{ qt}} - 0.25 \text{ gal}$

73. $1 \text{ gal} = 4 \text{ qt}$

$= \dfrac{4 \text{ qt}}{1} \cdot \dfrac{2 \text{ pt}}{1 \text{ qt}} = 8 \text{ pt}$

$= \dfrac{8 \text{ pt}}{1} \cdot \dfrac{2 \text{ c}}{1 \text{ pt}} = 16 \text{ c}$

$= \dfrac{16 \text{ c}}{1} \cdot \dfrac{8 \text{ fl oz}}{1 \text{ c}} = 128 \text{ fl oz}$

75. $0.5 \text{ qt} = \dfrac{0.5 \text{ qt}}{1} \cdot \dfrac{1 \text{ gal}}{4 \text{ qt}} = 0.125 \text{ gal}$

$0.5 \text{ qt} = \dfrac{0.5 \text{ qt}}{1} \cdot \dfrac{2 \text{ pt}}{1 \text{ qt}} = 1 \text{ pt}$

$= 2 \text{ c}$

$= \dfrac{2 \text{ c}}{1} \cdot \dfrac{8 \text{ fl oz}}{1 \text{ c}} = 16 \text{ fl oz}$

77. $1 \text{ c} = 8 \text{ fl oz}$

$1 \text{ c} = \dfrac{1 \text{ c}}{1} \cdot \dfrac{1 \text{ pt}}{2 \text{ c}} = 0.5 \text{ pt}$

$= \dfrac{0.5 \text{ pt}}{1} \cdot \dfrac{1 \text{ qt}}{2 \text{ pt}} = 0.25 \text{ qt}$

$= \dfrac{0.25 \text{ qt}}{1} \cdot \dfrac{1 \text{ gal}}{4 \text{ qt}} = 0.0625 \text{ gal}$

79. $64 \text{ fl oz} = \dfrac{64 \text{ fl oz}}{1} \cdot \dfrac{1 \text{ c}}{8 \text{ fl oz}} = 8 \text{ c}$

$= \dfrac{8 \text{ c}}{1} \cdot \dfrac{1 \text{ pt}}{2 \text{ c}} = 4 \text{ pt}$

$= \dfrac{4 \text{ pt}}{1} \cdot \dfrac{1 \text{ qt}}{2 \text{ pt}} = 2 \text{ qt}$

$= \dfrac{2 \text{ qt}}{1} \cdot \dfrac{1 \text{ gal}}{4 \text{ qt}} = 0.5 \text{ gal}$

Section 7.3   Metric Units of Length

Section 7.3 Practice Exercises

1. (a) False
   (b) False
   (c) True

3. $2200 \text{ yd} = \dfrac{2200 \text{ yd}}{1} \cdot \dfrac{1 \text{ mi}}{1760 \text{ yd}} = 1.25 \text{ mi}$

5. $48 \text{ oz} = \dfrac{48 \text{ oz}}{1} \cdot \dfrac{1 \text{ lb}}{16 \text{ oz}} = 3 \text{ lb}$

7. $1 \text{ day} = \dfrac{1 \text{ day}}{1} \cdot \dfrac{24 \text{ hr}}{1 \text{ day}} \cdot \dfrac{60 \text{ min}}{1 \text{ hr}} = 1440 \text{ min}$

9. $3.5 \text{ lb} = \dfrac{3.5 \text{ lb}}{1} \cdot \dfrac{16 \text{ oz}}{1 \text{ lb}} = 56 \text{ oz}$

11. b, f, g

13. 3.2 cm or 32 mm

15. 2.1 cm or 21 mm

17. (a) 5 cm
   (b) 2 cm
   (c) $2(5) + 2(2) = 10 + 4 = 14 \text{ cm}$
   (d) $(5)(2) = 10 \text{ cm}^2$

19. a

21. d

23. d

25. $\dfrac{1 \text{ km}}{1000 \text{ m}}$

27. $\dfrac{1 \text{ m}}{100 \text{ cm}}$

29. $\dfrac{1 \text{ m}}{10 \text{ dm}}$

31. $2430 \text{ m} = \dfrac{2430 \text{ m}}{1} \cdot \dfrac{1 \text{ km}}{1000 \text{ m}} = 2.43 \text{ km}$

33. $103 \text{ dm} = \dfrac{103 \text{ dm}}{1} \cdot \dfrac{1 \text{ m}}{10 \text{ dm}} = 10.3 \text{ m}$

35. $50 \text{ m} = \dfrac{50 \text{ m}}{1} \cdot \dfrac{1000 \text{ mm}}{1 \text{ m}} = 50{,}000 \text{ mm}$

37. $4 \text{ km} = \dfrac{4 \text{ km}}{1} \cdot \dfrac{1000 \text{ m}}{1 \text{ km}} = 4000 \text{ m}$

39. $4.31 \text{ cm} = 43.1 \text{ mm}$

41. $3328 \text{ dm} = 0.3328 \text{ km}$

43. $345 \text{ mm} = 0.345 \text{ m}$

45. $0.25 \text{ km} = 250 \text{ m}$

47. $4003 \text{ cm} = 400.3 \text{ dm}$

49. $0.07 \text{ mm} = 0.007 \text{ cm}$

51. $20.91 \text{ m} = 2091 \text{ cm}$

53. $2538 \text{ m} = 2.538 \text{ km}$

55. $270 \text{ m} = 0.27 \text{ km}$

57. $2(12 \text{ cm}) + 2(40 \text{ cm}) = 24 \text{ cm} + 80 \text{ cm}$
$= 104 \text{ cm} = 1.04 \text{ m}$
No, she needs 1.04 m of molding.

59. $110 \text{ mm} = 0.11 \text{ m}$
$\dfrac{1.43 \text{ m}}{0.11 \text{ m}} = 13$
It will take 13 tiles.

61. $0.108 \text{ km} = 108 \text{ m}$
$\dfrac{108 \text{ m}}{4.5 \text{ m}} = 24$
There can be 24 parking spaces.

63. $30{,}000 \text{ mm}^2$
$= \dfrac{30{,}000 \text{ mm}^2}{1} \cdot \dfrac{1 \text{ cm}}{10 \text{ mm}} \cdot \dfrac{1 \text{ cm}}{10 \text{ mm}}$
$= 300 \text{ cm}^2$

65. $4.1 \text{ m}^2 = \dfrac{4.1 \text{ m}^2}{1} \cdot \dfrac{100 \text{ cm}}{1 \text{ m}} \cdot \dfrac{100 \text{ cm}}{1 \text{ m}}$
$= 41{,}000 \text{ cm}^2$

Section 7.4   Metric Units of Mass, Capacity, and Medical Applications

Section 7.4   Practice Exercises

1. Answers will vary.

3. $670 \text{ km} = 670{,}000 \text{ m}$
$= 67{,}000{,}000 \text{ cm}$
$= 670{,}000{,}000 \text{ mm}$

5. $2.5 \text{ cm} = 25 \text{ mm}$
$= 0.025 \text{ m}$
$= 0.000025 \text{ km}$

7. $1.35 \text{ mm} = 0.135 \text{ cm}$
$= 0.00135 \text{ m}$
$= 0.00000135 \text{ km}$

9. $539 \text{ g} = 0.539 \text{ kg}$

11. $2.5 \text{ kg} = 2500 \text{ g}$

13. $0.0334 \text{ g} = 33.4 \text{ mg}$

15. $90 \text{ hg} = 9 \text{ kg}$

17. $45 \text{ dg} = 4.5 \text{ g}$

19. $1.58 \text{ kg} = 1580 \text{ g}$
$= 158{,}000 \text{ cg}$
$= 1{,}580{,}000 \text{ mg}$

21. $170 \text{ g} = 0.17 \text{ kg}$
$= 17{,}000 \text{ cg}$
$= 170{,}000 \text{ mg}$

23. $42{,}500 \text{ cg} = 425{,}000 \text{ mg}$
$= 425 \text{ g}$
$= 0.425 \text{ kg}$

25. $325 \text{ mg} = 32.5 \text{ cg}$
$= 0.325 \text{ g}$
$= 0.000325 \text{ kg}$

27. $1 \text{ cL} < 1 \text{ L}$

29. $1 \text{ mL} = 1 \text{ cc}$

31. $1 \text{ cL} < 1 \text{ kL}$

33. cubic centimeter

35. $3200 \text{ mL} = 3.2 \text{ L}$

37. $7 \text{ L} = 700 \text{ cL}$

39. $42 \text{ mL} = 0.42 \text{ dL}$

41. $64 \text{ cc} = 64 \text{ mL}$

43. $0.04 \text{ L} = 40 \text{ mL} = 40 \text{ cc}$

45. $15 \text{ mL} = 1.5 \text{ cL}$
$= 0.015 \text{ L}$
$= 0.000015 \text{ kL}$

47. $35.5 \text{ cL} = 355 \text{ mL}$
$= 0.355 \text{ L}$
$= 0.000355 \text{ kL}$

49. $2 \text{ L} = 0.002 \text{ kL}$
$= 200 \text{ cL}$
$= 2000 \text{ mL}$

51. $0.0377 \text{ kL} = 37.7 \text{ L}$
$= 3770 \text{ cL}$
$= 37{,}700 \text{ mL}$

53. c

55. b

57. c, d

59. $112.014 \text{ m} = 11.2014 \text{ dam}$

61. $600 \text{ mg} = 0.6 \text{ g}$

63. $19 \text{ L} = 0.019 \text{ kL}$

65. $7(600 \text{ mg} + 500 \text{ mg} + 250 \text{ mg})$
$= 7(1350 \text{ mg})$
$= 9450 \text{ mg}$
$= 9.45 \text{ g}$
Stacy gets 9.45 g per week.

67. $\dfrac{\$74.25}{45 \text{ L}} = \$1.65 \text{ per L}$
The price is \$1.65 per liter.

69. $6(710 \text{ mL}) = 4260 \text{ mL} = 4.26 \text{ L}$
A 6-pack contains 4.26 L.

71. $1 \text{ qt} = \dfrac{1 \text{ qt}}{1} \cdot \dfrac{2 \text{ pt}}{1 \text{ qt}} \cdot \dfrac{2 \text{ c}}{1 \text{ pt}} = 4 \text{ c}$
$4(130 \text{ mg}) = 520 \text{ mg}$
520 mg of sodium per 1-qt bottle.

73. $1 \text{ wk} = \dfrac{1 \text{ wk}}{1} \cdot \dfrac{7 \text{ days}}{1 \text{ wk}} \cdot \dfrac{24 \text{ hr}}{1 \text{ day}} = 168 \text{ hr}$
$\dfrac{168 \text{ hr}}{8 \text{ hr}} = 21 \text{ doses}$
$21(250 \text{ mg}) = 5250 \text{ mg} = 5.25 \text{ g}$
5.25 g of the drug would be given in 1 wk.

75. $2 \text{ cc} = 2 \text{ mL}$

77. $2 \text{ cc} = 2 \text{ mL}$
$\dfrac{1 \text{ L}}{2 \text{ mL}} = \dfrac{1000 \text{ mL}}{2 \text{ mL}} = 500$
500 people can be vaccinated.

79. $0.2 \text{ mg}(48) = 9.6 \text{ mg}$
9.6 mg should be given to the patient.

81. **(a)** $20 \text{ mg}(20) = 400 \text{ mg}$
One dose is 400 mg.

(b) $5(4)(400 \text{ mg}) = 8000 \text{ mg}$ or 8 g
8 g would be given over a 5-day period.

83. $0.01 \text{ mg} = 10 \text{ mcg}$

85. $0.2 \text{ mg} = 200 \text{ mcg}$

87. $1000 \text{ mcg} = 1 \text{ mg}$

89. $\dfrac{45 \text{ mg}}{15 \text{ mg}} = 3$

3 ml should be given.

91. $\dfrac{18 \text{ g}}{1 \text{ dL}} = \dfrac{18 \text{ g}}{100 \text{ mL}} = 0.18 \text{ g/mL}$

$(20 \text{ mL})(0.18 \text{ g/mL}) = 3.6 \text{ g}$

93. $3300 \text{ kg} = \dfrac{3300 \text{ kg}}{1} \cdot \dfrac{1 \text{ metric ton}}{1000 \text{ kg}}$

$= 3.3 \text{ metric tons}$

95. 10.9 metric tons

$= \dfrac{10.9 \text{ metric tons}}{1} \cdot \dfrac{1000 \text{ kg}}{1 \text{ metric ton}}$

$= 10,900 \text{ kg}$

Problem Recognition Exercises: U.S. Customary and Metric Conversions

1. $36 \text{ c} = \dfrac{36 \text{ c}}{1} \cdot \dfrac{1 \text{ pt}}{2 \text{ c}} \cdot \dfrac{1 \text{ qt}}{2 \text{ pt}} = 9 \text{ qt}$

3. $\dfrac{3}{4} \text{ lb} = \dfrac{0.75 \text{ lb}}{1} \cdot \dfrac{16 \text{ oz}}{1 \text{ lb}} = 12 \text{ oz}$

5. $12 \text{ ft} = \dfrac{12 \text{ ft}}{1} \cdot \dfrac{1 \text{ yd}}{3 \text{ ft}} = 4 \text{ yd}$

7. $45 \text{ dm} = 4.5 \text{ m}$

9. $\dfrac{1}{2} \text{ mi} = \dfrac{0.5 \text{ mi}}{1} \cdot \dfrac{5280 \text{ ft}}{1 \text{ mi}} = 2640 \text{ ft}$

11. $8 \text{ pt} = \dfrac{8 \text{ pt}}{1} \cdot \dfrac{1 \text{ qt}}{2 \text{ pt}} = 4 \text{ qt}$

13. $21 \text{ m} = 0.021 \text{ km}$

15. $36 \text{ mL} = 36 \text{ cc}$

17. $4322 \text{ g} = 4.322 \text{ kg}$

19. $20 \text{ fl oz} = \dfrac{20 \text{ fl oz}}{1} \cdot \dfrac{1 \text{ c}}{8 \text{ fl oz}} = 2.5 \text{ c}$

21. $4 \text{ pt} = \dfrac{4 \text{ pt}}{1} \cdot \dfrac{1 \text{ qt}}{2 \text{ pt}} \cdot \dfrac{1 \text{ gal}}{4 \text{ qt}} = \dfrac{1}{2} \text{ gal}$

23. $5.46 \text{ kg} = 5460 \text{ g}$

25. $9.1 \text{ mi} = \dfrac{9.1 \text{ mi}}{1} \cdot \dfrac{1760 \text{ yd}}{1 \text{ mi}} = 16,016 \text{ yd}$

27. $1.62 \text{ tons} = \dfrac{1.62 \text{ tons}}{1} \cdot \dfrac{2000 \text{ lb}}{1 \text{ ton}} = 3240 \text{ lb}$

29. $60 \text{ hr} = \dfrac{60 \text{ hr}}{1} \cdot \dfrac{1 \text{ day}}{24 \text{ hr}} = 2.5 \text{ days}$

Section 7.5   Converting Between U.S. Customary and Metric Units

Section 7.5   Practice Exercises

1. Answers will vary.

3. $500 \text{ g} = 0.5 \text{ kg}$
$500 \text{ g} = 50,000 \text{ cg}$
d, f

5. $500 \text{ cg} = 5 \text{ g}$
$500 \text{ cg} = 5000 \text{ mg}$
b, e

7. 200 L = 0.2 kL
200 L = 200,000 mL
c, f

9. 200 mL = 200 cc
200 mL = 0.2 L
b, g

11. b

13. a

15. $2 \text{ in.} \approx \dfrac{2 \text{ in.}}{1} \cdot \dfrac{2.54 \text{ cm}}{1 \text{ in.}} \approx 5.1 \text{ cm}$

17. $8 \text{ m} \approx \dfrac{8 \text{ m}}{1} \cdot \dfrac{1 \text{ yd}}{0.914 \text{ m}} \approx 8.8 \text{ yd}$

19. $400 \text{ ft} \approx \dfrac{400 \text{ ft}}{1} \cdot \dfrac{0.305 \text{ m}}{1 \text{ ft}} - 122 \text{ m}$

21. $45 \text{ in} \approx \dfrac{45 \text{ in}}{1} \cdot \dfrac{1 \text{ ft}}{12 \text{ in}} \cdot \dfrac{0.305 \text{ m}}{1 \text{ ft}} \approx 1.1 \text{ m}$

23. $0.5 \text{ ft} \approx \dfrac{0.5 \text{ ft}}{1} \cdot \dfrac{12 \text{ in}}{1 \text{ ft}} \cdot \dfrac{2.54 \text{ cm}}{1 \text{ in}} \approx 15.2 \text{ cm}$

25. $6 \text{ lb} \approx \dfrac{6 \text{ lb}}{1} \cdot \dfrac{0.45 \text{ kg}}{1 \text{ lb}} = 2.7 \text{ kg}$

27. $10 \text{ g} \approx \dfrac{10 \text{ g}}{1} \cdot \dfrac{1 \text{ oz}}{28 \text{ g}} \approx 0.4 \text{ oz}$

29. $0.54 \text{ kg} \approx \dfrac{0.54 \text{ kg}}{1} \cdot \dfrac{1 \text{ lb}}{0.45 \text{ kg}} = 1.2 \text{ lb}$

31. $2.2 \text{ tons} \approx \dfrac{2.2 \text{ tons}}{1} \cdot \dfrac{2000 \text{ lb}}{1 \text{ ton}} = 4400 \text{ lb}$
$= \dfrac{4400 \text{ lb}}{1} \cdot \dfrac{0.45 \text{ kg}}{1 \text{ lb}} = 1980 \text{ kg}$

33. $6 \text{ qt} \approx \dfrac{6 \text{ qt}}{1} \cdot \dfrac{0.95 \text{ L}}{1 \text{ qt}} = 5.7 \text{ L}$

35. $120 \text{ mL} \approx \dfrac{120 \text{ mL}}{1} \cdot \dfrac{1 \text{ fl oz}}{30 \text{ mL}} = 4 \text{ fl oz}$

37. 960 cc = 960 mL
$= \dfrac{960 \text{ mL}}{1} \cdot \dfrac{1 \text{ fl oz}}{30 \text{ mL}}$
$= 32 \text{ fl oz}$

39. $2 \text{ lb} = \dfrac{2 \text{ lb}}{1} \cdot \dfrac{16 \text{ oz}}{1 \text{ lb}} = 32 \text{ oz}$
$\dfrac{\$3.19}{2 \text{ lb}} = \dfrac{\$3.19}{32 \text{ oz}} \approx \$0.100 \text{ per ounce}$
$354 \text{ g} \approx \dfrac{354 \text{ g}}{1} \cdot \dfrac{1 \text{ oz}}{28 \text{ g}} \approx 12.6 \text{ oz}$
$\dfrac{\$1.49}{354 \text{ g}} \approx \dfrac{\$1.49}{12.6 \text{ oz}} \approx \$0.118 \text{ per ounce}$
The box of sugar costs $0.100 per ounce, and the packets cost $0.118 per ounce. The 2-lb box is the better buy.

41. $18 \text{ mi} \approx \dfrac{18 \text{ mi}}{1} \cdot \dfrac{1.61 \text{ km}}{1 \text{ mi}} = 28.98 \text{ km}$
18 mi is about 28.98 km. Therefore the 30-km race is longer than 18 mi.

43. $97 \text{ lb} \approx \dfrac{97 \text{ lb}}{1} \cdot \dfrac{0.45 \text{ kg}}{1 \text{ lb}} = 43.65 \text{ kg}$
97 lb is approximately 43.65 kg.

45. $1 \text{ gal} = 4 \text{ qt} \approx \dfrac{4 \text{ qt}}{1} \cdot \dfrac{0.95 \text{ L}}{1 \text{ qt}} = 3.8 \text{ L}$
$3.8(\$1.90) = \$7.22$
The price is approximately $7.22 per gallon.

47. 2.54 cm = 1 in.
A hockey puck is 1 in. thick.

49. 99,790 g = 99.790 kg
$\approx \dfrac{99.790 \text{ kg}}{1} \cdot \dfrac{1 \text{ lb}}{0.45 \text{ kg}}$
$\approx 222 \text{ lb}$
Tony weighs about 222 lb.

51. $45 \text{ cc} = 45 \text{ mL} \approx \dfrac{45 \text{ mL}}{1} \cdot \dfrac{1 \text{ fl oz}}{30 \text{ mL}}$
$= 1.5 \text{ fl oz}$
45 cc is 1.5 fl oz.

53. $2(2.9 + 2.2 + 2.9 + 2.2 + 5.8 + 4.4) \text{ ft}$
$= 2(20.4) \text{ ft} = 40.8 \text{ ft}$

55. $F = \dfrac{9}{5}C + 32 = \dfrac{9}{5}(25) + 32 = 45 + 32 = 77$

$\quad$ 77°F

57. $C = \dfrac{5}{9}(F - 32) = \dfrac{5}{9}(68 - 32) = \dfrac{5}{9}(36) = 20$

$\quad$ 20°C

59. $F = \dfrac{9}{5}C + 32 = \dfrac{9}{5}(30) + 32 = 54 + 32 = 86$

$\quad$ 86°F

61. $F = \dfrac{9}{5}C + 32 = \dfrac{9}{5}(4000) + 32$

$\quad = 7200 + 32 = 7232$

$\quad$ 7232°F

63. $F = \dfrac{9}{5}C + 32 = \dfrac{9}{5}(35) + 32 = 63 + 32 = 95$

$\quad$ It is a hot day. The temperature is 95°F.

65. $F = \dfrac{9}{5}C + 32 = \dfrac{9}{5}(100) + 32 = 9(20) + 32$

$\quad = 180 + 32 = 212$

67. $5700 \text{ lb} \approx \dfrac{5700 \text{ lb}}{1} \cdot \dfrac{0.45 \text{ kg}}{1 \text{ lb}} = 2565 \text{ kg}$

$\quad = 2.565$ metric tons

The Navigator weighs approximately 2.565 metric tons.

69. 108 metric tons $= 108{,}000$ kg

$\quad \approx \dfrac{108{,}000 \text{ kg}}{1} \cdot \dfrac{1 \text{ lb}}{0.45 \text{ kg}}$

$\quad = 240{,}000$ lb

The average weight of the blue whale is approximately 240,000 lb.

Chapter 7   Review Exercises

Section 7.1

1. $48 \text{ in.} = \dfrac{48 \text{ in.}}{1} \cdot \dfrac{1 \text{ ft}}{12 \text{ in.}} = 4 \text{ ft}$

3. $2 \text{ mi} = \dfrac{2 \text{ mi}}{1} \cdot \dfrac{1760 \text{ yd}}{1 \text{ mi}} = 3520 \text{ yd}$

5. $7040 \text{ ft} = \dfrac{7040 \text{ ft}}{1} \cdot \dfrac{1 \text{ mi}}{5280 \text{ ft}} = 1\dfrac{1}{3} \text{ mi}$

7. $2 \text{ yd} = \dfrac{2 \text{ yd}}{1} \cdot \dfrac{3 \text{ ft}}{1 \text{ yd}} \cdot \dfrac{12 \text{ in.}}{1 \text{ ft}} = 72 \text{ in.}$

9. $\quad$ 3 ft 9 in.
$\quad$ + 5 ft 6 in.
$\quad \overline{}$
$\quad$ 8 ft 15 in. $= 8 \text{ ft} + 1 \text{ ft} + 3 \text{ in.}$
$\qquad\qquad\quad = 9 \text{ ft } 3 \text{ in.}$

11. 5'3" − 2'5" = $\quad$ 5 ft 3 in. = $\quad$ 4 ft 15 in.
$\qquad\qquad\qquad$ − 2 ft 5 in. $\quad$ − 2 ft $\;$ 5 in.
$\qquad\qquad\qquad \overline{} \qquad \overline{}$
$\qquad\qquad\qquad\qquad\qquad\qquad\quad$ 2 ft 10 in.
$\qquad\qquad\qquad\qquad\qquad\qquad\quad$ or 2'10"

13. $4 \times (5'3") = 4(5 \text{ ft } 3 \text{ in.})$
$\qquad\qquad\quad = 20 \text{ ft } 12 \text{ in.}$
$\qquad\qquad\quad = 21 \text{ ft or } 21'$

15. $(6 \text{ ft } 3 \text{ in.}) \div 3 = \dfrac{6 \text{ ft}}{3} + \dfrac{3 \text{ in.}}{3} = 2 \text{ ft } 1 \text{ in.}$

17. $\quad$ 1 ft $\;$ 4 in.
$\qquad$ 1 ft $\;$ 4 in.
$\qquad$ 2 ft
$\qquad$ 2 ft
$\qquad$ + $\qquad$ 10 in.
$\qquad \overline{}$
$\qquad$ 6 ft 18 in. $= 6 \text{ ft} + 1 \text{ ft} + 6 \text{ in.}$
$\qquad\qquad\qquad = 7 \text{ ft } 6 \text{ in. or } 7\dfrac{1}{2} \text{ ft}$

Section 7.2

19. $72 \text{ hr} = \dfrac{72 \text{ hr}}{1} \cdot \dfrac{1 \text{ day}}{24 \text{ hr}} = 3 \text{ days}$

21. $5 \text{ lb} = \dfrac{5 \text{ lb}}{1} \cdot \dfrac{16 \text{ oz}}{1 \text{ lb}} = 80 \text{ oz}$

23. $12 \text{ fl oz} = \dfrac{12 \text{ fl oz}}{1} \cdot \dfrac{1 \text{ c}}{8 \text{ fl oz}} = 1\dfrac{1}{2} \text{ c}$

25. $3500 \text{ lb} = \dfrac{3500 \text{ lb}}{1} \cdot \dfrac{1 \text{ ton}}{2000 \text{ lb}}$

$= 1.75 \text{ tons or } 1\dfrac{3}{4} \text{ tons}$

27. $1800 \text{ sec} = \dfrac{1800 \text{ sec}}{1} \cdot \dfrac{1 \text{ min}}{60 \text{ sec}} \cdot \dfrac{1 \text{ hr}}{60 \text{ min}}$

$= 0.5 \text{ hr}$

29. $12 \text{ oz} = \dfrac{12 \text{ oz}}{1} \cdot \dfrac{1 \text{ lb}}{16 \text{ oz}} = \dfrac{3}{4} \text{ lb}$

31. $2{:}24{:}30 = 2 \text{ hr} + 24 \text{ min} + 30 \text{ sec}$

$2 \text{ hr} = \dfrac{2 \text{ hr}}{1} \cdot \dfrac{60 \text{ min}}{1 \text{ hr}} = 120 \text{ min}$

$30 \text{ sec} = \dfrac{30 \text{ sec}}{1} \cdot \dfrac{1 \text{ min}}{60 \text{ sec}} = 0.5 \text{ min}$

$2 \text{ hr} + 24 \text{ min} + 30 \text{ sec}$
$= 120 \text{ min} + 24 \text{ min} + 0.5 \text{ min}$
$= 144.5 \text{ min}$

33. $\left(1\dfrac{1}{2} \text{ tons}\right) \div 8 = (3000 \text{ lb}) \div 8 = 375 \text{ lb}$

375 lb will go to each location.

Section 7.3

35. b

37. c

39. $52 \text{ cm} = 520 \text{ mm}$

41. $2.338 \text{ km} = 2338 \text{ m}$

43. $34 \text{ dm} = 3.4 \text{ m}$

45. $4 \text{ cm} = 0.04 \text{ m}$

47. $1.2 \text{ m} = 1200 \text{ mm}$

49. $3.851 \text{ km} - 163 \text{ m} = 3851 \text{ m} - 163 \text{ m}$
$= 3688 \text{ m}$
The difference is 3688 m.

Section 7.4

51. $6.1 \text{ g} = 610 \text{ cg}$

53. $3212 \text{ mg} = 3.212 \text{ g}$

55. $5 \text{ cg} = 50 \text{ mg}$

57. $300 \text{ mL} = 0.3 \text{ L}$

59. $830 \text{ cL} = 8.3 \text{ L}$

61. $225 \text{ cc} = 225 \text{ mL} = 22.5 \text{ cL}$

63. $125 \text{ cm} = 1.25 \text{ m}$
$\text{Perimeter} = 2(2 \text{ m}) + 2(1.25 \text{ m})$
$= 4 \text{ m} + 2.5 \text{ m}$
$= 6.5 \text{ m}$
$\text{Area} = (1.25 \text{ m})(2 \text{ m}) = 2.5 \text{ m}^2$

65. $3200 \text{ g} = 3.2 \text{ kg}$
$68 \text{ kg} - 3.2 \text{ kg} = 64.8 \text{ kg}$
The difference is 64.8 kg.

67. (a) $80(0.04 \text{ mg}) = 3.2 \text{ mg}$
3.2 mg should be prescribed.

(b) $3.2 \text{ mg}(2)(7) = 44.8 \text{ mg}$
44.8 mg would be taken in a week.

69. $3 \text{ cc} = 3 \text{ mL}$
$3 \text{ mL} - 1.8 \text{ mL} = 1.2 \text{ mL} = 1.2 \text{ cc}$
There are 1.2 cc or 1.2 mL of fluid left.

Section 7.5

71. $6.2 \text{ in.} \approx \dfrac{6.2 \text{ in.}}{1} \cdot \dfrac{2.54 \text{ cm}}{1 \text{ in.}} \approx 15.75 \text{ cm}$

73. $140 \text{ g} \approx \dfrac{140 \text{ g}}{1} \cdot \dfrac{1 \text{ oz}}{28 \text{ g}} = 5 \text{ oz}$

75. $3.4 \text{ ft} \approx \dfrac{3.4 \text{ ft}}{1} \cdot \dfrac{0.305 \text{ m}}{1 \text{ ft}} \approx 1.04 \text{ m}$

77. $120 \text{ km} \approx \dfrac{120 \text{ km}}{1} \cdot \dfrac{1 \text{ mi}}{1.61 \text{ km}} \approx 74.53 \text{ mi}$

79. $1.5 \text{ fl oz} \approx \dfrac{1.5 \text{ fl oz}}{1} \cdot \dfrac{30 \text{ mL}}{1 \text{ fl oz}} = 45 \text{ mL}$
$= 45 \text{ cc}$

81. $30 \text{ in.} \approx \dfrac{30 \text{ in.}}{1} \cdot \dfrac{2.54 \text{ cm}}{1 \text{ in.}} = 76.2 \text{ cm}$
$76.2 \text{ cm} - 38 \text{ cm} = 38.2 \text{ cm}$
The difference in height is 38.2 cm.

83. $(30 \text{ mL})(2)(7) = 420 \text{ mL} = 0.42 \text{ L}$
The total amount of cough syrup is approximately 0.42 L.

85. $C = \dfrac{5}{9}(F - 32)$

87. $F = \dfrac{9}{5}C + 32$

Chapter 7   Test

1. c, d, g, j

3. a, b, e

5. $11{,}000 \text{ lb} = \dfrac{11{,}000 \text{ lb}}{1} \cdot \dfrac{1 \text{ ton}}{2000 \text{ lb}} = 5.5 \text{ tons}$

7. $\dfrac{3}{4} \text{ c} = \dfrac{\frac{3}{4} \text{ c}}{1} \cdot \dfrac{8 \text{ fl oz}}{1 \text{ c}} = 6 \text{ fl oz}$
$6 \text{ oz} + 4 \text{ oz} = 10 \text{ oz of liquid}$

9. $2' + 30'' + 2' + 30'' = 4' + 60''$
$60 \text{ in.} = \dfrac{60 \text{ in.}}{1} \cdot \dfrac{1 \text{ ft}}{12 \text{ in.}} = 5 \text{ ft}$
$4'60'' = 4 \text{ ft} + 5 \text{ ft} = 9 \text{ ft or } 9'$

11.
$$
\begin{array}{r}
8 \text{ lb } 1 \text{ oz} = \quad 7 \text{ lb } 17 \text{ oz} \\
- 7 \text{ lb } 10 \text{ oz} \quad - 7 \text{ lb } 10 \text{ oz} \\
\hline
7 \text{ oz}
\end{array}
$$
He lost 7 oz.

13. $1:15:15 = 1 \text{ hr } 15 \text{ min } 15 \text{ sec}$
$1 \text{ hr} = 60 \text{ min}$
$15 \text{ sec} = 0.25 \text{ min}$
$1 \text{ hr } 15 \text{ min } 15 \text{ sec}$
$= 60 \text{ min} + 15 \text{ min} + 0.25 \text{ min}$
$= 75.25 \text{ min}$

15. c

17. $0.015 \text{ L} = 15 \text{ mL}$

19. $411 \text{ g} = 41{,}100 \text{ cg}$

21. $2 \text{ L} \approx \dfrac{2 \text{ L}}{1} \cdot \dfrac{1 \text{ qt}}{0.95 \text{ L}} \approx 2.1 \text{ qt}$

23. $4.5 \text{ km} \approx \dfrac{4.5 \text{ km}}{1} \cdot \dfrac{1 \text{ mi}}{1.61 \text{ km}} \approx 2.8 \text{ mi}$

25. $20 \text{ in.} \approx \dfrac{20 \text{ in.}}{1} \cdot \dfrac{2.54 \text{ cm}}{1 \text{ in.}} = 50.8 \text{ cm}$
$38 \text{ in.} \approx \dfrac{38 \text{ in.}}{1} \cdot \dfrac{2.54 \text{ cm}}{1 \text{ in.}} = 96.52 \text{ cm}$
50.8 cm tall and 96.52-cm wingspan

27. $C = \dfrac{5}{9}(F - 32) = \dfrac{5}{9}(375 - 32)$
$= \dfrac{5}{9}(343) \approx 190.6$
190.6°C

29. $(0.1 \text{ mg})(70)(4) = 28 \text{ mg}$

Chapter 1–7   Cumulative Review Exercises

1. (a) $2499 \approx 2000$
(b) $42{,}099 \approx 42{,}100$

3. $\text{Area} = (10 \text{ cm})(18 \text{ cm}) = 180 \text{ cm}^2$

5. (a) Ford Motor Company spends the most. That amount is $7400 million or $7,400,000,000.

(b)
$$\begin{array}{r} 4620 \\ -\,4318 \\ \hline 302 \end{array}$$
The difference between IBM and Motorola is $302 million or $302,000,000.

(c)
$$\begin{array}{r} 7400 \\ 6200 \\ 4620 \\ 4379 \\ +\,4318 \\ \hline 26{,}917 \end{array}$$
The total amount spent is $26,917 million or $26,917,000,000.

7. The number 32,542 is not divisible by 3 because the sum of the digits (16) is not divisible by 3.

9. Area $= \dfrac{1}{2}(54 \text{ in.})(20 \text{ in.}) = 540 \text{ in.}^2$

11. The LCD of $\dfrac{1}{2}, \dfrac{6}{5},$ and $\dfrac{3}{10}$ is 10.

13.
$$\begin{aligned} 6\frac{2}{3}+2\frac{5}{6} &= \frac{20}{3}+\frac{17}{6} \\ &= \frac{40}{6}+\frac{17}{6} \\ &= \frac{57}{6} \\ &= \frac{19}{2} \text{ or } 9\frac{1}{2} \end{aligned}$$

15. $6\dfrac{2}{3} \div 2\dfrac{5}{6} = \dfrac{20}{3} \div \dfrac{17}{6} = \dfrac{20}{3} \cdot \dfrac{\cancel{6}^{2}}{17} = \dfrac{40}{17} \text{ or } 2\dfrac{6}{17}$

17. $\dfrac{1}{3} = 0.\overline{3}$

19. $1.25 = 1\dfrac{1}{4} = \dfrac{5}{4}$

21. $\dfrac{3}{8} = 0.375$

23. (a) $\dfrac{6}{5}$

(b) $6+5=11; \ \dfrac{6}{11}$

25.
$$\begin{aligned} \frac{18}{5} &= \frac{x}{25} \\ 5x &= (18)(25) \\ 5x &= 450 \\ \frac{5x}{5} &= \frac{450}{5} \\ x &= 90 \text{ cars} \end{aligned}$$

27.
$$\frac{6}{8} \ \blacklozenge \ \frac{2}{3}$$
$$(6)(3) \ \blacklozenge \ (8)(2)$$
$$18 \neq 16$$
No, because $\dfrac{6}{8} \neq \dfrac{2}{3}$.

29.
$$\begin{aligned} \frac{3}{4} &= \frac{4}{x} \\ 3x &= (4)(4) \\ 3x &= 16 \\ \frac{3x}{3} &= \frac{16}{3} \\ x &= \frac{16}{3} \text{ yd} \end{aligned}$$

31. $x = 0.45(60)$
$x = 27$ people

33.
$$\begin{aligned} 2100 &= 0.14(x) \\ \frac{2100}{0.14} &= \frac{0.14x}{0.14} \\ 15{,}000 &= x \end{aligned}$$
$15,000 in sales

35. $5800 \text{ g} = 5.8 \text{ kg}$

37. $72 \text{ in.} \approx \dfrac{72 \text{ in.}}{1} \cdot \dfrac{2.54 \text{ cm}}{1 \text{ in.}} \approx 182.9 \text{ cm}$

39. $3\dfrac{1}{2} \text{ qt} = \dfrac{3\frac{1}{2} \text{ qt}}{1} \cdot \dfrac{2 \text{ pt}}{1 \text{ qt}} = 7 \text{ pt}$

Chapter 8   Geometry

Chapter Opener Puzzle

1. f

2. i

3. d

4. g

5. b

6. a

7. h

8. e

9. j

10. c

Section 8.1   Lines and Angles

Section 8.1   Practice Exercises

1. Answers will vary.

3. A line extends forever in both directions. A line segment is a portion of a line between and including two endpoints.

5. The single arrowhead indicates that the figure is a ray.

7. The dot represents a point.

9. The double arrowheads indicate that the figure is a line.

11.
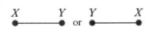

13.
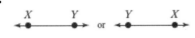

15.

17.

19. 20°

21. 90°

23. 148°

25. Right

27. Obtuse

29. Acute

31. Straight

33. $90° - 80° = 10°$

35. $90° - 27° = 63°$

37. $90° - 29.5° = 60.5°$

39. $90° - 89° = 1°$

41. $180° - 80° = 100°$

43. $180° - 127° = 53°$

45. $180° - 37.4° = 142.6°$

47. $180° - 179° = 1°$

49. No, because the sum of two angles that are both greater than 90° will be more than 180°.

51. Yes. For two angles to add to 90°, the angles themselves must both be less than 90°.

53. A 90° angle; $90° + 90° = 180°$

55.

57.

59. $m(\angle a) = 41°$ since $\angle a$ and the given angle are vertical angles.
$m(\angle b) = 180° - 41° = 139°$ since $\angle b$ is the supplement to the given angle.
$m(\angle c) = 180° - 41° = 139°$ since $\angle c$ is the supplement to the given angle.

61. $m(\angle a) = 180° - (42° + 112°)$
$= 180° - 154° = 26°$
since $\angle a$ together with the two given angles form a straight angle.
$m(\angle b) = 112°$ since $\angle b$ and the 112° angle are vertical angles.
$m(\angle c) = 26°$ since $\angle c$ and $\angle a$ are vertical angles.
$m(\angle d) = 42°$ since $\angle d$ and the 42° angle are vertical angles.

63. The two lines are perpendicular.

65. Vertical angles

67. a, c or b, h or e, g or f, d

69. a, e or f, b

71. $m(\angle a) = 180° - 125° = 55°$ since $\angle a$ and the given angle are supplementary.
$m(\angle b) = 125°$ since $\angle b$ and the given angle are vertical angles.
$m(\angle c) = 55°$ since $\angle c$ and $\angle a$ are vertical angles.
$m(\angle d) = 55°$ since $\angle d$ and $\angle a$ are corresponding angles.
$m(\angle e) = 125°$ since $\angle e$ and $\angle b$ are corresponding angles.
$m(\angle f) = 55°$ since $\angle f$ and $\angle a$ are alternate exterior angles.
$m(\angle g) = 125°$ since $\angle g$ and $\angle e$ are vertical angles.

73. True

75. True

77. False

79. True

81. True

83. $22° + 48° = 70°$

85. $69° + 21° = 90°$

87. **(a)** $180° - (42° + 90°) = 180° - 132° = 48°$
(b) $48°$
(c) $42° + 90° = 132°$

89. $180°$, since $30 \sec = \dfrac{1}{2} \min$

91. $120°$, since $20 \sec = \dfrac{1}{3} \min$

Section 8.2   Triangles and the Pythagorean Theorem

Section 8.2   Practice Exercises

1. Triangles

3. Yes

5. No

7. No

9. $90° + 36° + m(\angle a) = 180°$
$126° + m(\angle a) = 180°$
$m(\angle a) = 180° - 126°$
$m(\angle a) = 54°$

11. $62° + m(\angle b) + 40° = 180°$
$m(\angle b) + 102° = 180°$
$m(\angle b) = 180° - 102°$
$m(\angle b) = 78°$

13. $m(\angle b) = 180° - 100° = 80°$
$m(\angle a) + 40° + 80° = 180°$
$m(\angle a) + 120° = 180°$
$m(\angle a) = 180° - 120°$
$m(\angle a) = 60°$

15. $m(\angle a) = 40°$
$40° + m(\angle b) + 68° = 180°$
$m(\angle b) + 108° = 180°$
$m(\angle b) = 180° - 108°$
$m(\angle b) = 72°$

17. c, f

19. b, d

21. b, c, e

23. $\sqrt{49} = 7$

25. $7^2 = 49$

27. $4^2 = 16$

29. $\sqrt{16} = 4$

31. $\sqrt{36} = 6$

33. $6^2 = 36$

35. $9^2 = 81$

37. $\sqrt{81} = 9$

39. $a^2 + b^2 = c^2$
$3^2 + 4^2 = c^2$
$9 + 16 = c^2$
$25 = c^2$
$\sqrt{25} = c$
$5 = c$
$c = 5$ m

41. $a^2 + b^2 = c^2$
$9^2 + b^2 = 15^2$
$81 + b^2 = 225^2$
$b^2 = 225 - 81$
$b^2 = 144$
$b = \sqrt{144}$
$b = 12$ yd

43. $a^2 + b^2 = c^2$
$24^2 + b^2 = 26^2$
$576 + b^2 = 676$
$b^2 = 676 - 576$
$b^2 = 100$
$b = \sqrt{100}$
$b = 10$
Leg = 10 ft

45. $a^2 + b^2 = c^2$
$32^2 + 24^2 = c^2$
$1024 + 576 = c^2$
$1600 = c^2$
$\sqrt{1600} = c$
$40 = c$
Hypotenuse = 40 in.

47. $a^2 + b^2 = c^2$
$16^2 + 12^2 = c^2$
$256 + 144 = c^2$
$400 = c^2$
$\sqrt{400} = c$
$20 = c$
The brace is 20 in. long.

49. $a^2 + b^2 = c^2$
$12^2 + b^2 = 15^2$
$144 + b^2 = 225$
$b^2 = 225 - 144$
$b^2 = 81$
$b = \sqrt{81}$
$b = 9$
The height is 9 km.

51. The distances traveled form the legs of a right triangle.

$$24^2 + 7^2 = c^2$$
$$576 + 49 = c^2$$
$$625 = c^2$$
$$\sqrt{625} = c$$
$$25 = c$$

The car is 25 mi from the starting point.

53.
$$a^2 + b^2 = c^2$$
$$a^2 + 6^2 = 10^2$$
$$a^2 + 36 = 100$$
$$a^2 = 100 - 36$$
$$a^2 = 64$$
$$a = \sqrt{64}$$
$$a = 8$$
$$P = 8\text{ m} + 6\text{ m} + 10\text{ m} = 24\text{ m}$$

55.
$$a^2 + b^2 = c^2$$
$$a^2 + 12^2 = 13^2$$
$$a^2 + 144 = 169$$
$$a^2 = 169 - 144$$
$$a^2 = 25$$
$$a = \sqrt{25}$$
$$a = 5$$
$$P = 5\text{ km} + 12\text{ km} + 13\text{ km} = 30\text{ km}$$

57. The right triangle has legs $a = 4$ in. and $b = 11$ in. $- 8$ in. $= 3$ in.

$$a^2 + b^2 = c^2$$
$$4^2 + 3^2 = c^2$$
$$16 + 9 = c^2$$
$$25 = c^2$$
$$\sqrt{25} = c$$
$$5 = c$$
$$c = 5\text{ in.}$$
$$P = 5\text{ in.} + 11\text{ in.} + 4\text{ in.} + 8\text{ in.} = 28\text{ in.}$$

59. The unknown side c is the hypotenuse of a right triangle with legs
$a = 18 - (7 + 5) = 18 - 12 = 6$ ft and
$b = 20 - (6 + 6) = 20 - 12 = 8$ ft.

$$a^2 + b^2 = c^2$$
$$6^2 + 8^2 = c^2$$
$$36 + 64 = c^2$$
$$100 = c^2$$
$$\sqrt{100} = c$$
$$10 = c$$
$$P = 10 + 7 + 6 + 18 + 20 + 5 + 6 = 72\text{ ft}$$

61. $\sqrt{10}$ is between 3 and 4; 3.162

63. $\sqrt{116}$ is between 10 and 11; 10.770

65. $\sqrt{5}$ is between 2 and 3; 2.236

67. $\sqrt{427.75} \approx 20.682$

69. $\sqrt{1,246,000} \approx 1116.244$

71. $\sqrt{0.49} = 0.7$

73. $\sqrt{0.56} \approx 0.748$

75.
$$20^2 + b^2 = 29^2$$
$$400 + b^2 = 841$$
$$b^2 = 841 - 400$$
$$b^2 = 441$$
$$b = \sqrt{441}$$
$$b = 21\text{ ft}$$

77. $5^2 + 10^2 = c^2$

$\quad 25 + 100 = c^2$

$\quad\quad 125 = c^2$

$\quad\quad \sqrt{125} = c$

$\quad\quad 11.180 \approx c$

Hypotenuse = 11.180 mi

79. $12^2 + b^2 = 22^2$

$\quad 144 + b^2 = 484$

$\quad\quad b^2 = 484 - 144$

$\quad\quad b^2 = 340$

$\quad\quad b = \sqrt{340}$

$\quad\quad b \approx 18.439$

Leg = 18.439 in.

81. $1^2 + 1^2 = c^2$

$\quad 1 + 1 = c^2$

$\quad\quad 2 = c^2$

$\quad\quad \sqrt{2} = c$

$\quad\quad 1.41 \approx c$

The diagonal length is 1.41 ft.

83. $25^2 + 25^2 = c^2$

$\quad 625 + 625 = c^2$

$\quad\quad 1250 = c^2$

$\quad\quad \sqrt{1250} = c$

$\quad\quad 35.36 \approx c$

The diagonal length is 35.36 ft.

Section 8.3   Quadrilaterals, Perimeter, and Area

Section 8.3   Practice Exercises

1.

$P = l + w + l + w = l + l + w + w = 2l + 2w$

3. An isosceles triangle has two sides of equal length.

5. An acute triangle has all acute angles.

7. An obtuse triangle has an obtuse angle.

9. A quadrilateral is a polygon with four sides.

11. A trapezoid has one pair of opposite sides that are parallel.

13. A rectangle has four right angles.

15. $P = 2l + 2w = 2(25 \text{ cm}) + 2(15 \text{ cm})$
$\quad = 50 \text{ cm} + 30 \text{ cm} = 80 \text{ cm}$

17. $P = 4s = 4(65 \text{ mm}) = 260 \text{ mm}$

19. $P = 1.8 \text{ m} + 3 \text{ m} + 2 \text{ m} + 3.9 \text{ m} = 10.7 \text{ m}$

21. $P = 3$ ft   8 in.

$\quad\quad 2$ ft 10 in.

$\quad\quad + 4$ ft

$\quad\quad\overline{\quad 9 \text{ ft } 18 \text{ in.}} = 9 \text{ ft} + 1 \text{ ft} + 6 \text{ in.}$

$\quad\quad\quad\quad\quad\quad\quad = 10 \text{ ft } 6 \text{ in.}$

23. $P = 2l + 2w = 2(2 \text{ ft}) + 2(6 \text{ in.})$
$\quad = 4 \text{ ft} + 12 \text{ in.} = 4 \text{ ft} + 1 \text{ ft} = 5 \text{ ft}$

25. $x = 300 \text{ mm} + 250 \text{ mm} = 550 \text{ mm}$
$\quad y = 7.5 \text{ dm} - 4.5 \text{ dm} = 3 \text{ dm}$
$\quad P = 550 \text{ mm} + 3 \text{ dm} + 300 \text{ mm} + 4.5 \text{ dm}$
$\quad\quad\quad\quad\quad\quad\quad\quad + 250 \text{ mm} + 7.5 \text{ dm}$
$\quad = 1100 \text{ mm} + 15 \text{ dm} = 11 \text{ dm} + 15 \text{ dm}$
$\quad = 26 \text{ dm or } 2600 \text{ mm}$

27. The total length of the unknown vertical sides is 80 ft. The total length of the unknown horizontal sides is 40 ft + 20 ft = 60 ft.
$P = 80 \text{ ft} + 60 \text{ ft} + 40 \text{ ft} + 20 \text{ ft} + 80 \text{ ft}$
$\quad = 280 \text{ ft}$
280 ft of rain gutters are needed.

29. $A = s^2 = (24 \text{ yd})^2 = 576 \text{ yd}^2$

31. $A = \dfrac{1}{2}bh = \dfrac{1}{2}(12 \text{ m})(9 \text{ m}) = 54 \text{ m}^2$

33. $A = \dfrac{1}{2}(a+b)h = \dfrac{1}{2}(47 \text{ in.} + 35 \text{ in.})16 \text{ in.}$

$\quad = \dfrac{1}{2}(82 \text{ in.})16 \text{ in.} = 656 \text{ in.}^2$

35. $A = bh = (4.6 \text{ ft})(4 \text{ ft}) = 18.4 \text{ ft}^2$

37. $A = bh = (5 \text{ ft } 6 \text{ in.})(2 \text{ ft } 3 \text{ in.})$

$\quad = (5.5 \text{ ft})(2.25 \text{ ft}) = 12.375 \text{ ft}^2$

39. Area of outer rectangle:

$\quad A = lw = (18.4 \text{ ft})(12.8 \text{ ft}) = 235.52 \text{ ft}^2$

Area of smaller removed rectangle:

$\quad A = lw = (3.1 \text{ ft})(5.8 \text{ ft}) = 17.98 \text{ ft}^2$

Area of shaded region

$\quad = 235.52 \text{ ft}^2 - 17.98 \text{ ft}^2 = 217.54 \text{ ft}^2$

41. Area of outer rectangle:

$\quad A = lw = (22 \text{ mm})(16 \text{ mm}) = 352 \text{ mm}^2$

Area of removed triangle:

$\quad A = \dfrac{1}{2}bh = \dfrac{1}{2}(16 \text{ mm})(9 \text{ mm}) = 72 \text{ mm}^2$

Area of shaded region

$\quad = 352 \text{ mm}^2 - 72 \text{ mm}^2 = 280 \text{ mm}^2$

43. Area of outer rectangle:

$\quad A = lw = (8 \text{ in.})(10 \text{ in.}) = 80 \text{ in.}^2$

Area of one small removed rectangle:

$\quad A = lw = (5 \text{ in.})(2 \text{ in.}) = 10 \text{ in.}^2$

Area of shaded region

$\quad = 80 \text{ in.}^2 - 2(10 \text{ in.}^2)$

$\quad = 80 \text{ in.}^2 - 20 \text{ in.}^2 = 60 \text{ in.}^2$

45. The area to be tiled is a trapezoid:

$\quad A = \dfrac{1}{2}(a+b)h = \dfrac{1}{2}(8 \text{ ft} + 10 \text{ ft})1.5 \text{ ft}$

$\quad = \dfrac{1}{2}(18 \text{ ft})1.5 \text{ ft} = 13.5 \text{ ft}^2$

The area to be carpeted is a rectangle with the tiled area removed.

$\quad A = lw - 13.5 \text{ ft}^2 = (22 \text{ ft})(18 \text{ ft}) - 13.5 \text{ ft}^2$

$\quad = 396 \text{ ft}^2 - 13.5 \text{ ft}^2 = 382.5 \text{ ft}^2$

The area to be carpeted is 382.5 ft^2.

The area to be tiled is 13.5 ft^2.

47. Area of one triangle:

$\quad A = \dfrac{1}{2}bh = \dfrac{1}{2}(3 \text{ yd})(1.5 \text{ yd}) = 2.25 \text{ yd}^2$

Area of rectangle:

$\quad A = lw = (4 \text{ yd})(3 \text{ yd}) = 12 \text{ yd}^2$

Total area $= 12 \text{ yd}^2 + 2(2.25 \text{ yd}^2)$

$\quad = 12 \text{ yd}^2 + 4.5 \text{ yd}^2$

$\quad = 16.5 \text{ yd}^2$

The area of the sign is 16.5 yd^2.

49. (a) $A = lw = (23 \text{ ft})(21 \text{ ft}) = 483 \text{ ft}^2$

The area is 483 ft^2.

(b) $\dfrac{483 \text{ ft}^2}{250 \text{ ft}^2} \approx 1.932$

They will need 2 paint kits.

51. $A = (3s)^2 = 9s^2$

The area is increased by 9 times.

53. False

55. True

Section 8.4   Circles, Circumference, and Area

Section 8.4   Practice Exercises

1. Circumference $= 2\pi r$ and πd

Area $= \pi r^2$

Similarities: All formulas involve π, and both can be found using the radius.

Differences: One involves $2r$, the other involves r^2.

3. $A = lw = (42 \text{ cm})(30 \text{ cm}) = 1260 \text{ cm}^2$

5. $A = \dfrac{1}{2}bh = \dfrac{1}{2}(42 \text{ cm})(30 \text{ cm}) = 630 \text{ cm}^2$

7. Yes. Since a rectangle is a special type of parallelogram (one that contains four right angles), the area formula for a parallelogram applies to a rectangle.

9. $d = 2r = 2(6 \text{ in.}) = 12 \text{ in.}$

11. $d = 2r = 2\left(\dfrac{3}{2} \text{ m}\right) = 3 \text{ m}$

13. $r = \dfrac{d}{2} = \dfrac{8 \text{ in.}}{2} = 4 \text{ in.}$

15. $r = \dfrac{d}{2} = \dfrac{16.6 \text{ m}}{2} = 8.3 \text{ m}$

17. c

19. π is the circumference divided by the diameter. That is, $\pi = \dfrac{C}{d}$.

21. (a) $C = 2\pi r = 2\pi(2 \text{ m}) = 4\pi \text{ m}$

   (b) $C = 2\pi r \approx 2(3.14)(2 \text{ m}) = 12.56 \text{ m}$

23. (a) $C = \pi d = \pi(20 \text{ cm}) = 20\pi \text{ cm}$

   (b) $C = \pi d \approx (3.14)(20 \text{ cm}) = 62.8 \text{ cm}$

25. (a) $C = 2\pi r = 2\pi(2.1 \text{ cm}) = 4.2\pi \text{ cm}$

   (b) $C = 2\pi r \approx 2(3.14)(2.1 \text{ cm})$
   $= 13.188 \text{ cm}$

27. (a) $C = 2\pi r = 2\pi\left(2\dfrac{1}{2} \text{ km}\right) = 2\pi(2.5 \text{ km})$
   $= 5\pi \text{ km}$

   (b) $C = 2\pi r \approx 2(3.14)\left(2\dfrac{1}{2} \text{ km}\right)$
   $= 15.7 \text{ km}$

29. $C = \pi d = \pi(6 \text{ cm}) \approx (3.14)(6 \text{ cm})$
   $= 18.84 \text{ cm}$

31. $C = \pi d = \pi(4.5 \text{ in.}) \approx (3.14)(4.5 \text{ in.})$
   $= 14.13 \text{ in.}$

33. $C = \pi d = \pi(2.2 \text{ cm}) \approx (3.14)(2.2 \text{ cm})$
   $= 6.908 \text{ cm}$

35. (a) $A = \pi r^2 \approx \pi(7 \text{ m})^2 = 49\pi \text{ m}^2$

   (b) $A = \pi r^2 \approx \left(\dfrac{22}{7}\right)(7 \text{ m})^2$
   $= \left(\dfrac{22}{7}\right)(49 \text{ m}^2) = 154 \text{ m}^2$

37. $r = \dfrac{d}{2} = \dfrac{42 \text{ in.}}{2} = 21 \text{ in.}$

   (a) $A = \pi r^2 \approx \pi(21 \text{ in.})^2 = 441\pi \text{ in.}^2$

   (b) $A = \pi r^2 \approx \left(\dfrac{22}{7}\right)(21 \text{ in.})^2$
   $= \left(\dfrac{22}{7}\right)(441 \text{ in.}^2) = 1386 \text{ in.}^2$

39. $r = \dfrac{d}{2} = \dfrac{25 \text{ mm}}{2} = 12.5 \text{ mm}$

   (a) $A = \pi r^2 \approx \pi(12.5 \text{ mm})^2$
   $= 156.25\pi \text{ mm}^2$

   (b) $A = \pi r^2 \approx (3.14)(12.5 \text{ mm})^2$
   $= (3.14)(156.25 \text{ mm}^2) \approx 491 \text{ mm}^2$

41. (a) $A = \pi r^2 \approx \pi(6.2 \text{ ft})^2 = 38.44\pi \text{ ft}^2$

   (b) $A = \pi r^2 \approx (3.14)(6.2 \text{ ft})^2$
   $= (3.14)(38.44 \text{ ft}^2) \approx 121 \text{ ft}^2$

43. $A = \dfrac{1}{2}(a+b)h - 2\pi r^2$

   $\approx \dfrac{1}{2}(5 \text{ ft} + 4 \text{ ft})(2 \text{ ft}) - 2(3.14)(1 \text{ ft})^2$

   $= \dfrac{1}{2}(9 \text{ ft})(2 \text{ ft}) - 2(3.14)(1 \text{ ft}^2)$

   $= 9 \text{ ft}^2 - 6.28 \text{ ft}^2$

   $= 2.72 \text{ ft}^2$

45. $A = s^2 - \pi\left(\dfrac{d}{2}\right)^2$

$\approx (16 \text{ in.})^2 - (3.14)\left(\dfrac{16 \text{ in.}}{2}\right)^2$

$= (16 \text{ in.})^2 - (3.14)(8 \text{ in.})^2$

$= 256 \text{ in.}^2 - (3.14)(64 \text{ in.}^2)$

$= 256 \text{ in.}^2 - 200.96 \text{ in.}^2$

$= 55.04 \text{ in.}^2$

47. $A = \dfrac{1}{2}bh + \dfrac{1}{2}\pi\left(\dfrac{d}{2}\right)^2$

$\approx \dfrac{1}{2}(4 \text{ in.})(6 \text{ in.}) + \dfrac{1}{2}(3.14)\left(\dfrac{4 \text{ in.}}{2}\right)^2$

$= \dfrac{1}{2}(4 \text{ in.})(6 \text{ in.}) + \dfrac{1}{2}(3.14)(2 \text{ in.})^2$

$= \dfrac{1}{2}(4 \text{ in.})(6 \text{ in.}) + \dfrac{1}{2}(3.14)(4 \text{ in.}^2)$

$= 12 \text{ in.}^2 + 6.28 \text{ in.}^2$

$= 18.28 \text{ in.}^2$

49. Area of outer circle $= \pi r^2$

$\approx (3.14)(10 \text{ mm})^2$

$= (3.14)(100 \text{ mm}^2)$

$= 314 \text{ mm}^2$

Area of inner circle $= \pi r^2$

$\approx (3.14)(8 \text{ mm})^2$

$= (3.14)(64 \text{ mm}^2)$

$= 200.96 \text{ mm}^2$

Area $= 314 \text{ mm}^2 - 200.96 \text{ mm}^2$

$= 113.04 \text{ mm}^2$

51. $A = lw + 3\left(\dfrac{1}{2}\pi\left(\dfrac{d}{2}\right)^2\right)$

$\approx (18 \text{ in.})(10 \text{ in.}) + 3\left(\dfrac{1}{2}(3.14)\left(\dfrac{6 \text{ in.}}{2}\right)^2\right)$

$= (18 \text{ in.})(10 \text{ in.}) + 3\left(\dfrac{1}{2}(3.14)(3 \text{ in.})^2\right)$

$= (18 \text{ in.})(10 \text{ in.}) + 3\left(\dfrac{1}{2}(3.14)(9 \text{ in.}^2)\right)$

$= 180 \text{ in.}^2 + 42.39 \text{ in.}^2$

$= 222.39 \text{ in.}^2$

53. $C = \pi d \approx (3.14)(5.3 \text{ mi}) = 16.642 \text{ mi}$

55. $A = \pi r^2$

$\approx (3.14)(30 \text{ ft})^2$

$= (3.14)(900 \text{ ft}^2)$

$= 2826 \text{ ft}^2$

57. (a) $A = \pi\left(\dfrac{d}{2}\right)^2 \approx 3.14\left(\dfrac{32 \text{ mi}}{2}\right)^2$

$= 3.14(16 \text{ mi})^2 = 3.14\left(256 \text{ mi}^2\right)$

$\approx 804 \text{ mi}^2$

(b) $A = \pi\left(\dfrac{d}{2}\right)^2 \approx 3.14\left(\dfrac{10 \text{ mi}}{2}\right)^2$

$= 3.14(5 \text{ mi})^2 = 3.14\left(25 \text{ mi}^2\right)$

$\approx 79 \text{ mi}^2$

59. (a) $C = \pi d \approx (3.14)(26 \text{ in.}) = 81.64 \text{ in.}$

(b) $\dfrac{12,000 \text{ in.}}{81.64 \text{ in.}} \approx 147 \text{ times}$

61. (a) $C = \pi d \approx (3.14)(6.75 \text{ in.}) = 21 \text{ in.}$

(b) $25(21 \text{ in.}) = 525 \text{ in.}$

$\dfrac{525 \text{ in.}}{1} \cdot \dfrac{1 \text{ ft}}{12 \text{ in.}} \approx 43.8 \text{ ft}$

63. $A = \pi r^2 = \pi(5.1 \text{ ft})^2 \approx 81.7128 \text{ ft}^2$

$C = 2\pi r = 2\pi(5.1 \text{ ft}) \approx 32.0442 \text{ ft}$

65. $A = \pi \left(\dfrac{d}{2}\right)^2 = \pi \left(\dfrac{103.24 \text{ mm}}{2}\right)^2$

$\approx 8371.1644 \text{ mm}^2$

$C = \pi d = \pi (103.24 \text{ mm}) \approx 324.3380 \text{ mm}$

Problem Recognition Exercises: Area, Perimeter, and Circumference

1. $A = s^2 = (5 \text{ ft})^2 = 25 \text{ ft}^2$

$P = 4s = 4(5 \text{ ft}) = 20 \text{ ft}$

3. $A = lw = (4 \text{ m})(3 \text{ m}) = 12 \text{ m}^2$ or

$A = lw = (400 \text{ cm})(300 \text{ cm})$

$= 120{,}000 \text{ cm}^2$

$P = 2l + 2w = 2(4 \text{ m}) + 2(3 \text{ m})$

$= 8 \text{ m} + 6 \text{ m} = 14 \text{ m}$ or

$P = 2l + 2w = 2(400 \text{ cm}) + 2(300 \text{ cm})$

$= 800 \text{ cm} + 600 \text{ cm} = 1400 \text{ cm}$

5. $A = bh = (1 \text{ yd})\left(\dfrac{1}{3} \text{ yd}\right) = \dfrac{1}{3} \text{ yd}^2$ or

$A = bh = (3 \text{ ft})(1 \text{ ft}) = 3 \text{ ft}^2$

$P = 2a + 2b = 2(1 \text{ yd}) + 2\left(\dfrac{1}{2} \text{ yd}\right)$

$= 2 \text{ yd} + 1 \text{ yd} = 3 \text{ yd}$ or

$P = 2l + 2w = 2(6 \text{ ft}) + 2(1.5 \text{ ft})$

$= 12 \text{ ft} + 3 \text{ ft} = 9 \text{ ft}$

7. $A = \dfrac{1}{2}bh = \dfrac{1}{2}(3 \text{ yd})(4 \text{ yd}) = 6 \text{ yd}^2$

$c^2 = a^2 + b^2 = 3^2 + 4^2 = 9 + 16 = 25$

$c = \sqrt{25} = 5 \text{ yd}$

$P = a + b + c = 3 \text{ yd} + 4 \text{ yd} + 5 \text{ yd} = 12 \text{ yd}$

9. $A = \dfrac{1}{2}(a + b)h = \dfrac{1}{2}(14 \text{ m} + 8 \text{ m})(4 \text{ m})$

$= \dfrac{1}{2}(22 \text{ m})(4 \text{ m}) = 44 \text{ m}^2$

$P = 14 \text{ m} + 5 \text{ m} + 8 \text{ m} + 5 \text{ m} = 32 \text{ m}$

11. $A = \pi \left(\dfrac{d}{2}\right)^2 = 3.14 \left(\dfrac{6 \text{ yd}}{2}\right)^2 = 3.14(3\text{yd})^2$

$= 3.14(9 \text{ yd}^2) = 28.26 \text{ yd}^2$

$C = \pi d = 3.14(6 \text{ yd}) = 18.84 \text{ yd}$

13. $A = \pi r^2 = \dfrac{22}{7}(7 \text{ cm})^2 = \dfrac{22}{7}(49 \text{ cm}^2)$

$= 154 \text{ cm}^2$

$C = 2\pi r = 2\left(\dfrac{22}{7}\right)(7 \text{ cm}) = 44 \text{ cm}$

Section 8.5   Volume

Section 8.5   Practice Exercises

1. Answers will vary.

3. $C = 2\pi r \approx 2(3.14)(4 \text{ in.}) = 25.12 \text{ in.}$

$A = \pi r^2 \approx (3.14)(4 \text{ in.})^2$

$= (3.14)(16 \text{ in.}^2) = 50.24 \text{ in.}^2$

5. $A = lw - 2\pi \left(\dfrac{d}{2}\right)^2$

$\approx (24 \text{ cm})(12 \text{ cm}) - 2(3.14)\left(\dfrac{8 \text{ cm}}{2}\right)^2$

$= (24 \text{ cm})(12 \text{ cm}) - 2(3.14)(4 \text{ cm})^2$

$= (24 \text{ cm})(12 \text{ cm}) - 2(3.14)(16 \text{ cm}^2)$

$= 288 \text{ cm}^2 - 100.48 \text{ cm}^2$

$= 187.52 \text{ cm}^2$

7. b, d

9. $A = s^2 = (1 \text{ ft})^2 = 1 \text{ ft}^2$
$V = s^3 = (1 \text{ ft})^3 = 1 \text{ ft}^3$

11. $A = s^2 = (1 \text{ km})^2 = 1 \text{ km}^2$
$V = s^3 = (1 \text{ km})^3 = 1 \text{ km}^3$

13. $V = s^3 = (1.4 \text{ cm})^3 = 2.744 \text{ cm}^3$

15. $V = lwh = (8 \text{ ft})(12 \text{ ft})(6 \text{ in.})$
$= (8 \text{ ft})(12 \text{ ft})(0.5 \text{ ft}) = 48 \text{ ft}^3$

17. $V = \pi r^2 h \approx (3.14)(2 \text{ mm})^2 (1 \text{ mm})$
$= (3.14)(4 \text{ mm}^2)(1 \text{ mm}) = 12.56 \text{ mm}^3$

19. $V = \dfrac{4}{3}\pi r^3 \approx \dfrac{4}{3}(3.14)(9 \text{ yd})^3$
$= \dfrac{4}{3}(3.14)(729 \text{ yd}^3) = 3052.08 \text{ yd}^3$

21. $V = \dfrac{1}{3}\pi r^2 h$
$\approx \dfrac{1}{3}(3.14)(5 \text{ cm})^2 (9 \text{ cm})$
$= \dfrac{1}{3}(3.14)(25 \text{ cm}^2)(9 \text{ cm})$
$= 235.5 \text{ cm}^3$

23. $V = \dfrac{1}{2} \cdot \left(\dfrac{4}{3}\pi r^3 \right) \approx \dfrac{1}{2} \cdot \dfrac{4}{3}(3.14)\left(\dfrac{12 \text{ ft}}{2} \right)^3$
$= \dfrac{1}{2} \cdot \dfrac{4}{3}(3.14)(6 \text{ ft})^3$
$= \dfrac{1}{2} \cdot \dfrac{4}{3}(3.14)(216 \text{ ft}^3) = 452.16 \text{ ft}^3$

25. $r = \dfrac{d}{2} = \dfrac{8.2 \text{ in.}}{2} = 4.1 \text{ in.}$
$V = \dfrac{4}{3}\pi r^3 \approx \dfrac{4}{3}(3.14)(4.1 \text{ in.})^3$
$= \dfrac{4}{3}(3.14)(68.921 \text{ in.}^3) \approx 289 \text{ in.}^3$

27. $V = \dfrac{1}{3}\pi r^2 h \approx \dfrac{1}{3}(3.14)\left(\dfrac{10 \text{ ft}}{2} \right)^2 (12 \text{ ft})$
$= \dfrac{1}{3}(3.14)(5 \text{ ft})^2 (12 \text{ ft})$
$= \dfrac{1}{3}(3.14)(25 \text{ ft}^2)(12 \text{ ft}) = 314 \text{ ft}^3$

29. $r = \dfrac{d}{2} = \dfrac{6 \text{ in.}}{2} = 3 \text{ in.} = 0.25 \text{ ft}$
$V = \pi r^2 h \approx (3.14)(0.25 \text{ ft})^2 (50 \text{ ft})$
$= (3.14)(0.0625 \text{ ft}^2)(50 \text{ ft}) \approx 10 \text{ ft}^3$

31. **(a)** $r = \dfrac{d}{2} = \dfrac{27 \text{ ft}}{2} = 13.5 \text{ ft}$
$V = \pi r^2 h \approx 3.14(13.5 \text{ ft})^2 (54 \text{ in.})$
$= 3.14(13.5 \text{ ft})^2 (4.5 \text{ ft})$
$= 3.14(182.25 \text{ ft}^2)(4.5 \text{ ft})$
$\approx 2575 \text{ ft}^3$

(b) $\dfrac{2575 \text{ ft}^3}{0.1337 \text{ ft}^3} \approx 19,260 \text{ gal}$

33. $R = \dfrac{d}{2} = \dfrac{6 \text{ mm}}{2} = 3 \text{ mm}$
$r = \dfrac{d}{2} = \dfrac{2 \text{ mm}}{2} = 1 \text{ mm}$
$V = \pi R^2 h - \pi r^2 h$
$\approx (3.14)(3 \text{ mm})^2 (20 \text{ mm})$
$\qquad - (3.14)(1 \text{ mm})^2 (20 \text{ mm})$
$\approx 565.2 \text{ mm}^3 - 62.8 \text{ mm}^3 \approx 502 \text{ mm}^3$

35. $r = \dfrac{d}{2} = \dfrac{2 \text{ ft}}{2} = 1 \text{ ft}$
$V = \pi r^2 h + \dfrac{4}{3}\pi r^3$
$\approx (3.14)(1 \text{ ft})^2 (9 \text{ ft}) + \dfrac{4}{3}(3.14)(1 \text{ ft})^3$
$= (3.14)(1 \text{ ft}^2)(9 \text{ ft}) + \dfrac{4}{3}(3.14)(1 \text{ ft}^3)$
$\approx 28.26 \text{ ft}^3 + 4.19 \text{ ft}^3 \approx 32 \text{ ft}^3$

37. $V = \dfrac{1}{3}\pi r^2 h + \dfrac{4}{3}\pi r^3$

$\approx \dfrac{1}{3}(3.14)(2 \text{ in.})^2(5.5 \text{ in.})$

$\qquad\qquad + \dfrac{4}{3}(3.14)(2 \text{ in.})^3$

$= \dfrac{1}{3}(3.14)(4 \text{ in.}^2)(5.5 \text{ in.})$

$\qquad\qquad + \dfrac{4}{3}(3.14)(8 \text{ in.}^3)$

$\approx 23 \text{ in.}^3 + 33 \text{ in.}^3 \approx 56 \text{ in.}^3$

39. $V = s^3 - lwh$

$= (1 \text{ ft})^3 - (10 \text{ in.})(10 \text{ in.})(1 \text{ ft})$

$= (1 \text{ ft})^3 - \left(\dfrac{10}{12} \text{ ft}\right)\left(\dfrac{10}{12} \text{ ft}\right)(1 \text{ ft})$

$= 1 \text{ ft}^3 - \dfrac{100}{144} \text{ ft}^3 = \dfrac{44}{144} \text{ ft}^3$

$= \dfrac{11}{36} \text{ ft}^3$ or approximately 0.306 ft^3

or

$V = s^3 - lwh$

$= (1 \text{ ft})^3 - (10 \text{ in.})(10 \text{ in.})(1 \text{ ft})$

$= (12 \text{ in.})^3 - (10 \text{ in.})(10 \text{ in.})(12 \text{ in.})$

$= 1728 \text{ in.}^3 - 1200 \text{ in.}^3 = 528 \text{ in.}^3$

41. $V = s^3 - \pi r^2 h$

$\approx (5 \text{ in.})^3 - (3.14)(1 \text{ in.})^2(5 \text{ in.})$

$= 125 \text{ in.}^3 - 15.7 \text{ in.}^3 = 109.3 \text{ in.}^3$

43. 6 in. = 0.5 ft

$V = (40 \text{ ft})(15 \text{ ft})(0.5 \text{ ft})$

$\qquad + (35 \text{ ft} - 15 \text{ ft})(40 \text{ ft} - 25 \text{ ft})(0.5 \text{ ft})$

$= (40 \text{ ft})(15 \text{ ft})(0.5 \text{ ft})$

$\qquad\qquad + (20 \text{ ft})(15 \text{ ft})(0.5 \text{ ft})$

$= 300 \text{ ft}^3 + 150 \text{ ft}^3 = 450 \text{ ft}^3$

45. $V = \dfrac{1}{3}\pi r^2 h \approx \dfrac{1}{3}(3.14)(3 \text{ in.})^2(9 \text{ in.})$

$= \dfrac{1}{3}(3.14)(9 \text{ in.}^2)(9 \text{ in.}) = 84.78 \text{ in.}^3$

47. $V - \pi r^2 h \approx (3.14)(20 \text{ cm})^2(40 \text{ cm})$

$= (3.14)(400 \text{ cm}^2)(40 \text{ cm})$

$= 50,240 \text{ cm}^3$

Chapter 8   Review Exercises

Section 8.1

1. d

3. c

5. The measure of an acute angle is between $0°$ and $90°$.

7. The measure of a straight angle is $180°$.

9. (a) $90° - 33° = 57°$

(b) $180° - 33° = 147°$

11. $m(\angle ABE) = 90° - 30° = 60°$

13. $m(\angle ABG) = 180° - 5° = 175°$

15. b

17. a, c

19. $m(\angle b) = 118°$ since $\angle a$ and $\angle b$ are supplementary angles.

21. $m(\angle d) = 62°$ since $\angle d$ and $\angle a$ are vertical angles.

23. $m(\angle f) = 118°$ since $\angle f$ and the given angle are supplementary angles.

Section 8.2

25. $66° + 74° + x = 180°$
$140° + x = 180°$
$x = 180° - 140°$
$x = 40°$
$m(\angle x) = 40°$

27. An obtuse triangle has one obtuse angle.

29. A right triangle has a right (90°) angle.

31. An isosceles triangle has two sides of equal length and two angles of equal measure.

33. $\sqrt{25} = 5$

35. $\sqrt{100} = 10$

37. The sum of the squares of the legs of a right triangle equals the square of the hypotenuse.

39. $c^2 = a^2 + b^2$
$c^2 = 12^2 + 16^2$
$c^2 = 144 + 256$
$c^2 = 400$
$c = \sqrt{400}$
$c = 20$ ft

Section 8.3

41. They both have sides of equal length, but a square also has four right angles.

43. A square is a rectangle with four sides of equal length.

45. $P = 29$ cm $+ 10$ cm $+ 21$ cm $+ 30$ cm
$= 90$ cm

47. $P = 2l + 2w = 2(16$ mi$) + 2(12$ mi$)$
$= 32$ mi $+ 24$ mi $= 56$ mi

49. $P = 2l + 2w$
120 ft $= 36$ ft $+ 2w$
120 ft $- 36$ ft $= 2w$
84 ft $= 2w$
$\dfrac{84 \text{ ft}}{2} = \dfrac{2w}{2}$
42 ft $= w$

51. $A = \dfrac{1}{2}bh = \dfrac{1}{2}(8$ in.$)(5$ in.$) = 20$ in.^{2}

53. $A = (150$ ft $- 2 \cdot 12$ ft$)(80$ ft $- 2 \cdot 12$ ft$)$
$= (150$ ft $- 24$ ft$)(80$ ft $- 24$ ft$)$
$= (126$ ft$)(56$ ft$) = 7056$ ft^2

Section 8.4

55. $d = 2r = 2(45$ mm$) = 90$ mm

57. $r = \dfrac{d}{2} = \dfrac{45 \text{ mm}}{2} = 22.5$ mm

59. $C = 2\pi r = 2\pi(8$ m$) \approx 2(3.14)(8$ m$)$
$= 50.24$ m

$A = \pi r^2 = \pi(8$ m$)^2 \approx (3.14)(8$ m$)^2$
$= (3.14)(64$ m$^2) = 200.96$ m^2

61. $C = \pi d \approx \left(\dfrac{22}{7}\right)(140$ in.$) = 440$ in.

$r = \dfrac{d}{2} = \dfrac{140 \text{ in.}}{2} = 70$ in.

$A = \pi r^2 \approx \left(\dfrac{22}{7}\right)(70$ in.$)^2$

$= \left(\dfrac{22}{7}\right)(4900$ in.$^2) = 15,400$ in.^{2}

63. $A = lw - \dfrac{1}{2}(\pi r^2)$

$\approx (20$ in.$)(8$ in.$) - \dfrac{1}{2}(3.14)(4$ in.$)^2$

$= (20$ in.$)(8$ in.$) - \dfrac{1}{2}(3.14)(16$ in.$^2)$

$= 160$ in.$^2 - 25.12$ in.$^2 = 134.88$ in.^{2}

65. $A = s^2 + \dfrac{1}{2}(\pi r^2)$

$\approx (2 \text{ yd})^2 + \dfrac{1}{2}(3.14)\left(\dfrac{2 \text{ yd}}{2}\right)^2$

$= (2 \text{ yd})^2 + \dfrac{1}{2}(3.14)(1 \text{ yd})^2$

$= 4 \text{ yd}^2 + \dfrac{1}{2}(3.14)(1 \text{ yd}^2)$

$= 4 \text{ yd}^2 + 1.57 \text{ yd}^2 = 5.57 \text{ yd}^2$

Section 8.5

67. $r = \dfrac{d}{2} = \dfrac{6 \text{ ft}}{2} = 3 \text{ ft}$

$V = \pi r^2 h \approx (3.14)(3 \text{ ft})^2 (8 \text{ ft})$

$= (3.14)(9 \text{ ft}^2)(8 \text{ ft}) = 226.08 \text{ ft}^3$

69. $V = \dfrac{1}{3}\pi r^2 h \approx \dfrac{1}{3}(3.14)(3 \text{ km})^2 (4 \text{ km})$

$= \dfrac{1}{3}(3.14)(9 \text{ km}^2)(4 \text{ km}) = 37.68 \text{ km}^3$

71. $r = \dfrac{d}{2} = \dfrac{6 \text{ in.}}{2} = 3 \text{ in.}$

$V = \dfrac{4}{3}\pi r^3 \approx \dfrac{4}{3}(3.14)(3 \text{ in.})^3$

$= \dfrac{4}{3}(3.14)(27 \text{ in.}^3) \approx 113 \text{ in.}^3$

73. $V = (54 \text{ in.})(10 \text{ in.})(50 \text{ in.}) = 27{,}000 \text{ in.}^3$

$5 \text{ ft} = \dfrac{5 \text{ ft}}{1} \cdot \dfrac{12 \text{ in.}}{1 \text{ ft}} = 60 \text{ in.}$

$60 \text{ in.} - 50 \text{ in.} = 10 \text{ in.}$

$V = (10 \text{ in.})(15 \text{ in.})(10 \text{ in.}) = 1500 \text{ in.}^3$

Total volume $= 27{,}000 \text{ in.}^3 + 1500 \text{ in.}^3$

$= 28{,}500 \text{ in.}^3$

Chapter 8   Test

1. d

3. $90° - 16° = 74°$

5. $37° + 40° + x = 180°$
$77° + x = 180°$
$x = 180° - 77°$
$x = 103°$

7. $C = \pi d = \left(\dfrac{22}{7}\right)\left(\dfrac{5}{2} \text{ ft}\right) = \dfrac{55}{7} \text{ ft or } 7\dfrac{6}{7} \text{ ft}$

9. $A = \dfrac{1}{2}(a+b)h = \dfrac{1}{2}(4 \text{ ft} + 12 \text{ ft})(6 \text{ ft})$

$= \dfrac{1}{2}(16 \text{ ft})(6 \text{ ft}) = 48 \text{ ft}^2$

11. Obtuse

13. Acute

15. Right

17. Straight

19. $m(\angle x) = 180° - 55° = 125°$
$m(\angle y) = 55°$

21. $m(\angle S) = 90° - 41° = 49°$

23. $m(\angle A) + (180° - 120°)$
$+ (180° - 140°) = 180°$
$m(\angle A) + 60° + 40° = 180°$
$m(\angle A) + 100° = 180°$
$m(\angle A) = 180° - 100°$
$m(\angle A) = 80°$

25. $(80 \text{ m})^2 + (60 \text{ m})^2 = c^2$
$6400 \text{ m}^2 + 3600 \text{ m}^2 = c^2$
$10{,}000 \text{ m}^2 = c^2$
$\sqrt{10{,}000 \text{ m}^2} = c$
$100 \text{ m} = c$

27. c

29. b

31. e

33. $P = 2l + 2w$
$= 2(12 \text{ ft}) + 2(15 \text{ ft})$
$= 24 \text{ ft} + 30 \text{ ft}$
$= 54 \text{ ft}$

$54 \text{ ft} = \dfrac{54 \text{ ft}}{1} \cdot \dfrac{1 \text{ yd}}{3 \text{ ft}} = 18 \text{ yd}$

$\dfrac{18 \text{ yd}}{6 \text{ yd}} = 3$

3 rolls are needed.

35. $A = lw = (12 \text{ in.})(8 \text{ in.}) = 96 \text{ in.}^2$

$r = \dfrac{d}{2} = \dfrac{12 \text{ in.}}{2} = 6 \text{ in.}$

$A = \pi r^2$
$\approx (3.14)(6 \text{ in.})^2$
$= (3.14)(36 \text{ in.}^2)$
$= 113.04 \text{ in.}^2$

$113.04 \text{ in.}^2 - 96 \text{ in.}^2 = 17.04 \text{ in.}^2$
The area of the rectangular pizza is
96 in.^2 The area of the round pizza is
approximately 113.04 in.^2 The round
pizza is larger by about 17 in.^2

37. $V = lwh$
$= (18 \text{ in.})(5 \text{ in.})(14 \text{ in.})$
$= 1260 \text{ in.}^3$
The volume is 1260 in.^3

Chapters 1–8   Cumulative Review Exercises

1.
$$\begin{array}{r} 3835 \\ 21\overline{)\,80{,}535} \\ -63 \\ \hline 17\,5 \\ -16\,8 \\ \hline 73 \\ -63 \\ \hline 105 \\ -105 \\ \hline 0 \end{array}$$

3. $21 \div 0$ is undefined.

5.
$$\begin{array}{r} 1{,}275{,}000 \\ +\,1{,}236{,}000 \\ \hline 2{,}511{,}000 \end{array}$$

7. $\dfrac{1}{4}$ is used, so $\dfrac{3}{4}$ is left.

$\dfrac{3}{4}(14 \text{ oz}) = \dfrac{3}{4} \cdot \dfrac{14}{1} \text{ oz} = \dfrac{42}{4} \text{ oz} = 10\dfrac{1}{2} \text{ oz}$

There is $10\dfrac{1}{2}$ oz left.

9. $\dfrac{1}{3} \div 6 = \dfrac{1}{3} \div \dfrac{6}{1} = \dfrac{1}{3} \cdot \dfrac{1}{6} = \dfrac{1}{18}$

11. $6 = 2 \cdot 3$
$4 = 2 \cdot 2$
$10 = 2 \cdot 5$
LCM: $2 \cdot 2 \cdot 3 \cdot 5 = 60$

13. $\dfrac{13}{6} - \dfrac{3}{4} - \dfrac{3}{10} = \dfrac{130}{60} - \dfrac{45}{60} - \dfrac{18}{60} = \dfrac{67}{60}$

15. $5\dfrac{1}{9} = \dfrac{5 \cdot 9 + 1}{9} = \dfrac{46}{9}$

17. $\dfrac{\$26.98}{2} = \$13.49$

Geraldo will save the cost of one shirt
which is $13.49.

19. $0.\overline{2} = \dfrac{2}{9}$

21. $\dfrac{2\frac{1}{2}}{3\frac{3}{4}} = \dfrac{\frac{5}{2}}{\frac{15}{4}} = \dfrac{5}{2} \div \dfrac{15}{4} = \dfrac{\cancel{5}^{1}}{\cancel{2}_{1}} \cdot \dfrac{\cancel{4}^{2}}{\cancel{15}_{3}} = \dfrac{2}{3}$

23. $\dfrac{25}{7} = \dfrac{60}{x}$

$25x = (7)(60)$

$25x = 420$

$\dfrac{25x}{25} = \dfrac{420}{25}$

$x = 16.8$

17 pizzas

25. $\dfrac{\$8590}{2.5 \text{ hr}} = \$3436 \text{ per hour}$

27. $46.8 = 0.65(x)$

$\dfrac{46.8}{0.65} = \dfrac{0.65x}{0.65}$

$72 = x$

29. $\$180 - \$150 = \$30$

$\$30 = x(\$150)$

$\dfrac{\$30}{\$150} = \dfrac{\$150x}{\$150}$

$0.2 = x$

20% markup

31.
$$
\begin{array}{r}
8 \text{ in.} \\
1 \text{ ft } 9 \text{ in.} \\
16 \text{ in.} \\
1 \text{ ft } 3 \text{ in.} \\
2 \text{ ft} \\
+ 1 \text{ yd} \\
\hline
1 \text{ yd } 4 \text{ ft } 36 \text{ in.}
\end{array}
$$
$\phantom{1 \text{ yd } 4 \text{ ft } 36 \text{ in.}} = 3 \text{ ft} + 4 \text{ ft} + 3 \text{ ft}$

$\phantom{1 \text{ yd } 4 \text{ ft } 36 \text{ in.}} = 10 \text{ ft}$

33. $\dfrac{1}{2} \text{ c} = \dfrac{\frac{1}{2} \text{ c}}{1} \cdot \dfrac{8 \text{ fl oz}}{1 \text{ c}} = 4 \text{ fl oz}$

$4 \text{ fl oz} + 6 \text{ fl oz} = 10 \text{ fl oz}$

$6 \text{ fl oz} = \dfrac{6 \text{ fl oz}}{1} \cdot \dfrac{1 \text{ c}}{8 \text{ fl oz}} = \dfrac{3}{4} \text{ c}$

$\dfrac{1}{2} \text{ c} + \dfrac{3}{4} \text{ c} = 1\dfrac{1}{4} \text{ c}$

There is a total of $1\dfrac{1}{4}$ c or 10 fl oz of liquid.

35. $F = \dfrac{9}{5}C + 32 = \dfrac{9}{5}(5) + 32 = 9 + 32 = 41^{\circ}F$

37. $P = 1.75 \text{ ft} + 2 \text{ ft} + 2 \text{ ft} + 1.75 \text{ ft} + 3.5 \text{ ft}$

$ = 11 \text{ ft}$

39. $A = bh = (3 \text{ yd})(3 \text{ ft}) = (3 \text{ yd})(1 \text{ yd}) = 3 \text{ yd}^2$

or

$A = bh = (3 \text{ yd})(3 \text{ ft}) = (9 \text{ ft})(3 \text{ ft}) = 27 \text{ ft}^2$

Chapter 9   Introduction to Statistics

Chapter Opener Puzzle

		¹p			²m	³e	a	n
		r		⁴m		v		
	⁵m	o	d	e		e		
		b		d		n		
⁶s	t	a	t	i	s	t	i	c	s
		b		a				
		i		n				
		l						
		i						
		t						
		y						

Section 9.1   Tables, Bar Graphs, Pictographs, and Line Graphs

Section 9.1   Practice Exercises

1. Answers will vary.

3. Since Mt. Everest is the highest mountain, look at the 2nd column to see that Mt. Everest is in Asia.

5. Heights are in the 3rd column. Mt. Aconcagua is 22,834 ft – 20,320 ft = 2514 ft higher than Denali.

7. 25.1 – 21.5 = 3.6 yr

9. 24.3 – 21.5 = 2.8 yr

11. Men

13.

	Dog	Cat	Neither
Boy	4	1	3
Girl	3	4	5

15. (a) Students had the best access to a computer in 2007. Only four students were sharing a computer.

(b)
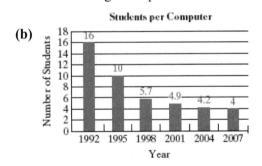

17.
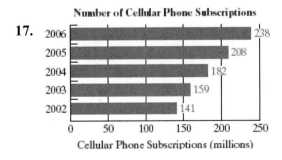

19. **(a)** One icon represents 100 servings sold.

  (b) The length of the "bar" for Saturday is given by $4\frac{1}{2}$ ice cream cones. There were about 450 servings sold on Saturday.

  (c) The "bar" containing $2\frac{3}{4}$ ice cream cones corresponds to Sunday.

21. **(a)** Each book icon represents 1 billion dollars. The "bar" of greatest length corresponds to Barnes & Noble/B. Dalton. There are $4\frac{1}{2}$ book icons, so Barnes & Noble/B. Dalton has approximately $4.5 billion in book sales.

  (b) There are 10 full book icons, and a total of about $1\frac{1}{2}$ partial icons. There is approximately $11.5 billion in book sales.

23. In 1920, 55.6% of men and 7.2% of women were in the labor force. 55.6% − 7.2% = 48.4%

25. The trend for women over 65 in the labor force shows a slight increase.

27. For example: 18%

29. In January 2007, the most hybrids were sold. There were 14,900 sold.

31. $12,400 - 8000 = 4400$

33. Between 2005 and 2006

35.

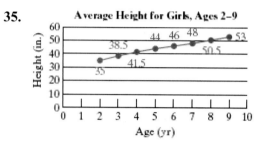

37. There are 14 servings per container. Each serving has 8 g of fat, so there are $8 \times 14 = 112$ g of fat in one container.

39. $$8\text{ g} = (0.13)x$$
$$\frac{8\text{ g}}{0.13} = \frac{0.13x}{0.13}$$
$$61.5g \approx x$$

The daily value of fat is approximately 61.5 g.

Section 9.2   Frequency Distributions and Histograms

Section 9.2   Practice Exercises

1. Answers will vary.

3. $14 + 18 + 24 + 10 + 6 = 72$
There are 72 data.

5. The 9–12 category had 24 data values, which is the highest frequency.

7.

Class Intervals (Age Group)	Tally	Frequency (Number of Professors)
56–58	||	2
59–61	|	1
62–64	|	1
65–67	⦚⦚⦚⦚ ||	7
68–70	⦚⦚⦚⦚	5
71–73	||||	4

(a) The class of 65–67 has the most values.

(b) 2 + 1 + 1 + 7 + 5 + 4 = 20

(c) $\dfrac{5}{20} = 0.25$ or 25%

Of the professors, 25% retire when they are 68 to 70 years old.

9.

Class Intervals (Amount Purchased)	Tally	Frequency (Number of Customers)
8.0–9.9	IIII	4
10.0–11.9	I	1
12.0–13.9	IIII|	5
14.0–15.9	IIII	4
16.0–17.9		0
18.0–19.9	II	2

(a) The 12.0–13.9 class has the highest frequency.

(b) 4 + 1 + 5 + 4 + 0 + 2 = 16
There are 16 data values represented in the table.

(c) $\dfrac{2}{16} = 0.125$ or 12.5%
Of the customers, 12.5% purchase 18 to 19.9 gal of gas.

11. The class widths are not the same.

13. There are too few classes.

15. The class intervals overlap. For example, it is unclear whether the data value 12 should be placed in the first class or the second class.

17.

Class Interval (Height, in.)	Frequency (No. of Students)
62–63	2
64–65	3
66–67	4
68–69	4
70–71	4
72–73	3

19.

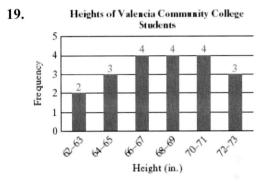

Heights of Valencia Community College Students

21.

Class	Tally	Frequency
20–39	IIII I	6
40–59	IIII IIII I	11
60–79	IIII	4
80–99	I	1
100–119		0
120–139		0
140–159		0
160–179	I	1

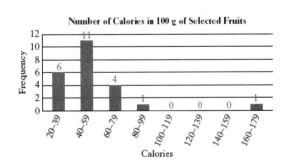

Number of Calories in 100 g of Selected Fruits

Section 9.3   Circle Graphs

Section 9.3   Practice Exercises

1. Answers will vary.

3. Total number of traffic fatalities
$= 16,000 + 11,520 + 10,880 + 9600$
$\qquad\qquad\qquad + 6400 + 9600$
$= 64,000$ fatalities

5. $11,520 - 10,880 = 640$
640 more people died in the 25–34 age group than in the 35–44 age group.

7. $\dfrac{16,000}{64,000} = 0.25$
25% of the deaths were from the 15–24 age group.

9. $\dfrac{16,000}{6400} = 2.5$
There are 2.5 times as many deaths in the 15–24 age group than in the 55–64 age group.

11. Total viewers (in millions)
$= 3 + 5.4 + 4.2 + 3.8 + 3.6 = 20$
There are 20 million viewers represented.

13. *The Young and the Restless* has 5.4 million viewers and *Guiding Light* has 3 million viewers.
$\dfrac{5.4}{3} = 1.8$
There are 1.8 times as many viewers who watch *The Young and the Restless* as *Guiding Light*.

15. *General Hospital* has 3.6 million viewers.
$\dfrac{3.6}{20} = 0.18$
Of the viewers, 18% watch *General Hospital*.

17. The store carries 8000 CDs. From the graph, 12% are musica Latina. Find 12% of 8000.

$x = (0.12)(8000) = 960$
There are 960 Latina CDs.

19. From the graph, 5% of the CDs are jazz and 3% are classical. This accounts for 8% of the CDs. Find 8% of 8000.
$x = (0.08)(8000) = 640$
There are 640 CDs that are either classical or jazz.

21. From the graph, 25% were played in Louisiana. Find 25% of 36.
$x = (0.25)(36) = 9$
There were 9 Super Bowls played in Louisiana.

23. From the graph, 6% were played in Georgia. Find 6% of 36.
$x = (0.06)(36) \approx 2$
There were 2 Super Bowls played in Georgia.

25.

27.

29.

31.

33.

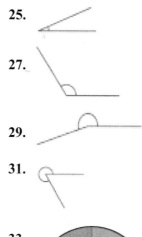

35. (a) Total Expenses = 9000 + 600 + 2400 = $12,000

	Expenses	Percent	Number of Degrees
Tuition	$9000	$\frac{9000}{12,000} = 0.75$ or 75%	$(0.75)(360°) = 270°$
Books	600	$\frac{600}{12,000} = 0.05$ or 5%	$(0.05)(360°) = 18°$
Housing	2400	$\frac{2400}{12,000} = 0.20$ or 20%	$(0.20)(360°) = 72°$

(b)

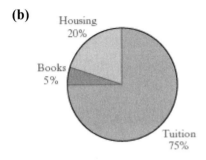

37.

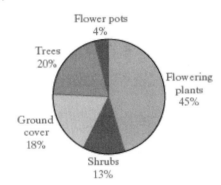

Section 9.4   Mean, Median, and Mode

Section 9.4 Practice Exercises

1. Answers will vary.

3. Mean $= \dfrac{4+6+5+10+4+5+8}{7} = \dfrac{42}{7} = 6$

5. Mean $= \dfrac{0+5+7+4+7+2+4+3}{8} = \dfrac{32}{8}$
$= 4$

7. Mean $= \dfrac{10+13+18+20+15}{5} = \dfrac{76}{5} = 15.2$

9. Mean $= \dfrac{11.0+9.1+8.3+7.9+7.5}{5} = \dfrac{43.8}{5}$
$= 8.76$ in.

11. Mean $= \dfrac{5.5+6.0+5.8+5.8+6.0+5.6}{6}$
$= \dfrac{34.7}{6} \approx 5.8$ hr

13. (a)

$$
\begin{array}{r}
360 \\
370 \\
380 \\
400 \\
400 \\
+\ 470 \\
\hline
2380
\end{array}
$$

$\text{Mean} = \dfrac{2380}{6} \approx 397 \text{ Cal}$

(b)

$$
\begin{array}{r}
310 \\
325 \\
350 \\
390 \\
440 \\
+\ 500 \\
\hline
2315
\end{array}
$$

$\text{Mean} = \dfrac{2315}{6} \approx 386 \text{ Cal}$

(c)

$$
\begin{array}{r}
397 \\
-\ 386 \\
\hline
11
\end{array}
$$

There is only an 11-Cal difference in the means.

15. (a) $\text{Mean} = \dfrac{98 + 80 + 78 + 90}{4} = \dfrac{346}{4}$
$= 86.5$

Zach's mean test score was 86.5%.

(b) $\text{Mean} = \dfrac{98 + 80 + 78 + 90 + 59}{5} = \dfrac{405}{5}$
$= 81$

The mean of all five tests was 81%.

(c) The low score of 59% decreased Zach's average by 86.5% − 81% = 5.5%.

17. Arrange the numbers in order from least to greatest.
13   14   16   <u>17</u>   19   20   22
Median = 17

19. Arrange the numbers in order from least to greatest.
100   109   <u>110   111</u>   118   123

$\text{Median} = \dfrac{110 + 111}{2} = \dfrac{221}{2} = 110.5$

21. Arrange the numbers in order from least to greatest.
40   40   <u>50   55</u>   55   58

$\text{Median} = \dfrac{50 + 55}{2} = \dfrac{105}{2} = 52.5$

23. Arrange the numbers in order from least to greatest.
3.82   3.87   <u>3.93</u>   4.09   4.10
Median = 3.93 deaths per 100

25. Arrange the numbers in order from least to greatest.
43   46   52   55   <u>56   60</u>   61   62   64   69

$\text{Median} = \dfrac{56 + 60}{2} = \dfrac{116}{2} = 58 \text{ years old}$

27. Arrange the numbers in order from least to greatest.
42.4, 45.4, 46.5, 48.3, <u>51.7</u>, 56.4, 71.2, 86.8, 91.6
Median = 51.7 million passengers

29. The data value 4 appears most often. The mode is 4.

31. No data value occurs most often. There is no mode.

33. There are 2 modes: 21 and 24.

35. $600

37. 5.2%

39. These data are bimodal: $2.49 and $2.51

41. $\text{Mean} = \dfrac{92 + 98 + 43 + 98 + 97 + 85}{6} = \dfrac{513}{6}$
$= 85.5\%$

Arrange the numbers in order from least to greatest.
43   85   <u>92   97</u>   98   98

$\text{Median} = \dfrac{92 + 97}{2} = \dfrac{189}{2} = 94.5\%$

The median gave Jonathan a better overall score.

43. Mean $= \dfrac{\begin{array}{c}312 + 225 + 221 + 256 + 308 \\ + 280 + 147\end{array}}{7}$

$= \dfrac{1749}{7} \approx \$250$

Arrange the numbers in order from least to greatest.

147 221 225 <u>256</u> 280 308 312

Median = $256

There is no mode.

45. Mean $= \dfrac{\begin{array}{c}850 + 835 + 839 + 829 + 850 + 850 \\ + 850 + 847 + 1850 + 825\end{array}}{10}$

$= \dfrac{9425}{10} = \$942,500$

Arrange the numbers in order from least to greatest.

825 829 835 839 <u>847 850</u> 850 850 850 1850

Median $= \dfrac{847 + 850}{2} = \dfrac{1697}{2} = \$848,500$

Mode = $850,000

47.

Age (yr)	Number of Students	Product
16	2	(16)(2) = 32
17	9	(17)(9) = 153
18	6	(18)(6) = 108
19	3	(19)(3) = 57
Total:	20	350

Mean $= \dfrac{350}{20} = 17.5$

The mean age is approximately 17.5 years.

49.

Number of Students in Each Class	Number of Classes	Product
18	5	(18)(5) = 90
20	6	(20)(6) = 120
25	15	(25)(15) = 375
30	10	(30)(10) = 300
35	8	(35)(8) = 280
Total:	44	1165

Mean $= \dfrac{1165}{44} \approx 26$

The weighted mean is about 26 students initially enrolled in each class.

51.

Grade	Credit-Hours	Product
B+ = 3.5	3	(3.5)(3) = 10.5
A = 4.0	4	(4.0)(4) = 16.0
A = 4.0	1	(4.0)(1) = 4.0
B = 3.0	3	(3.0)(3) = 9.0
Total:	11	39.5

GPA $= \dfrac{39.5}{11} \approx 3.59$

53.

Grade	Credit-Hours	Product
C+ = 2.5	5	(2.5)(5) = 12.5
A = 4.0	4	(4.0)(4) = 16.0
D = 1.0	3	(1.0)(3) = 3.0
A = 4.0	1	(4.0)(1) = 4.0
Total:	13	35.5

GPA $= \dfrac{35.5}{13} \approx 2.73$

Section 9.5   Introduction to Probability

Section 9.5   Practice Exercises

1. Answers will vary.

3. Mean $= \dfrac{13+16+22+25+10}{5} = \dfrac{86}{5} = 17.2$

Arrange the numbers in order from least to greatest.
10   13   <u>16</u>   22   25
Median = 16
There is no mode.

5. Mean $= \dfrac{8+9+10+7+8+8+11+10}{8}$

$= \dfrac{71}{8} = 8.875$

Arrange the numbers in order from least to greatest.
7   8   8   <u>8</u>   <u>9</u>   10   10   11
Median $= \dfrac{8+9}{2} = \dfrac{17}{2} = 8.5$
The mode is 8.

7. Mean $= \dfrac{20+20+18+17+19+5}{6} = \dfrac{99}{6}$

$= 16.5$
The mean is 16.5.
Arrange the numbers in order from least to greatest.
5   17   <u>18</u>   <u>19</u>   20   20
Median $= \dfrac{18+19}{2} = \dfrac{37}{2} = 18.5$
The median is 18.5.
The mode is 20.

9. {1, 2, 3, 4, 5, 6, 7, 8, 9, 10}

11. The sample space consists of all possible sums of the numbers of dots.
{2, 3, 4, 5, 6, 7, 8, 9, 10, 11, 12}

13. in 3 ways {1, 3, 5}

15. c, d, g, h

17. The event can occur in 2 ways:
The die lands as a 1 or 2.
The sample space has 6 elements:
1, 2, 3, 4, 5, and 6.
$\dfrac{2}{6} = \dfrac{1}{3}$

19. The event can occur in 3 ways:
The die lands as a 2, 4, or 6.
The sample space has 6 elements.
$\dfrac{3}{6} = \dfrac{1}{2}$

21. There are 5 black socks. There are 8 socks in the drawer.
$\dfrac{5}{8}$

23. There is 1 blue sock. There are 8 socks in the drawer.
$\dfrac{1}{8}$

25. The probability is 1 because a number from 1–6 will definitely come up.

27. An impossible event is one in which the probability is 0.

29. There are 12 + 40 = 52 cards.
$\dfrac{12}{52} = \dfrac{3}{13}$

31. There are 7 + 5 + 4 = 16 marbles in the jar. There are 5 red and 7 yellow marbles.
$\dfrac{5+7}{16} = \dfrac{12}{16} = \dfrac{3}{4}$

33. The total number of vacationers is
$14 + 13 + 18 + 28 + 11 + 30 + 6 = 120$.

  (a) 18 vacationers stayed 4 days.
  $$\frac{18}{120} = \frac{3}{20}$$

  (b) 14 vacationers stay 2 days and
  13 vacationers stay 3 days.
  $$\frac{14 + 13}{120} = \frac{27}{120} = \frac{9}{40}$$

  (c) 30 vacationers stay 7 days and
  6 vacationers stay 8 days.
  $$\frac{30 + 6}{120} = \frac{36}{120} = 0.30 \text{ or } 30\%$$

35. (a) 21 cars are manufactured in America.
  $$\frac{21}{60} = \frac{7}{20}$$

  (b) There are $21 + 9 = 30$ cars
  manufactured in a country other than
  Japan.
  $$\frac{30}{60} = 0.50 \text{ or } 50\%$$

37. The total number of students is
$1 + 6 + 11 + 7 + 3 + 1 = 29$.

  (a) A total of $1 + 6 = 7$ students are early.
  $$\frac{7}{29}$$

  (b) A total of $7 + 3 + 1 = 11$ students are
  late.
  $$\frac{11}{29}$$

  (c) A total of $1 + 6 + 11 = 18$ students
  arrive on time or early.
  $$\frac{18}{29} \approx 0.62 \text{ or } 62\%$$

39. The probability of losing is the same as
the probability of not winning.
$$1 - \frac{2}{11} = \frac{11}{11} - \frac{2}{11} = \frac{9}{11}$$

41. $100\% - 1.2\% = 98.8\%$

Chapter 9   Review Exercises

Section 9.1

1. From the 2nd column, Godiva has the
most calories.

3. Blue Bell has 70 mg of sodium. Edy's
Grand has 35 mg of sodium.
$$\frac{70}{35} = 2$$
Blue Bell has 2 times more sodium than
Edy's Grand.

5. The bar corresponding to 1970 has height
374. The average size of the farms in 1970
was 374 acres.

7. The average size in 1990 was 430 acres.
The average size in 1980 was 426 acres.
$430 - 426 = 4$
The difference is 4 acres.

9. 1 icon represents 50 tornados.

11. The June bar has 4 icons which represents
200 tornados.

13. The number of liver transplants was
greatest in 2005.

15. Increasing

17.

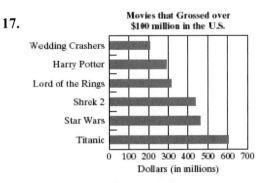

Section 9.2

19.

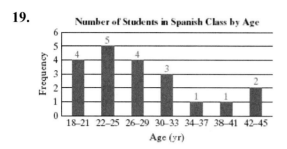

Section 9.3

21. $\dfrac{16}{24} = \dfrac{2}{3}$ of the subs contain beef.

Section 9.4

25. Mean $= \dfrac{\begin{array}{c}800 + 1000 + 1200 + 1300 + 900 \\ + 1200 + 1000\end{array}}{7}$

$= \dfrac{7400}{7} \approx 1060$

The mean daily calcium intake is approximately 1060 mg.

27. The mode is 4.

29. {blue, green, brown, black, gray, white}

31. a, c, d, e, g

23. (a)

Education Level	No. of People	Percent	No. of Degrees
Grade School	10	$\dfrac{10}{200} = 0.05$ or 5%	$(0.05)(360°) = 18°$
High School	50	$\dfrac{50}{200} = 0.25$ or 25%	$(0.25)(360°) = 90°$
Some College	60	$\dfrac{60}{200} = 0.30$ or 30%	$(0.30)(360°) =$ $108°$
Four-year degree	40	$\dfrac{40}{200} = 0.20$ or 20%	$(0.20)(360°) = 72°$
Post-graduate	40	$\dfrac{40}{200} = 0.20$ or 20%	$(0.20)(360°) = 72°$

Percent by Education Level

(b)

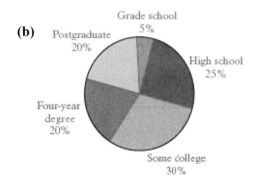

Chapter 9   Test

1.

World's Major Producers of Primary Energy (Quadrillions of Btu)

United States	72
Russia	43
China	35
Saudi Arabia	43
Canada	18

0 10 20 30 40 50 60 70 80

3. $1000

5. The "bar" with 5 icons corresponds to February.

7. $2.5 - 0.77 = 1.73$ in.

9.

Number of Minutes Used Monthly	Tally	Frequency
51–100	卌 I	6
101–150	II	2
151–200	III	3
201–250	II	2
251–300	IIII	4
301–350	III	3

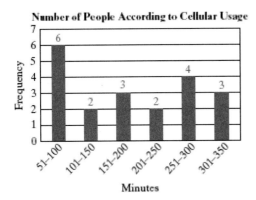

Number of People According to Cellular Usage

11. 20% have tile on their living room floor.
Find 20% of 200.
$x = (0.20)(200) = 40$
40 people would have tile.

13.
```
   19,340
   18,510
   22,834
   20,320
   16,864
    7,310
+ 29,035
────────
  134,213
```
$$\text{Mean} = \frac{134,213}{7} \approx 19,173 \text{ ft}$$

15. There is no mode.

17. (a) $\{1, 2, 3, 4, 5, 6, 7, 8\}$

(b) $\dfrac{1}{8}$

(c) There are 4 ways of obtaining an even number: 2, 4, 6, or 8.
$$\frac{4}{8} = \frac{1}{2}$$

(d) There are 2 ways of obtaining number less than 3 : 1 or 2.
$$\frac{2}{8} = \frac{1}{4}$$

19.

Grade	Number of Credit-Hours	Product
B = 3.0	4	(3.0)(4) = 12.0
A = 4.0	3	(4.0)(3) = 12.0
C = 2.0	3	(2.0)(3) = 6.0
A = 4.0	1	(4.0)(1) = 4.0
Total:	11	34.0

$$\text{GPA} = \frac{34.0}{11} \approx 3.09$$

Chapters 1–9   Cumulative Review Exercises

1. **(a)** Millions
   (b) Ten-thousands
   (c) Hundreds

3.
$$\begin{array}{r} 700 \\ \times\ 1\ 200 \\ \hline 840{,}000 \end{array}$$

5. $\dfrac{3}{8}$

7. $\dfrac{105}{96} \div \dfrac{7}{16} = \dfrac{\overset{15}{\cancel{105}}}{\underset{6}{\cancel{96}}} \cdot \dfrac{\overset{1}{\cancel{16}}}{\underset{1}{\cancel{7}}} = \dfrac{15}{6} = \dfrac{5}{2}$

9. $\dfrac{97}{102} - \dfrac{63}{102} = \dfrac{34}{102} = \dfrac{1}{3}$

11. $\dfrac{1}{2} + \dfrac{5}{3} - \dfrac{1}{6} = \dfrac{3}{6} + \dfrac{10}{6} - \dfrac{1}{6} = \dfrac{12}{6} = 2$

13. $13.28 + 0.27 = 13.55$
   $9.51 - 0.17 = 9.34$
   $14.35 + 0.10 = 14.45$
   $18.09 + 0.09 = 18.18$
   $21.63 - 0.37 = 21.26$

15. $68.412 \times 0.1 = 6.8412$

17. **(a)**
$$\begin{array}{r} 3{,}700{,}000 \\ -\ 3{,}000{,}000 \\ \hline 700{,}000 \end{array} \text{ km}^2 \text{ or } 0.7 \text{ million km}^2$$

   (b) $\dfrac{700{,}000}{3{,}700{,}000} = 0.189$ or 18.9%

19. $\dfrac{4}{50} = \dfrac{10}{x}$
   $4x = (50)(10)$
   $4x = 500$
   $\dfrac{4x}{4} = \dfrac{500}{4}$
   $x = 125$ min or 2 hr 5 min

21. $95 = (0.78)x$
   $\dfrac{95}{0.78} = \dfrac{0.78x}{0.78}$
   $122 \approx x$
   122 people

23. $78 = x(120)$
   $\dfrac{78}{120} = \dfrac{120x}{120}$
   $0.65 = x$
   65%

25. $2 \text{ ft} = \dfrac{2 \text{ ft}}{1} \cdot \dfrac{12 \text{ in.}}{1 \text{ ft}} = 24 \text{ in.}$
   2 ft 5 in. = 24 in. + 5 in. = 29 in.

27.
$$\begin{array}{r} 3 \text{ yd } 2 \text{ ft} \\ +\ 5 \text{ yd } 2 \text{ ft} \\ \hline 8 \text{ yd } 4 \text{ ft} = 8 \text{ yd} + 1 \text{ yd} + 1 \text{ ft} \\ = 9 \text{ yd } 1 \text{ ft} \end{array}$$

29. $\dfrac{16 \text{ lb } 12 \text{ oz}}{4} = \dfrac{16 \text{ lb}}{4} + \dfrac{12 \text{ oz}}{4}$
   $= 4 \text{ lb} + 3 \text{ oz}$
   $= 4 \text{ lb } 3 \text{ oz}$

31. Right

33. Area $= bh = (4 \text{ ft})(2 \text{ ft}) = 8 \text{ ft}^2$

35.

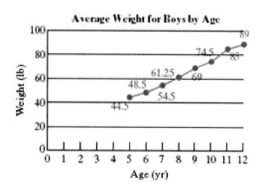

37. $30 - 3(5-2)^2 = 30 - 3(3)^2 = 30 - 3 \cdot 9$
   $= 30 - 27 = 3$

39. $\dfrac{1}{4}$

Chapter 10   Real Numbers

Chapter Opener Puzzle

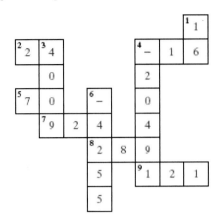

Section 10.1   Real Numbers and the Real Number Line

Section 10.1 Practice Exercises

1. Answers will vary.

3. -86 m

5. $3800

7. $-$500

9. -14 lb

11. 140,000

13. ![number line]

15. ![number line]

17. ![number line]

19. ![number line]

21. ![number line]

23. ![number line]

25. Rational

27. Rational

29. Rational

31. Irrational

33. Irrational

35. Rational

37. $0 > -3$

39. $-8 > -9$

41. $-9.1 < 2.2$

43. $-3.35 < -3.3$

45. $-\dfrac{2}{3} = -\dfrac{4}{6} > -\dfrac{5}{6}$

47. $\dfrac{7}{8} > -\dfrac{1}{9}$

49. $0 < \dfrac{1}{10}$

51. $-\dfrac{6}{5} < -\dfrac{5}{5} = -1$

53. $|-2| = 2$

55. $|4.5| = 4.5$

57. $\left|-\dfrac{5}{2}\right| = \dfrac{5}{2}$

59. $|0| = 0$

61. $|-3.2| = 3.2$

63. $|21| = 21$

65. **(a)** -8

　　 (b) $|-12| = 12$
　　　　 $|-8| = 8$
　　　　 $|-12|$ is greater.

67. **(a)** 7.8

　　 (b) $|5.2| = 5.2$
　　　　 $|7.8| = 7.8$
　　　　 $|7.8|$ is greater.

69. $|-5|$

71. Neither, they are equal.

73. -5

75. 12

77. $\dfrac{1}{6}$

79. $-\dfrac{2}{11}$

81. -8.1

83. 1.14

85. -6

87. $-(-2)$

89. $|7|$

91. $|-3|$

93. $-|14|$

95. $-|-30|$

97. $-|2| = -2$

99. $-|-5.3| = -5.3$

101. $-(-15) = 15$

103. $|-4.7| = 4.7$

105. $-\left|-\dfrac{12}{17}\right| = -\dfrac{12}{17}$

107. $-\left(-\dfrac{3}{8}\right) = \dfrac{3}{8}$

Section 10.2   Addition of Real Numbers

Section 10.2   Practice Exercises

1. Answers will vary.

3. $-\dfrac{2}{3} = -\dfrac{8}{12} > -\dfrac{11}{12}$

5. $|6| = |-6|$

7. $-10 = -|-10| < 10$

9. $2 + (-4) = -2$

11. $-3 + 5 = 2$

13. $-4 + (-4) = -8$

15. $-3 + 9 = 6$

17. $0 + (-7) = -7$

19. $-1 + (-3) = -4$

21. To add two numbers with the same sign, add their absolute values and apply the common sign.

23. $12 + 3 = 15$

25. $-40 + (-33) = -(40 + 33) = -73$

27. $-100 + (-24) = -(100 + 24) = -124$

29. $44 + 45 = 89$

31. $75 + (-23) = 75 - 23 = 52$

33. $-34 + 12 = -(34 - 12) = -22$

35. $-90 + 66 = -(90 - 66) = -24$

37. $78 + (-33) = 78 - 33 = 45$

39. $2 + (-2) = 2 - 2 = 0$

41. $-1.3 + 1.3 = 1.3 - 1.3 = 0$

43. $12 + (-3) = 9$

45. $-23 + (-3) = -26$

47. $4 + (-45) = -41$

49. $(-103) + (-47) = -150$

51. $0 + (-17) = -17$

53. $-19 + (-22) = -41$

55. $6 + (-12) + 8 = -6 + 8 = 2$

57. $-33 + (-15) + 18 = -48 + 18 = -30$

59. $7 + (-3) + 6 = 4 + 6 = 10$

61. $-10 + (-3) + 5 = -13 + 5 = -8$

63. $-18 + (-5) + 23 = -23 + 23 = 0$

65. $4 + (-12) + (-30) + 16 + 10$
$\qquad = -8 + (-30) + 16 + 10$
$\qquad = -38 + 16 + 10$
$\qquad = -22 + 10$
$\qquad = -12$

67. $23.9 + 2.1 = 26$

69. $-34.2 + (-4.1) = -(34.2 + 4.1) = -38.3$

71. $-\dfrac{3}{4} + \left(-\dfrac{5}{4}\right) = -\left(\dfrac{3}{4} + \dfrac{5}{4}\right) = -\dfrac{8}{4} = -2$

73. $-\dfrac{7}{8} + \left(-\dfrac{1}{4}\right) = -\left(\dfrac{7}{8} + \dfrac{1}{4}\right) = -\left(\dfrac{7}{8} + \dfrac{2}{8}\right) = -\dfrac{9}{8}$

75. $34.8 + (-45) = -(45 - 34.8) = -10.2$

77. $-23.1 + 24.5 = 24.5 - 23.1 = 1.4$

79. $\dfrac{3}{8} + \left(-\dfrac{3}{16}\right) = \dfrac{6}{16} + \left(-\dfrac{3}{16}\right) = \dfrac{6}{16} - \dfrac{3}{16} = \dfrac{3}{16}$

81. $\left(-\dfrac{5}{6}\right) + \dfrac{1}{4} = \left(-\dfrac{10}{12}\right) + \dfrac{3}{12} = -\left(\dfrac{10}{12} - \dfrac{3}{12}\right)$
$\qquad = -\dfrac{7}{12}$

83. $\left(-\dfrac{1}{7}\right) + \left(-\dfrac{2}{5}\right) = -\left(\dfrac{1}{7} + \dfrac{2}{5}\right) = -\left(\dfrac{5}{35} + \dfrac{14}{35}\right)$
$\qquad = -\left(\dfrac{19}{35}\right) = -\dfrac{19}{35}$

85. Sum, added to, increased by, more than, plus, total

87. $89 + (-11) = 89 - 11 = 78$

89. $-2 + (-4) + 14 + 20 = -6 + 14 + 20$
$\qquad\qquad\qquad\qquad = 8 + 20 = 28$

91. $-12 + (-4.5) = -16.5$

93. $-\dfrac{1}{3} + 2 = -\dfrac{1}{3} + \dfrac{6}{3} = \dfrac{5}{3}$

95. $-\dfrac{1}{5} + 1 = -\dfrac{1}{5} + \dfrac{5}{5} = \dfrac{4}{5}$

97. $14°F - 20°F = -(20°F - 14°F) = -6°F$

99. $\$23.89 - \$40.00 = -(\$40.00 - \$23.89)$
$\qquad\qquad\qquad\quad = -\$16.11$

101. $\$570.32 - \$250.00 = \$320.32$

103. For example: $-6 + (-8)$

105. For example: $5 + (-5)$

107. $-900 + 334 = -566$

109. $-103.4 + (-229.1) = -332.5$

111. $891 + 12 + (-223) + (-341) = 339$

Section 10.3   Subtraction of Real Numbers

Section 10.3 Practice Exercises

1. Answers will vary.

3. $-34 + (-13) = -(34 + 13) = -47$

5. $-\dfrac{5}{9} + \dfrac{7}{12} = -\dfrac{20}{36} + \dfrac{21}{36} = \dfrac{21}{36} - \dfrac{20}{36} = \dfrac{1}{36}$

7. $\begin{aligned} -\dfrac{5}{9} + \left(-\dfrac{7}{12}\right) &= -\dfrac{20}{36} + \left(-\dfrac{21}{36}\right) \\ &= -\left(\dfrac{20}{36} + \dfrac{21}{36}\right) = -\dfrac{41}{36} \end{aligned}$

9. $\begin{aligned} \left(-\dfrac{1}{2}\right) &+ 6.5 + (-8) + 2 + (-4) \\ &= (-0.5) + 6.5 + (-8) + 2 + (-4) \\ &= 6 + (-8) + 2 + (-4) \\ &= -2 + 2 + (-4) = 0 + (-4) = -4 \end{aligned}$

11. $2 - 9 = 2 + (-9) = -7$

13. $4 - (-3) = 4 + 3 = 7$

15. $-3 - 15 = -3 + (-15) = -18$

17. $-11 - (-13) = -11 + 13 = 2$

19. $35 - (-17) = 35 + 17 = 52$

21. $-24 - 9 = -24 + (-9) = -33$

23. $50 - 62 = 50 + (-62) = -12$

25. $-17 - (-25) = -17 + 25 = 8$

27. $-8 - (-8) = -8 + 8 = 0$

29. $120 - (-41) = 120 + 41 = 161$

31. $-15 - 19 = -15 + (-19) = -34$

33. $3 - 25 = 3 + (-25) = -22$

35. $-13 - 13 = -13 + (-13) = -26$

37. $24 - 25 = 24 + (-25) = -1$

39. $-6 - (-38) = -6 + 38 = 32$

41. $-48 - (-33) = -48 + 33 = -15$

43. Minus, difference, decreased, less than, subtract from

45. $14 - 23 = 14 + -23 = -9$

47. $5 - 12 = 5 + (-12) = -7$

49. $105 - 110 = 105 + (-110) = -5$

51. $320 - (-20) = 320 + 20 = 340$

53. $-35 - 24 = -35 + (-24) = -59$

55. $-34 - 21 = -34 + (-21) = -55$

57. $5.2 - 13.5 = 5.2 + (-13.5) = -8.3$

59. $-2.3 - 1.9 = -2.3 + (-1.9) = -4.2$

61. $-3.6 - (-9.1) = -3.6 + 9.1 = 5.5$

63. $5.5 - (-2.8) = 5.5 + 2.8 = 8.3$

65. $\dfrac{2}{3} - \left(-\dfrac{1}{6}\right) = \dfrac{2}{3} + \dfrac{1}{6} = \dfrac{4}{6} + \dfrac{1}{6} = \dfrac{5}{6}$

67. $-\dfrac{3}{10} - \left(-\dfrac{7}{10}\right) = -\dfrac{3}{10} + \dfrac{7}{10} = \dfrac{4}{10} = \dfrac{2}{5}$

69. $\begin{aligned} \dfrac{3}{14} - \dfrac{12}{7} &= \dfrac{3}{14} + \left(-\dfrac{12}{7}\right) = \dfrac{3}{14} + \left(-\dfrac{24}{14}\right) \\ &= -\dfrac{21}{14} = -\dfrac{3}{2} \end{aligned}$

71. $-\dfrac{1}{2} - \dfrac{5}{4} = -\dfrac{1}{2} + \left(-\dfrac{5}{4}\right) = -\dfrac{2}{4} + \left(-\dfrac{5}{4}\right) = -\dfrac{7}{4}$

73. $\begin{aligned} 2 + 5 - (-3) - 10 &= 2 + 5 + 3 + (-10) \\ &= 7 + 3 + (-10) \\ &= 10 + (-10) \\ &= 0 \end{aligned}$

75. $\begin{aligned} -5 + 6 &+ (-7) - 4 - (-9) \\ &= -5 + 6 + (-7) + (-4) + 9 \\ &= 1 + (-7) + (-4) + 9 \\ &= -6 + (-4) + 9 = -10 + 9 = -1 \end{aligned}$

77. $[-2-(-6)]^2 = [-2+6]^2 = [4]^2 = 16$

79. $[-5-(-6)]^3 = [-5+6]^3 = [1]^3 = 1$

81. $25-13-(-40) = 25+(-13)+40$
$= 12+40 = 52$

83. $5.5-\left(\dfrac{1}{2}-\dfrac{1}{5}\right) = 5.5-\left[\dfrac{1}{2}+\left(-\dfrac{1}{5}\right)\right]$
$= 5.5-\left[\dfrac{5}{10}+\left(-\dfrac{2}{10}\right)\right]$
$= 5.5-\dfrac{3}{10} = 5.5+(-0.3)$
$= 5.2$

85. $6000°F - (-423°F) = 6000°F + 423°F$
$= 6423°F$
The temperature difference is $6423°F$.

87. $17,476.55+1786.84-2342.47-754.32$
$\qquad\qquad\quad +321.63+1597.28$
$= 19,263.39-2342.47-754.32$
$\qquad\qquad\quad +321.63+1597.28$
$= 16,920.92-754.32+321.63+1597.28$
$= 16,166.60+321.63+1597.28$
$= 16,488.23+1597.28$
$= 18,085.51$
The balance was \$18,085.51.

89. $0.14-(-0.04) = 0.14+0.04 = 0.18$
The difference is 0.18 point.

91. $-\$320-\$55 = -\$320+(-\$55) = -\$375$
His new balance is $-\$375$.

93. $29,029-(-35,798) = 29,029+35,798$
$= 64,827$ ft

95. The range is $3°-(-8°) = 11°$.

97. For example, $4-10$

99. To find each number, subtract 4 from the previous number.
$-7-4 = -7+(-4) = -11$
$-11-4 = -11+(-4) = -15$
$-15-4 = -15+(-4) = -19$

101. To find each number, subtract $\dfrac{1}{3}$ from the previous number.
$-\dfrac{2}{3}-\dfrac{1}{3} = -\dfrac{2}{3}+\left(-\dfrac{1}{3}\right) = -\dfrac{3}{3} = -1$
$-1-\dfrac{1}{3} = -\dfrac{3}{3}+\left(-\dfrac{1}{3}\right) = -\dfrac{4}{3}$
$-\dfrac{4}{3}-\dfrac{1}{3} = -\dfrac{4}{3}+\left(-\dfrac{1}{3}\right) = -\dfrac{5}{3}$

103. Positive, since $a-b = a+(-b)$ and b is negative so $-b$ is positive.

105. Positive, since $|a|$ and $|b|$ are both positive

107. Negative

109. Negative, since a is positive

111. $-190-223 = -413$

113. $-23.24-(-90.01) = 66.77$

115. $89.2-(-23.6) = 112.8$

Problem Recognition Exercises: Addition and Subtraction of Real Numbers

1. $-7-5 = -7+(-5) = -(7+5) = -12$

3. $-7+(-5) = -(7+5) = -12$

5. $10-(-45) = 10+45 = 55$

7. $10+(-45) = -(45-10) = -35$

9. $-31.2-(-52.6) = -31.2+52.6$
$= 52.6-31.2 = 21.4$

11. $-31.2 - 52.6 = -31.2 + (-52.6)$
$= -(31.2 + 52.6) = -83.8$

13. $-19.5 + 21.5 = 21.5 - 19.5 = 2$

15. $-19.5 + (-21.5) = -(19.5 + 21.5) = -41$

17. $|-12 + 8| = |-4| = 4$

19. $|-12 - 8| = |-20| = 20$

21. $\dfrac{1}{8} - \dfrac{5}{4} = \dfrac{1}{8} - \dfrac{10}{8} = -\left(\dfrac{10}{8} - \dfrac{1}{8}\right)$
$= -\dfrac{9}{8} \text{ or } -1\dfrac{1}{8}$

23. $\dfrac{1}{8} + \left(-\dfrac{5}{4}\right) = \dfrac{1}{8} + \left(-\dfrac{10}{8}\right) = -\left(\dfrac{10}{8} - \dfrac{1}{8}\right)$
$= -\dfrac{9}{8} \text{ or } -1\dfrac{1}{8}$

25. $-\dfrac{7}{9} - \dfrac{1}{6} = -\dfrac{14}{18} - \dfrac{3}{18} = -\left(\dfrac{14}{18} + \dfrac{3}{18}\right) = -\dfrac{17}{18}$

27. $-\dfrac{7}{9} - \left(-\dfrac{1}{6}\right) = -\dfrac{7}{9} + \dfrac{1}{6} = -\dfrac{14}{18} + \dfrac{3}{18}$
$= -\left(\dfrac{14}{18} - \dfrac{3}{18}\right) = -\dfrac{11}{18}$

29. $2\dfrac{1}{4} - 5\dfrac{1}{2} = \dfrac{9}{4} - \dfrac{11}{2} = \dfrac{9}{4} - \dfrac{22}{4}$
$= -\left(\dfrac{22}{4} - \dfrac{9}{4}\right) = -\dfrac{13}{4} \text{ or } -3\dfrac{1}{4}$

31. $-1\dfrac{2}{5} - 3\dfrac{1}{10} = -\dfrac{7}{5} - \dfrac{31}{10} = -\dfrac{14}{10} - \dfrac{31}{10}$
$= -\left(\dfrac{31}{10} + \dfrac{14}{10}\right) = -\dfrac{45}{10}$
$= -\dfrac{9}{2} \text{ or } -4\dfrac{1}{2}$

33. $-\dfrac{3}{4} + 3 = -\dfrac{3}{4} + \dfrac{12}{4} = \dfrac{12}{4} - \dfrac{3}{4} = \dfrac{9}{4} \text{ or } 2\dfrac{1}{4}$

35. $-2 + 0.001 = -(2 - 0.001) = -1.999$

37. $-56 + 56 = 56 - 56 = 0$

39. $-56 - 56 = -(56 + 56) = -112$

Section 10.4   Multiplication and Division of Real Numbers

Section 10.4   Practice Exercises

1. $-2 + (-4)$ is negative.
$-(-5)$ is positive.
$-|-10|$ is negative.
$(-3)(-6)$ is positive.
When multiplying two numbers with the same sign, the product is positive.

3. $14 - (-5) = 14 + 5 = 19$

5. $-33 + (-11) = -(33 + 11) = -44$

7. $23 - 12 + (-4) - (-10)$
$= 23 + (-12) + (-4) + 10$
$= 11 + (-4) + 10 = 7 + 10 = 17$

9. $-3(5) = -15$

11. $-12 \cdot 4 = -48$

13. $-15(-3) = 45$

15. $9(-8) = -72$

17. $(-1.2)(-3.2) = 3.84$

19. $-6(0.4) = -2.4$

21. $7(-1.1) = -7.7$

23. $-14 \cdot 0 = 0$

25. $\left(-\dfrac{2}{\cancel{3}_1}\right)\left(-\dfrac{\cancel{6}^2}{7}\right) = \dfrac{4}{7}$

27. $\dfrac{\cancel{3}^{\,1}}{\cancel{8}_{\,1}}\left(-\dfrac{\cancel{8}^{\,1}}{\cancel{21}_{\,7}}\right)=-\dfrac{1}{7}$

29. $6\cdot\left(-\dfrac{5}{12}\right)=\dfrac{\cancel{6}^{\,1}}{1}\cdot\left(-\dfrac{5}{\cancel{12}_{\,2}}\right)=-\dfrac{5}{2}$ or $-2\dfrac{1}{2}$

31. $\left(-2\dfrac{3}{5}\right)\left(-1\dfrac{2}{3}\right)=\left(-\dfrac{13}{\cancel{5}_{\,1}}\right)\left(-\dfrac{\cancel{5}^{\,1}}{3}\right)=\dfrac{13}{3}$ or $4\dfrac{1}{3}$

33. $\left(-\dfrac{8}{9}\right)\cdot 0=0$

35. $(-3.5)(-1.4)-4.9$

37. $-3(-1)=3$

39. $-5\cdot 3=-15$

41. $1.3(-3)=-3.9$

43. $(5)(-2)(4)(-10)=400$

45. $(-11)(-4)(-2)=-88$

47. $(24)(-2)(0)(-3)=0$

49. $(-1)(-1)(-1)(-1)(-1)(-1)=1$

51. $(-1)(-1)(-1)(-1)(-1)=-1$

53. $-10^2=-(10)(10)=-100$

55. $(-10)^2=(-10)(-10)=100$

57. $-3^3=-(3)(3)(3)=-27$

59. $(-3)^3=(-3)(-3)(-3)=-27$

61. $-0.2^3=-(0.2)(0.2)(0.2)=-0.008$

63. $\left(-\dfrac{2}{3}\right)^3=\left(-\dfrac{2}{3}\right)\left(-\dfrac{2}{3}\right)\left(-\dfrac{2}{3}\right)=-\dfrac{8}{27}$

65. $(-6)^2=(-6)(-6)=36$

67. $-(-6)^2=-(-6)(-6)=-36$

69. $\dfrac{-15}{5}=-3$

71. $\dfrac{56}{-8}=-7$

73. $\dfrac{-25}{-15}=\dfrac{5}{3}$

75. $\dfrac{-2}{-3}=\dfrac{2}{3}$

77. $\dfrac{13}{0}$ is undefined.

79. $\dfrac{0}{-2}=0$

81. $(-20)\div(-5)=4$

83. $-0.91\div-0.7=1.3$

85. $\left(\dfrac{8}{7}\right)\div\left(-\dfrac{4}{5}\right)=\left(\dfrac{\cancel{8}^{\,2}}{7}\right)\left(-\dfrac{5}{\cancel{4}_{\,1}}\right)=-\dfrac{10}{7}$

87. $\left(-\dfrac{1}{6}\right)\div 0$ is undefined.

89. $\dfrac{-5}{-8}=\dfrac{5}{8}$

91. $-18\div 24=\dfrac{-18}{24}=-\dfrac{3}{4}$

93. $-100\div 20=-5$

95. $-32\div(-64)=\dfrac{-32}{-64}=\dfrac{1}{2}$

97. $-52\div 13=-4$

99. $8+(-6)=2$

101. $8(-6) = -48$

103. $-9 - (-12) = -9 + 12 = 3$

105. $-36 \div (-12) = 3$

107. $(-5)(-4) = 20$

109. $0 + (-15) = -15$

111. $\dfrac{1}{3} \div \left(-\dfrac{5}{6}\right) = \dfrac{1}{\cancel{3}} \cdot \left(-\dfrac{\overset{2}{\cancel{6}}}{5}\right) = -\dfrac{2}{5}$

113. $\dfrac{1}{3} - \left(-\dfrac{5}{6}\right) = \dfrac{1}{3} + \dfrac{5}{6} = \dfrac{2}{6} + \dfrac{5}{6} = \dfrac{7}{6}$

115. $(-2)^{50}$ is positive and $(-2)^{51}$ is negative, so $(-2)^{50}$ is greater.

117. $(5)^{41}$ has one more factor of 5 than $(5)^{40}$, so $(5)^{41}$ is greater.

119. $a \cdot b$ is negative since signs are different.

121. $-a \div (b)$ is positive since $-a$ and b are both negative.

123. $(-413)(871) = -359{,}723$

125. $(-52.12)(-101.5) = 5290.18$

127. $5{,}945{,}308 \div (-9452) = -629$

Problem Recognition Exercises: Operations on Real Numbers

1. $15 - (-5) = 15 + 5 = 20$

3. $15 + (-5) = 10$

5. $-36(-2) = 72$

7. $\dfrac{-36}{-2} = 18$

9. $20(-4) = -80$

11. $-20(4) = -80$

13. $-5 - 9 - 2 = -5 + (-9) + (-2) = -14 + (-2)$
$= -16$

15. $10 + (-3) + (-12) = 7 + (-12) = -5$

17. $(-1)(-2)(-3)(-4) = 2(-3)(-4)$
$= -6(-4) = 24$

19. $(-1)(-2)(-3)(4) = 2(-3)(4) = -6(4)$
$= -24$

21. $\dfrac{3}{5} \div \left(-\dfrac{10}{9}\right) = \dfrac{3}{5} \cdot \left(-\dfrac{9}{10}\right) = -\dfrac{27}{50}$

23. $\dfrac{3}{5} + \left(-\dfrac{10}{9}\right) = \dfrac{27}{45} + \left(-\dfrac{50}{45}\right) = -\dfrac{23}{45}$

25. $-\dfrac{2}{3} + \left(-\dfrac{7}{9}\right) = -\dfrac{6}{9} + \left(-\dfrac{7}{9}\right) = -\dfrac{13}{9}$

27. $\left(-2\dfrac{1}{4}\right)\left(1\dfrac{4}{9}\right) = -\dfrac{\cancel{9}}{4} \cdot \left(\dfrac{13}{\cancel{9}}\right)$
$= -\dfrac{13}{4}$ or $-3\dfrac{1}{4}$

29. $41.5 - (-13.6) = 41.5 + 13.6 = 55.1$

31. $-60.41 - 33.50 = -60.41 + (-33.50)$
$= -93.91$

33. $\dfrac{-12}{-11} = \dfrac{12}{11}$

35. $\dfrac{0}{-8} = 0$

37. $42 \div (-0.002) = -21{,}000$

39. $-44 - (-44) = -44 + 44 = 0$

Section 10.5   Order of Operations

Section 10.5   Practice Exercises

1. Answers will vary.

3. $\left(-\dfrac{2}{9}\right) \div \left(\dfrac{8}{27}\right) = \left(-\dfrac{\cancel{2}^{1}}{\cancel{9}}\right) \cdot \left(\dfrac{\cancel{27}^{3}}{\cancel{8}}\right) = -\dfrac{3}{4}$

5. $-2.8(-1.1) = 3.08$

7. $(-1)(-5)(-8)(3) = -120$

9. $5 + 2(3-5) = 5 + 2(-2) = 5 + (-4) = 1$

11. $-8 - 6^2 = -8 - 36 = -8 + (-36) = -44$

13. $4 + (3-8)^2 = 4 + (-5)^2 = 4 + 25 = 29$

15. $120 \div (-4)(5) = -30(5) = -150$

17. $-2.1 - 6 \div 5 = -2.1 - 1.2 = -3.3$

19. $\left[5.3 - (-2.7)\right]^2 = \left[5.3 + 2.7\right]^2 = 8^2 = 64$

21. $-2(3-6) + 10 = -2(-3) + 10 = 6 + 10 = 16$

23. $-16 \div (-4)(-5) = 4(-5) = -20$

25. $8 - (-3)(-2)^3 = 8 - (-3)(-8) = 8 - 24$
$= -16$

27. $12 + (14-16)^2 \div (-4) = 12 + (-2)^2 \div (-4)$
$= 12 + (4) \div (-4)$
$= 12 + (-1) = 11$

29. $-48 \div 12 \div (-2) = -4 \div (-2) = 2$

31. $90 \div (-3)(-1) \div (-6) = -30(-1) \div (-6)$
$= 30 \div (-6) = -5$

33. $\left|9^2 - (-7)^2\right| \div (-4) = |81 - 49| \div (-4)$
$= |32| \div (-4)$
$= 32 \div (-4) = -8$

35. $2 + 2^3 - |10 - 12| = 2 + 2^3 - |-2|$
$= 2 + 2^3 - 2 = 2 + 8 - 2$
$= 10 - 2 = 8$

37. $-6(48 \div 12)^2 = -6(4)^2 = -6(16) = -96$

39. $\left(-\dfrac{1}{2}\right) \cdot \dfrac{1}{3} \div \dfrac{1}{12} = -\dfrac{1}{6} \div \dfrac{1}{12} = -\dfrac{1}{\cancel{6}} \cdot \dfrac{\cancel{12}^{2}}{1} = -2$

41. $\dfrac{1}{6} + \left(-\dfrac{5}{\cancel{4}}\right) \cdot \dfrac{\cancel{4}^{1}}{3} = \dfrac{1}{6} + \left(-\dfrac{5}{3}\right) = \dfrac{1}{6} + \left(-\dfrac{10}{6}\right)$
$= -\dfrac{9}{6} = -\dfrac{3}{2}$

43. $\left(-\dfrac{2}{3}\right)^2 - \left(\dfrac{5}{21}\right) \div \dfrac{15}{7} = \dfrac{4}{9} - \dfrac{5}{21} \div \dfrac{15}{7}$
$= \dfrac{4}{9} - \dfrac{\cancel{5}^{1}}{\cancel{21}_{3}} \cdot \dfrac{\cancel{7}^{1}}{\cancel{15}_{3}}$
$= \dfrac{4}{9} - \dfrac{1}{9} = \dfrac{3}{9} = \dfrac{1}{3}$

45. $\dfrac{5}{2} \cdot \left(\dfrac{3}{2}\right)^2 + \left(-\dfrac{1}{2}\right)^3 = \dfrac{5}{2} \cdot \dfrac{9}{4} + \left(-\dfrac{1}{8}\right)$
$= \dfrac{45}{8} + \left(-\dfrac{1}{8}\right) = \dfrac{44}{8} = \dfrac{11}{2}$

47. $21 - \left[4 - (5-8)\right] = 21 - \left[4 - (-3)\right]$
$= 21 - \left[4 + 3\right] = 21 - 7$
$= 14$

49. $-17 - 2\left[18 \div (-3)\right] = -17 - 2[-6]$
$= -17 + 12 = -5$

51. $4 + 2\left[9 + (-4 + 12)\right] = 4 + 2\left[9 + 8\right]$
$= 4 + 2\left[17\right] = 4 + 34$
$= 38$

53. $2^2 - \left|-3 + 9\right| = 2^2 - \left|6\right| = 4 - 6 = -2$

55. $\dfrac{\left|3 + (-5)\right|}{4 - (3)(-2)} = \dfrac{\left|-2\right|}{4 - (3)(-2)} = \dfrac{2}{4 - (3)(-2)}$
$= \dfrac{2}{4 - (-6)} = \dfrac{2}{4 + 6} = \dfrac{2}{10} = \dfrac{1}{5}$

57. $\dfrac{13 - (2)(4)}{-1 - 2^2} = \dfrac{13 - (2)(4)}{-1 - 4} = \dfrac{13 - 8}{-1 - 4} = \dfrac{5}{-5}$
$= -1$

59. $\dfrac{1 - 4(3 - 5)}{5^2 - 2^2} = \dfrac{1 - 4(-2)}{5^2 - 2^2} = \dfrac{1 - 4(-2)}{25 - 4}$
$= \dfrac{1 + 8}{25 - 4} = \dfrac{9}{21} = \dfrac{3}{7}$

61. $\dfrac{6 - 3^2}{(5 - 2)^2} = \dfrac{6 - 3^2}{3^2} = \dfrac{6 - 9}{9} = \dfrac{-3}{9} = -\dfrac{1}{3}$

63. $\dfrac{-8° + (-11°) + (-4°) + 1° + 9° + 4° + (-5°)}{7}$
$= \dfrac{-14°}{7} = -2°$

65.

$\dfrac{-8 + (-8) + (-6) + (-5) + (-2) + (-2) + 3 + 3 + 0 + (-4)}{10}$
$= \dfrac{-29}{10} = -2.9$

67. $F = \dfrac{9}{5}(-89.6) + 32 = 1.8(-89.6) + 32$
$= -161.28 + 32 = -129.28°\,F$

69. $\left(\dfrac{1}{2}\right)^2 \div 0.05 + \left(-\dfrac{\overset{1}{\cancel{6}}}{\underset{1}{\cancel{4}}} \cdot \dfrac{\overset{2}{\cancel{8}}}{\underset{1}{\cancel{3}}}\right)$

$= \left(\dfrac{1}{2}\right)^2 \div 0.05 + (-2)$

$= \dfrac{1}{4} \div 0.05 + (-2)$

$= 0.25 \div 0.05 + (-2)$

$= 5 + (-2) = 3$

71. $2\left(\dfrac{7}{8} - \dfrac{1}{4}\right) - (-1.5)^2 = 2\left(\dfrac{7}{8} - \dfrac{2}{8}\right) - (-1.5)^2$

$= 2\left(\dfrac{5}{8}\right) - (-1.5)^2$

$= 2\left(\dfrac{5}{8}\right) - (2.25)$

$= \dfrac{10}{8} - 2.25$

$= 1.25 - 2.25 = -1$

Chapter 10   Review Exercises

Section 10.1

1. $-64{,}599$

3. $15°$

5–7.

9. Opposite: 4
   absolute value: 4

11. Opposite: -3.5
    absolute value: 3.5

13. (a) $-(-9) = 9$

    (b) $-|-9| = -9$

15. $-\dfrac{5}{6} > -1$

17. $|3| > -|3|$

19. $-2.8 < -2$

Section 10.2

21. $6 + (-2) = 4$

23. $-3 + -2 = -5$

25. To add two numbers with the same sign, add their absolute values and apply the common sign.

27. $35 + (-22) = 35 - 22 = 13$

29. $-29 + (-41) = -(29 + 41) = -70$

31. $-6.5 + (-4.16) = -(6.5 + 4.16) = -10.66$

33. $\left(-\dfrac{1}{5}\right) + \left(-\dfrac{7}{10}\right) = \left(-\dfrac{2}{10}\right) + \left(-\dfrac{7}{10}\right)$
$$= -\left(\dfrac{2}{10} + \dfrac{7}{10}\right) = -\dfrac{9}{10}$$

35. $23 + (-35) = -(35 - 23) = -12$

37. $-5 + (-13) + 20 = -18 + 20 = 2$

39. $-12 + 3 = -9$

41. $-3 + (-10) + 12 + 14 + (-10)$
$$= -13 + 12 + 14 + (-10)$$
$$= -1 + 14 + (-10)$$
$$= 13 + (-10) = 3$$

Section 10.3

43. 1. Leave the first number (the minuend) unchanged.
     2. Change the subtraction sign to an addition sign.
     3. Add the opposite of the second number (the subtrahend).

45. $19 - 44 = 19 + (-44) = -25$

47. $-289 - 130 = -289 + (-130) = -419$

49. $3.8 - 4.5 = 3.8 + (-4.5) = -0.7$

51. $0 - \left(-\dfrac{20}{21}\right) = 0 + \dfrac{20}{21} = \dfrac{20}{21}$

53. For example: 23 minus negative 6.

55. For example: Subtract -7 from -25.

57. $-\$40 + \$132 = \$92$
    Sam's balance is now \$92.

Section 10.4

59. $6(-3) = -18$

61. $\dfrac{-900}{-60} = 15$

63. $-2.8 \div 0.04 = -70$

65. $\left(-\dfrac{\overset{1}{\cancel{2}}}{\cancel{3}}\right)\left(-\dfrac{\overset{7}{\cancel{21}}}{\cancel{8}}\right) = \dfrac{7}{4}$

67. $\left(-\dfrac{1}{5}\right) \div 0$ is undefined.

69. $(-1)(-8)(2)(1)(-2) = -32$

71. $(-6)^2 = (-6)(-6) = 36$

73. $\left(-\dfrac{3}{4}\right)^3 = \left(-\dfrac{3}{4}\right)\left(-\dfrac{3}{4}\right)\left(-\dfrac{3}{4}\right) = -\dfrac{27}{64}$

75. $(-1)^{10} = 1$

77. Negative

79. $-45 \div (-15) = 3$

81. $30(-5) = -150$

Section 10.5

83. $28 \div (-7) \cdot 3 - (-1) = -4 \cdot 3 - (-1)$
$$= -12 - (-1) = -12 + 1$$
$$= -11$$

85. $\left|10 - (-3)^2\right| \cdot (-11) + 4 = \left|10 - 9\right| \cdot (-11) + 4$
$$= \left|1\right| \cdot (-11) + 4$$
$$= 1 \cdot (-11) + 4$$
$$= -11 + 4 = -7$$

87. $18 - (-5)^2 + 14 \div 2 = 18 - 25 + 14 \div 2$
$$= 18 - 25 + 7 = 0$$

89. $\left(-\dfrac{3}{8}\right)^2 - \left(-\dfrac{1}{2}\right)^3 = \dfrac{9}{64} - \left(-\dfrac{1}{8}\right) = \dfrac{9}{64} + \dfrac{1}{8}$
$$= \dfrac{9}{64} + \dfrac{8}{64} = \dfrac{17}{64}$$

91. $\dfrac{3 - \left|2 + (-7)\right|}{3^2 - 5^2} = \dfrac{3 - \left|-5\right|}{3^2 - 5^2} = \dfrac{3 - 5}{3^2 - 5^2}$
$$= \dfrac{3 - 5}{9 - 25} = \dfrac{-2}{-16} = \dfrac{1}{8}$$

Chapter 10   Test

1. (a) $-\$220$

   (b) 26

3. $-3, -\dfrac{3}{5}, 0, 4, -1, \dfrac{4}{7}$

5. $-5 < -2$

7. $0 > -2.4$

9. $-|-9| < 9$

11. $9 + (-14) = -5$

13. $-4 - (-13) = -4 + 13 = 9$

15. $-1.5 + 2.1 = 0.6$

17. $-\dfrac{2}{3} - \dfrac{4}{7} = -\dfrac{14}{21} - \dfrac{12}{21} = -\dfrac{26}{21}$ or $-1\dfrac{5}{21}$

19. $6(-12) = -72$

21. $\dfrac{-24}{-12} = 2$

23. $\dfrac{-44}{0}$ is undefined.

25. $\dfrac{3}{10} \div \left(-\dfrac{4}{5}\right) = \dfrac{3}{10} \cdot \left(-\dfrac{5}{4}\right) = -\dfrac{15}{40} = -\dfrac{3}{8}$

27. (a) Positive

   (b) Negative

29. $-3(-7) = 21$

31. $18 - (-4) = 18 + 4 = 22$

33. $-8.1 + 5 = -3.1$

35. $-14 + 22 - (-5) + (-10) = 8 + 5 + (-10)$
$$= 13 + (-10) = 3$$

37. $-20 \div (-2)^2 + (-14) = -20 \div (4) + (-14)$
$$= -5 + (-14) = -19$$

39. $-\dfrac{2}{15} + \left(-\dfrac{\overset{4}{\cancel{20}}}{\underset{3}{\cancel{21}}} \cdot \dfrac{\overset{1}{\cancel{7}}}{\underset{1}{\cancel{8}}}\right) = -\dfrac{2}{15} + \left(-\dfrac{4}{3}\right)$

$$= -\dfrac{2}{15} + \left(-\dfrac{20}{15}\right) = -\dfrac{22}{15}$$

41. $16 - 2\big[5 - (1-4)\big] = 16 - 2\big[5 - (-3)\big]$
$$= 16 - 2\big[5 + 3\big]$$
$$= 16 - 2\big[8\big] = 16 - 16$$
$$= 0$$

43. $\dfrac{4° + (-3°) + (-1°) + 5° + (-2°) + 0° + 4°}{7}$
$$= \dfrac{7°}{7} = 1°$$

Chapters 1–10   Cumulative Review Exercises

1.
$$\begin{array}{r} 3490 \\ + 123 \\ \hline 3613 \end{array}$$

3.
$$\begin{array}{r} 23 \\ 34 \\ 98 \\ + 22 \\ \hline 177 \end{array}$$

5. $720 = 2 \cdot 2 \cdot 2 \cdot 2 \cdot 3 \cdot 3 \cdot 5$ or $2^4 \cdot 3^2 \cdot 5$

7. Harold answered $14 - 3 = 11$ questions correctly. He got $\dfrac{11}{14}$ of the quiz correct.

9. $16 = 2 \cdot 2 \cdot 2 \cdot 2$
$40 = 2 \cdot 2 \cdot 2 \cdot 5$
$10 = 2 \cdot 5$
LCM: $2 \cdot 2 \cdot 2 \cdot 2 \cdot 5 = 80$

11. $3\dfrac{3}{5} + 2\dfrac{13}{15} = \dfrac{18}{5} + \dfrac{43}{15} = \dfrac{54}{15} + \dfrac{43}{15}$
$$= \dfrac{97}{15} \text{ or } 6\dfrac{7}{15}$$

13. (a) 34.230

(b) 9.0

15.
$$\begin{array}{r} 204.55 \\ \times \quad 2.4 \\ \hline 81\,820 \\ 409\,100 \\ \hline 490.920 \end{array}$$

17. $\dfrac{2\frac{1}{3}\text{ m}}{5\frac{5}{6}\text{ m}} = \dfrac{\frac{7}{3}}{\frac{35}{6}} = \dfrac{7}{3} \div \dfrac{35}{6} = \dfrac{\overset{1}{\cancel{7}}}{\underset{1}{\cancel{3}}} \cdot \dfrac{\overset{2}{\cancel{6}}}{\underset{5}{\cancel{35}}} = \dfrac{2}{5}$

19. $3\dfrac{1}{2} \cdot 6 = \dfrac{7}{2} \cdot \dfrac{6}{1} = \dfrac{42}{2} = 21$ mi

21. $(0.32)(600) = 192$

23. $15 = (0.06)x$
$$\dfrac{15}{0.06} = \dfrac{0.06x}{0.06}$$
$$250 = x$$

25. $2\text{ ft} = \dfrac{2\text{ ft}}{1} \cdot \dfrac{12\text{ in.}}{1\text{ ft}} = 24$ in.
2 ft 4 in. = 24 in. + 4 in. = 28 in.

27. 60 mL = 0.06 L

29. $6^2 + 8^2 = c^2$
$36 + 64 = c^2$
$100 = c^2$
$\sqrt{100} = c$
$10 = c$
The distance is 10 mi.

31. $A = s^2 = \left(2\dfrac{1}{4}\text{ m}\right)^2 = \left(\dfrac{9}{4}\text{ m}\right)^2 = \dfrac{81}{16}\text{ m}^2$
$$= 5\dfrac{1}{16}\text{ m}^2$$

33.

Number of miles	Tally	Frequency
3	||	2
4	⍫	5
5	|	1
6	||	2

35. $\text{Mean} = \dfrac{4+4+4+3+6+4+6+5+3+4}{10}$

$= \dfrac{43}{10} = 4.3 \text{ mi}$

37. $43 - (-12) = 43 + 12 = 55$

39. $(-4)^2 - 6^2 = 16 - 36 = -20$

Chapter 11　Solving Equations

Chapter Opener Puzzle

$3x$	$=$	5	$+$	1
4	$+$	x	$=$	8
$-3a$	$=$	-1	$+$	10
-2	$=$	y	$-$	10
$5t$	$-$	1	$=$	-11
2	$-$	$4x$	$=$	6

Section 11.1　Properties of Real Numbers

Section 11.1　Practice Exercises

1. $4x = 6$

$\dfrac{x}{5} = 3$

3. $8p$

5. $t + 4$

7. $v - 6$

9. $\dfrac{4}{n}$

11. $2g$

13. (a) $-6x = -6(\ ) = -6(2) = -12$

(b) $-6x = -6(\ ) = -6(-5) = 30$

15. (a) $3p + 5q = 3(\ ) + 5(\ ) = 3(2) + 5\left(-\dfrac{1}{5}\right)$

$= 6 + (-1) = 5$

(b) $3p + 5q = 3(\ ) + 5(\ ) = 3(-5) + 5(0)$

$= -15 + 0 = -15$

17. (a) $-a^2 = -(\ )^2 = -(-7)^2 = -49$

(b) $-a^2 = -(\ )^2 = -(7)^2 = -49$

19. (a) $-4(r - s)^2 = -4[(\ ) - (\ )]^2$

$= -4[(8) - (6)]^2 = -4(2)^2$

$= -4(4) = -16$

(b) $-4(r - s)^2 = -4[(\ ) - (\ )]^2$

$= -4[(3) - (-1)]^2$

$= -4(3 + 1)^2 = -4(4)^2$

$= -4(16) = -64$

21. $y(x - 4) = (\ )[(\ ) - 4]$

$= \left(\dfrac{2}{3}\right)[(-2) - 4] = \dfrac{2}{3}(-6) = -4$

23. $z^2 - x + 6 = (\ )^2 - (\ ) + 6 = (4)^2 - (-2) + 6$

$= 16 + 2 + 6 = 24$

25. $bc \div a = (\ )(\ ) \div (\ ) = (-3)(-2) \div (12)$

$= 6 \div 12 = \dfrac{1}{2}$

27. $b^2 - c^2 = (\ )^2 - (\ )^2 = (-3)^2 - (-2)^2$

$= 9 - 4 = 5$

29. $P = 2l + 2w = 2(\ ) + 2(\ )$
$\quad = 2(6 \text{ in.}) + 2(2.3 \text{ in.})$
$\quad = 12 \text{ in.} + 4.6 \text{ in.} = 16.6 \text{ in.}$

31. $A = \pi r^2 = (\ )(\ )^2 = \left(\dfrac{22}{7}\right)\left(\dfrac{7}{2} \text{ m}\right)^2$

$\quad = \left(\dfrac{\overset{11}{\cancel{22}}}{\underset{1}{\cancel{7}}}\right)\left(\dfrac{\overset{7}{\cancel{49}}}{\underset{2}{\cancel{4}}} \text{ m}^2\right) = \dfrac{77}{2} \text{ m}^2$

33. $5 + w = w + 5$

35. $-\dfrac{1}{3} + b = b + \left(-\dfrac{1}{3}\right)$ or $b - \dfrac{1}{3}$

37. $r(2) = 2r$

39. $t(-s) = -st$

41. $xy = yx$

43. $7 - p = 7 + (-p) = -p + 7$

45. $-2(6b) = (-2 \cdot 6)b = -12b$

47. $3 + (8 + t) = (3 + 8) + t = 11 + t$

49. $-4.2 + (2.5 + r) = (-4.2 + 2.5) + r$
$\quad\quad = -1.7 + r$

51. $3(6x) = (3 \cdot 6)x = 18x$

53. $-\dfrac{4}{7}\left(-\dfrac{7}{4}d\right) = \left[-\dfrac{4}{7} \cdot \left(-\dfrac{7}{4}\right)\right]d = 1d = d$

55. $-9 + (-12 + h) = \left(-9 + (-12)\right) + h$
$\quad\quad = -21 + h$

57. $4(x + 8) = 4(x) + 4(8) = 4x + 32$

59. $-2(p + 4) = -2(p) + (-2)(4)$
$\quad\quad = -2p + (-8) = -2p - 8$

61. $-10(t - 3) = -10(t) + (-10)(-3)$
$\quad\quad = -10t + 30$

63. $-5(-2 + x) = -5(-2) + (-5)(x)$
$\quad\quad = 10 + (-5x) = 10 - 5x$

65. $4(a + 4b - c) = 4[a + 4b + (-c)]$
$\quad\quad = 4(a) + 4(4b) + 4(-c)$
$\quad\quad = 4a + 16b + (-4c)$
$\quad\quad = 4a + 16b - 4c$

67. $4\left(\dfrac{2}{3} + g\right) = 4\left(\dfrac{2}{3}\right) + 4(g) = \dfrac{8}{3} + 4g$

69. $-(3 - n) = -[3 + (-n)] = -1[3 + (-n)]$
$\quad\quad = -1(3) + (-1)(-n) = -3 + n$

71. $-(-a - 8) = -[-a + (-8)] = -1[-a + (-8)]$
$\quad\quad = -1(-a) + (-1)(-8) = a + 8$

73. $-(3x + 9 - 5y)$
$\quad\quad = -[3x + 9 + (-5y)]$
$\quad\quad = -1[3x + 9 + (-5y)]$
$\quad\quad = -1(3x) + (-1)(9) + (-1)(-5y)$
$\quad\quad = -3x + (-9) + 5y = -3x - 9 + 5y$

75. $-(-5q - 2s - 3t)$
$\quad\quad = -[-5q + (-2s) + (-3t)]$
$\quad\quad = -1[-5q + (-2s) + (-3t)]$
$\quad\quad = -1(-5q) + (-1)(-2s) + (-1)(-3t)$
$\quad\quad = 5q + 2s + 3t$

77. $6(2x) = (6 \cdot 2)x = 12x$

79. $6(2 + x) = 6(2) + 6(x) = 12 + 6x$

81. $-6(-1 - k) = -6\left(-1 + (-k)\right)$
$\quad\quad = -6(-1) + (-6)(-k) = 6 + 6k$

83. $-6 + (-1 - k) = -6 + \left(-1 + (-k)\right)$
$\quad\quad = \left(-6 + (-1)\right) + (-k)$
$\quad\quad = -7 + (-k) = -7 - k$

85. $-8 + (4 - p) = -8 + [4 + (-p)]$
$\quad\quad = (-8 + 4) + (-p)$
$\quad\quad = -4 + (-p) = -4 - p$

87. $-8(4 - p) = -8[4 + (-p)]$
$\quad\quad = -8(4) + (-8)(-p) = -32 + 8p$

89. $8\left(\dfrac{1}{2}a\right) = \left(8 \cdot \dfrac{1}{2}\right)a = 4a$

91. $8\left(\dfrac{1}{2}+a\right)=8\left(\dfrac{1}{2}\right)+8(a)=4+8a$

95. $\dfrac{5}{9}(9y)=\left(\dfrac{5}{9}\cdot 9\right)y=5y$

93. $\dfrac{5}{9}(9+y)=\dfrac{5}{9}(9)+\dfrac{5}{9}(y)=5+\dfrac{5}{9}y$

Section 11.2   Simplifying Expressions

Section 11.2   Practice Exercises

1. Answers will vary.

3. $6(p+3)=6(p)+6(3)=6p+18$

5. $4(-6q)=[4\cdot(-6)]q=-24q$

7. $13+(-4-h)=13+[-4+(-h)]$
$=[13+(-4)]+(-h)$
$=9+(-h)=9-h$

9. $2a$: variable term
$5b^2$: variable term
6: constant term

11. 8: constant term
$9a$: variable term

13. $4pq$: variable term
$-9p$: variable term

15. $10h^2$: variable term
-15: constant term
$-4h$: variable term

17. $6, -4$

19. $-14, 12$

21. $1, -1$

23. $5, -8, -3$

25. *Like* terms

27. Unlike terms

29. *Like* terms

31. Unlike terms

33. Unlike terms

35. *Like* terms

37. $6rs+8rs=(6+8)rs=14rs$

39. $-4h+12h=(-4+12)h=8h$

41. $4x^2+9-x^2=4x^2-x^2+9$
$=(4-1)x^2+9=3x^2+9$

43. $10x-12y-4x-3y=10x-4x-12y-3y$
$=6x-15y$

45. $-6k-9k+12k=-3k$

47. $-8uv+6u+12uv=-8uv+12uv+6u$
$=4uv+6u$

49. $6-14m-15-2m=-14m-2m+6-15$
$=-16m-9$

51. $18-3a+5b-6a+2$
$=-3a-6a+5b+18+2$
$=-9a+5b+20$

53. $-5p^2+6p-p^2+7-8p$
$=-5p^2-p^2+6p-8p+7$
$=-6p^2-2p+7$

55. $\dfrac{1}{2}y+\dfrac{3}{2}y-\dfrac{5}{6}=\dfrac{4}{2}y-\dfrac{5}{6}=2y-\dfrac{5}{6}$

57. $\dfrac{3}{4}a + 3 - \dfrac{1}{8}a + 6 = \dfrac{3}{4}a - \dfrac{1}{8}a + 3 + 6$

$\qquad\qquad\qquad = \dfrac{6}{8}a - \dfrac{1}{8}a + 3 + 6$

$\qquad\qquad\qquad = \dfrac{5}{8}a + 9$

59. $2.3x^2 + 4.1x - 5.3x^2 - 6x$

$\qquad = 2.3x^2 - 5.3x^2 + 4.1x - 6x$

$\qquad = -3x^2 - 1.9x$

61. $4.4 - 0.9a + 3.2 = -0.9a + 4.4 + 3.2$

$\qquad\qquad\qquad\;\; = -0.9a + 7.6$

63. $5(t - 6) + 2 = 5[t + (-6)] + 2$

$\qquad\qquad\quad = 5(t) + 5(-6) + 2$

$\qquad\qquad\quad = 5t + (-30) + 2$

$\qquad\qquad\quad = 5t + (-28)$

$\qquad\qquad\quad = 5t - 28$

65. $-3(2x + 1) - 13 = -3(2x) + (-3)(1) - 13$

$\qquad\qquad\qquad\quad = -6x + (-3) - 13$

$\qquad\qquad\qquad\quad = -6x + (-16)$

$\qquad\qquad\qquad\quad = -6x - 16$

67. $4 + 6(y - 3) = 4 + 6[y + (-3)]$

$\qquad\qquad\quad = 4 + 6(y) + 6(-3)$

$\qquad\qquad\quad = 4 + 6y + (-18)$

$\qquad\qquad\quad = 6y + 4 + (-18)$

$\qquad\qquad\quad = 6y + (-14)$

$\qquad\qquad\quad = 6y - 14$

69. $21 - 7(3 - q) = 21 + (-7)[3 + (-q)]$

$\qquad\qquad\quad = 21 + (-7)(3) + (-7)(-q)$

$\qquad\qquad\quad = 21 + (-21) + (7q)$

$\qquad\qquad\quad = 0 + 7q$

$\qquad\qquad\quad = 7q$

71. $-3 - (2n + 1) = -3 + (-1)(2n + 1)$

$\qquad\qquad\quad = -3 + (-1)(2n) + (-1)(1)$

$\qquad\qquad\quad = -3 + (-2n) + (-1)$

$\qquad\qquad\quad = -2n + (-3) + (-1)$

$\qquad\qquad\quad = -2n + (-4)$

$\qquad\qquad\quad = -2n - 4$

73. $-2(a + 3b) - (4a - 5b)$

$\qquad = -2(a + 3b) + (-1)[4a + (-5b)]$

$\qquad = -2(a) + (-2)(3b) + (-1)(4a) + (-1)(-5b)$

$\qquad = -2a + (-6b) + (-4a) + (5b)$

$\qquad = -2a + (-4a) + (-6b) + (5b)$

$\qquad = -6a + (-b)$

$\qquad = -6a - b$

75. $10(x + 5) - 3(2x + 9)$

$\qquad = 10(x + 5) + (-3)(2x + 9)$

$\qquad = 10(x) + 10(5) + (-3)(2x) + (-3)(9)$

$\qquad = 10x + 50 + (-6x) + (-27)$

$\qquad = 10x + (-6x) + 50 + (-27)$

$\qquad = 4x + 23$

77. $-(12z + 1) + 2(7z - 5)$

$\qquad = -1(12z + 1) + 2[7z + (-5)]$

$\qquad = -1(12z) + (-1)(1) + 2(7z) + 2(-5)$

$\qquad = -12z + (-1) + 14z + (-10)$

$\qquad = -12z + 14z + (-1) + (-10)$

$\qquad = 2z + (-11)$

$\qquad = 2z - 11$

79. $3(w + 3) - (4w + y) - 3y$

$\qquad = 3(w + 3) + (-1)(4w + y) + (-3y)$

$\qquad = 3(w) + 3(3) + (-1)(4w) + (-1)(y) + (-3y)$

$\qquad = 3w + 9 + (-4w) + (-y) + (-3y)$

$\qquad = 3w + (-4w) + (-y) + (-3y) + 9$

$\qquad = -w + (-4y) + 9$

$\qquad = -w - 4y + 9$

81. $20a - 4(b + 3a) - 5b$

$\qquad = 20a + (-4)(b + 3a) + (-5b)$

$\qquad = 20a + (-4)(b) + (-4)(3a) + (-5b)$

$\qquad = 20a + (-4b) + (-12a) + (-5b)$

$\qquad = 20a + (-12a) + (-4b) + (-5b)$

$\qquad = 8a + (-9b)$

$\qquad = 8a - 9b$

83. $6 - (3m - n) - 2(m + 8) + 5n$

$\qquad = 6 + (-1)[3m + (-n)] + (-2)(m + 8) + 5n$

$\qquad = 6 + (-1)(3m) + (-1)(-n) + (-2)(m) + (-2)(8) + 5n$

$\qquad = 6 + (-3m) + n + (-2m) + (-16) + 5n$

$\qquad = -3m + (-2m) + n + 5n + 6 + (-16)$

$\qquad = -5m + 6n + (-10)$

$\qquad = -5m + 6n - 10$

85. $15+2(w-4)-(2w-5z)+7z$
$=15+2[w+(-4)]+(-1)[2w+(-5z)]+7z$
$=15+2(w)+2(-4)+(-1)(2w)+(-1)(-5z)+7z$
$=15+2w+(-8)+(-2w)+5z+7z$
$=2w+(-2w)+5z+7z+15+(-8)$
$=0+12z+7$
$=12z+7$

87. $6\left(\dfrac{1}{2}x-\dfrac{2}{3}\right)-4\left(\dfrac{5}{2}x+\dfrac{3}{4}\right)$

$=6\left[\dfrac{1}{2}x+\left(-\dfrac{2}{3}\right)\right]+(-4)\left(\dfrac{5}{2}x+\dfrac{3}{4}\right)$

$=6\left(\dfrac{1}{2}x\right)+6\left(-\dfrac{2}{3}\right)+(-4)\left(\dfrac{5}{2}x\right)+(-4)\left(\dfrac{3}{4}\right)$

$=3x+(-4)+(-10x)+(-3)$
$=3x+(-10x)+(-4)+(-3)$
$=-7x+(-7)$
$=-7x-7$

89. $\dfrac{2}{3}(9y+6)-\dfrac{3}{2}(18y-16)$

$=\dfrac{2}{3}(9y+6)+\left(-\dfrac{3}{2}\right)[18y+(-16)]$

$=\dfrac{2}{3}(9y)+\dfrac{2}{3}(6)+\left(-\dfrac{3}{2}\right)(18y)+\left(-\dfrac{3}{2}\right)(-16)$

$=6y+4+(-27y)+24$
$=6y+(-27y)+4+24$
$=-21y+28$

91. $10(0.2q-3)-100(0.04q-0.5)$
$=10[0.2q+(-3)]+(-100)[0.04q+(-0.5)]$
$=10(0.2q)+10(-3)+(-100)(0.04q)+(-100)(-0.5)$
$=2q+(-30)+(-4q)+50$
$=2q+(-4q)+(-30)+50$
$=-2q+20$

93. $100(1.04a-2.1b)-10(21.1a+0.3b)$
$=100[1.04a+(-2.1b)]+(-10)(21.1a+0.3b)$
$=100(1.04a)+100(-2.1b)+(-10)(21.1a)$
$\qquad\qquad\qquad\qquad +(-10)(0.3b)$
$=104a+(-210b)+(-211a)+(-3b)$
$=104a+(-211a)+(-210b)+(-3b)$
$=-107a+(-213b)$
$=-107a-213b$

Section 11.3   Addition and Subtraction Properties of Equality

Section 11.3 Practice Exercises

1. Answers will vary.

3. $-10a+3b-3a+13b$
$\qquad\qquad =-10a-3a+3b+13b$
$\qquad\qquad =-13a+16b$

5. $-(-8h+2k-13)$
$\qquad =-1[-8h+2k+(-13)]$
$\qquad =-1(-8h)+(-1)(2k)+(-1)(-13)$
$\qquad =8h+(-2k)+13$
$\qquad =8h-2k+13$

7. $5z-8(z-3)-20$
$\qquad =5z+(-8)[z+(-3)]+(-20)$
$\qquad =5z+(-8)(z)+(-8)(-3)+(-20)$
$\qquad =5z+(-8z)+24+(-20)$
$\qquad =-3z+4$

9. $\qquad 5x+3=-2$
$\qquad 5(-1)+3 \; \blacklozenge \; -2$
$\qquad\quad -5+3 \; \blacklozenge \; -2$
$\qquad\qquad\quad -2=-2$
-1 is a solution.

11. $\quad 10=p-16$
$\qquad 10 \; \blacklozenge \; 26-16$
$\qquad 10=10$
26 is a solution.

13. $-z + 8 = 20$

$-12 + 8 \blacklozenge 20$

$-4 \neq 20$

12 is not a solution.

15. $6m - 3 = -6$

$6\left(-\dfrac{1}{2}\right) - 3 \blacklozenge -6$

$-3 - 3 \blacklozenge -6$

$-6 = -6$

$-\dfrac{1}{2}$ is a solution.

17. $13 = 13 + 6t$

$13 \blacklozenge 13 + 6(0)$

$13 \blacklozenge 13 + 0$

$13 = 13$

0 is a solution.

19. $25 = -5q - 5$

$25 \blacklozenge -5(4) - 5$

$25 \blacklozenge -20 - 5$

$25 \neq -25$

4 is not a solution.

21. $13 + (-13) = 0$

23. $7 + (-7) = 0$

25. $3.2 + (-3.2) = 0$

27. $g - 23 = 14$

$g - 23 + 23 = 14 + 23$

$g + 0 = 37$

$g = 37$

29. $-4 + k = 12$

$-4 + 4 + k = 12 + 4$

$0 + k = 16$

$k = 16$

31. $-18 = n - 3$

$-18 + 3 = n - 3 + 3$

$-15 = n + 0$

$-15 = n$

33. $-\dfrac{5}{6} + p = \dfrac{1}{3}$

$-\dfrac{5}{6} + \dfrac{5}{6} + p = \dfrac{2}{6} + \dfrac{5}{6}$

$0 + p = \dfrac{7}{6}$

$p = \dfrac{7}{6}$

35. $k - 4.3 = -1.2$

$k - 4.3 + 4.3 = -1.2 + 4.3$

$k + 0 = 3.1$

$k = 3.1$

37. $13 = -21 + w$

$13 + 21 = -21 + 21 + w$

$34 = 0 + w$

$34 = w$

39. $52 - 52 = 0$

41. $18 - 18 = 0$

43. $100 - 100 = 0$

45. $x + 34 = 6$

$x + 34 - 34 = 6 - 34$

$x + 0 = -28$

$x = -28$

47. $17 + b = 20$

$17 - 17 + b = 20 - 17$

$0 + b = 3$

$b = 3$

49. $-32 = t + 14$

$-32 - 14 = t + 14 - 14$

$-46 = t + 0$

$-46 = t$

51. $8.2 = 21.8 + m$

$8.2 - 21.8 = 21.8 - 21.8 + m$

$-13.6 = 0 + m$

$-13.6 = m$

53.
$$a + \frac{3}{5} = -\frac{7}{10}$$
$$a + \frac{3}{5} - \frac{3}{5} = -\frac{7}{10} - \frac{3}{5}$$
$$a + 0 = -\frac{7}{10} - \frac{6}{10}$$
$$a = -\frac{13}{10}$$

55.
$$21 = 14 + w$$
$$21 - 14 = 14 - 14 + w$$
$$7 = 0 + w$$
$$7 = w$$

57.
$$1 + p = 0$$
$$1 - 1 + p = 0 - 1$$
$$0 + p = -1$$
$$p = -1$$

59.
$$-34 + t = -40$$
$$-34 + 34 + t = -40 + 34$$
$$0 + t = -6$$
$$t = -6$$

61.
$$\frac{2}{3} = y - \frac{5}{12}$$
$$\frac{2}{3} + \frac{5}{12} = y - \frac{5}{12} + \frac{5}{12}$$
$$\frac{8}{12} + \frac{5}{12} = y + 0$$
$$\frac{13}{12} = y$$

63.
$$-2.5 = -1.1 + m$$
$$-2.5 + 1.1 = -1.1 + 1.1 + m$$
$$-1.4 = 0 + m$$
$$-1.4 = m$$

65.
$$w - 23 = -11$$
$$w - 23 + 23 = -11 + 23$$
$$w + 0 = 12$$
$$w = 12$$

67.
$$x + 21 = 16$$
$$x + 21 - 21 = 16 - 21$$
$$x + 0 = -5$$
$$x = -5$$

69.
$$-2 = a - 15$$
$$-2 + 15 = a - 15 + 15$$
$$13 = a + 0$$
$$13 = a$$

71.
$$4.01 + p = 3.22$$
$$4.01 - 4.01 + p = 3.22 - 4.01$$
$$0 + p = -0.79$$
$$p = -0.79$$

73.
$$t + \frac{3}{8} = 2$$
$$t + \frac{3}{8} - \frac{3}{8} = 2 - \frac{3}{8}$$
$$t + 0 = \frac{16}{8} - \frac{3}{8}$$
$$t = \frac{13}{8} \text{ or } 1\frac{5}{8}$$

75.
$$27 = z - 22$$
$$27 + 22 = z - 22 + 22$$
$$49 = z + 0$$
$$49 = z$$

77.
$$-70 = -55 + w$$
$$-70 + 55 = -55 + 55 + w$$
$$-15 = 0 + w$$
$$-15 = w$$

79.
$$10x - 9x - 11 = 15$$
$$x - 11 = 15$$
$$x - 11 + 11 = 15 + 11$$
$$x + 0 = 26$$
$$x = 26$$

81.
$$-13 + 15 = p + 5$$
$$2 = p + 5$$
$$2 - 5 = p + 5 - 5$$
$$-3 = p + 0$$
$$-3 = p$$

83.
$$4(k + 2) - 3k = -6 + 9$$
$$4(k) + 4(2) - 3k = -6 + 9$$
$$4k + 8 - 3k = -6 + 9$$
$$4k - 3k + 8 = -6 + 9$$
$$k + 8 = 3$$
$$k + 8 - 8 = 3 - 8$$
$$k + 0 = -5$$
$$k = -5$$

Section 11.4   Multiplication and Division Properties of Equality

Section 11.4   Practice Exercises

1. Answers will vary.

3.
$$p - 12 = 33$$
$$p - 12 + 12 = 33 + 12$$
$$p + 0 = 45$$
$$p = 45$$

5.
$$16 = h - 5$$
$$16 + 5 = h - 5 + 5$$
$$21 = h + 0$$
$$21 = h$$

7.
$$p - 6 = -19$$
$$p - 6 + 6 = -19 + 6$$
$$p + 0 = -13$$
$$p = -13$$

9.
$$n + \frac{1}{2} = -\frac{2}{3}$$
$$n + \frac{1}{2} - \frac{1}{2} = -\frac{2}{3} - \frac{1}{2}$$
$$n + 0 = -\frac{4}{6} - \frac{3}{6}$$
$$n = -\frac{7}{6}$$

11. $3 \cdot \dfrac{1}{3} = 1$

13. $-\dfrac{4}{7} \cdot \left(-\dfrac{7}{4}\right) = 1$

15. $-7 \div (-7) = 1$

17. $5.1 \div 5.1 = 1$

19.
$$14b = -42$$
$$\frac{14b}{14} = \frac{-42}{14}$$
$$b = -3$$

21.
$$-8k = 56$$
$$\frac{-8k}{-8} = \frac{56}{-8}$$
$$k = -7$$

23.
$$-t = -13$$
$$-1t = -13$$
$$\frac{-1t}{-1} = \frac{-13}{-1}$$
$$t = 13$$

25.
$$\frac{2}{3}m = 14$$
$$\frac{3}{2} \cdot \frac{2}{3}m = \frac{3}{2}\left(\frac{14}{1}\right)$$
$$m = 21$$

27.
$$\frac{b}{7} = -3$$
$$\frac{1}{7}b = -3$$
$$7\left(\frac{1}{7}b\right) = 7(-3)$$
$$b = -21$$

29.
$$-2.8 = -0.7t$$
$$\frac{-2.8}{-0.7} = \frac{-0.7t}{-0.7}$$
$$4 = t$$

31.
$$-\frac{u}{2} = -15$$
$$-\frac{1}{2}u = -15$$
$$-2\left(-\frac{1}{2}u\right) = -2(-15)$$
$$u = 30$$

33.
$$6 = -18w$$
$$\frac{6}{-18} = \frac{-18w}{-18}$$
$$-\frac{1}{3} = w$$

35.
$$1.3x = 5.33$$
$$\frac{1.3x}{1.3} = \frac{5.33}{1.3}$$
$$x = 4.1$$

37.

$$\frac{5}{4}k = -\frac{1}{2}$$

$$\frac{4}{5}\left(\frac{5}{4}k\right) = \frac{4}{5}\left(-\frac{1}{2}\right)$$

$$k = -\frac{2}{5}$$

39.

$$0 = \frac{3}{8}m$$

$$\frac{8}{3}(0) = \frac{8}{3}\left(\frac{3}{8}m\right)$$

$$0 = m$$

41.

$$-\frac{9}{4}x = -\frac{3}{5}$$

$$-\frac{4}{9}\left(-\frac{9}{4}x\right) = -\frac{4}{9}\left(-\frac{3}{5}\right)$$

$$x = \frac{4}{15}$$

43. $100 = 5k$

$$\frac{100}{5} = \frac{5k}{5}$$

$$20 = k$$

45. $31 = -p$

$$\frac{31}{-1} = \frac{-1p}{-1}$$

$$-31 = p$$

47.

$$3p = \frac{5}{2}$$

$$\frac{1}{3}(3p) = \frac{1}{3}\left(\frac{5}{2}\right)$$

$$p = \frac{5}{6}$$

49. $-4a = 0$

$$\frac{-4a}{-4} = \frac{0}{-4}$$

$$a = 0$$

51. If the operation between a number and a variable is subtraction, use the addition property to isolate the variable.

53. If the operation between a number and a variable is multiplication, use the division property to isolate the variable.

55. $4 + x = -12$

$$4 - 4 + x = -12 - 4$$

$$x = -16$$

57. $4y = -12$

$$\frac{4y}{4} = \frac{-12}{4}$$

$$y = -3$$

59. $q - 4 = -12$

$$q - 4 + 4 = -12 + 4$$

$$q = -8$$

61.

$$\frac{h}{4} = -12$$

$$\frac{1}{4}h = -12$$

$$4\left(\frac{1}{4}h\right) = 4(-12)$$

$$h = -48$$

63.

$$\frac{2}{3} + t = 1$$

$$\frac{2}{3} - \frac{2}{3} + t = 1 - \frac{2}{3}$$

$$t = \frac{3}{3} - \frac{2}{3}$$

$$t = \frac{1}{3}$$

65. $-9a = -12$

$$\frac{-9a}{-9} = \frac{-12}{-9}$$

$$a = \frac{4}{3}$$

67. $7 = r - 23$

$$7 + 23 = r - 23 + 23$$

$$30 = r$$

69.

$$-\frac{y}{3} = 5$$

$$-\frac{1}{3}y = 5$$

$$-3\left(-\frac{1}{3}y\right) = -3(5)$$

$$y = -15$$

71.
$$2p = \frac{5}{6}$$
$$\frac{1}{2}(2p) = \frac{1}{2}\left(\frac{5}{6}\right)$$
$$p = \frac{5}{12}$$

73.
$$-\frac{3}{7}x = \frac{9}{10}$$
$$-\frac{7}{3}\left(-\frac{3}{7}x\right) = -\frac{7}{3}\left(\frac{9}{10}\right)$$
$$x = -\frac{21}{10}$$

75.
$$t - 12.9 = 15$$
$$t - 12.9 + 12.9 = 15 + 12.9$$
$$t = 27.9$$

77.
$$5 + u = 3.2$$
$$5 - 5 + u = 3.2 - 5$$
$$u = -1.8$$

79.
$$50 = a + 72$$
$$50 - 72 = a + 72 - 72$$
$$-22 = a$$

81.
$$-1 = b - 16$$
$$-1 + 16 = b - 16 + 16$$
$$15 = b$$

83.
$$-12 = 30x$$
$$\frac{1}{30}(-12) = \frac{1}{30}(30x)$$
$$-\frac{2}{5} = x$$

85.
$$-6 = -\frac{1}{2}q$$
$$-2(-6) = -2\left(-\frac{1}{2}q\right)$$
$$12 = q$$

87.
$$5x - 2x = -15$$
$$3x = -15$$
$$\frac{3x}{3} = \frac{-15}{3}$$
$$x = -5$$

89.
$$3p + 4p = 25 - 4$$
$$7p = 21$$
$$\frac{7p}{7} = \frac{21}{7}$$
$$p = 3$$

91.
$$-2(a + 3) - 6a + 6 = 8$$
$$-2(a) + (-2)(3) + (-6a) + 6 = 8$$
$$-2a + (-6) + (-6a) + 6 = 8$$
$$-2a + (-6a) + (-6) + 6 = 8$$
$$-8a + 0 = 8$$
$$\frac{-8a}{-8} = \frac{8}{-8}$$
$$a = -1$$

Section 11.5   Solving Equations with Multiple Steps

Section 11.5   Practice Exercises

1. **Step 1:** Clear parentheses.
Step 2: Subtract $4x$ from both sides to collect the variable terms on the left. Combine like terms.
Step 3: Subtract 2 from both sides to collect the constants on the right. Simplify.
Step 4: Divide both sides by −11 to obtain a coefficient of 1 on the x-term. Simplify.

3.
$$\frac{1}{3}b = -4$$
$$3\left(\frac{1}{3}b\right) = 3(-4)$$
$$b = -12$$

5.
$$-\frac{3}{8} = w + \frac{1}{4}$$
$$-\frac{3}{8} - \frac{1}{4} = w + \frac{1}{4} - \frac{1}{4}$$
$$-\frac{3}{8} - \frac{2}{8} = w$$
$$-\frac{5}{8} = w$$

7.
$$-8h = 0$$
$$\frac{-8h}{-8} = \frac{0}{-8}$$
$$h = 0$$

9.
$$3m + 2 = 14$$
$$3m + 2 - 2 = 14 - 2$$
$$3m = 12$$
$$\frac{3m}{3} = \frac{12}{3}$$
$$m = 4$$

11.
$$-8c - 12 = 36$$
$$-8c - 12 + 12 = 36 + 12$$
$$-8c = 48$$
$$\frac{-8c}{-8} = \frac{48}{-8}$$
$$c = -6$$

13.
$$1 = -4z + 21$$
$$1 - 21 = -4z + 21 - 21$$
$$-20 = -4z$$
$$\frac{-20}{-4} = \frac{-4z}{-4}$$
$$5 = z$$

15.
$$9 = 12x - 7$$
$$9 + 7 = 12x - 7 + 7$$
$$16 = 12x$$
$$\frac{16}{12} = \frac{12x}{12}$$
$$\frac{4}{3} = x$$

17.
$$3.4 - 2d = 8.2$$
$$3.4 - 3.4 - 2d = 8.2 - 3.4$$
$$-2d = 4.8$$
$$\frac{-2d}{-2} = \frac{4.8}{-2}$$
$$d = -2.4$$

19.
$$-0.57 = 15h + 16.23$$
$$-0.57 - 16.23 = 15h + 16.23 - 16.23$$
$$-16.8 = 15h$$
$$\frac{-16.8}{15} = \frac{15h}{15}$$
$$-1.12 = h$$

21.
$$\frac{b}{3} - 12 = -9$$
$$\frac{b}{3} - 12 + 12 = -9 + 12$$
$$\frac{b}{3} = 3$$
$$\frac{1}{3}b = 3$$
$$3\left(\frac{1}{3}b\right) = 3(3)$$
$$b = 9$$

23.
$$-9 = \frac{w}{2} - 3$$
$$-9 + 3 = \frac{w}{2} - 3 + 3$$
$$-6 = \frac{w}{2}$$
$$-6 = \frac{1}{2}w$$
$$2(-6) = 2\left(\frac{1}{2}w\right)$$
$$-12 = w$$

25.
$$3x + \frac{1}{2} = \frac{5}{4}$$
$$3x + \frac{1}{2} - \frac{1}{2} = \frac{5}{4} - \frac{1}{2}$$
$$3x = \frac{5}{4} - \frac{2}{4}$$
$$3x = \frac{3}{4}$$
$$\frac{1}{3}(3x) = \frac{1}{3}\left(\frac{3}{4}\right)$$
$$x = \frac{1}{4}$$

27.
$$10 - y = 37$$
$$10 - 10 - y = 37 - 10$$
$$-y = 27$$
$$(-1)(-y) = -1(27)$$
$$y = -27$$

29.
$$8 + 4b = 2 + 2b$$
$$8 + 4b - 2b = 2 + 2b - 2b$$
$$8 + 2b = 2$$
$$8 - 8 + 2b = 2 - 8$$
$$2b = -6$$
$$\frac{2b}{2} = \frac{-6}{2}$$
$$b = -3$$

31.
$$7 - 5t = 3t - 2$$
$$7 - 5t - 3t = 3t - 3t - 2$$
$$7 - 8t = -2$$
$$7 - 7 - 8t = -2 - 7$$
$$-8t = -9$$
$$\frac{-8t}{-8} = \frac{-9}{-8}$$
$$t = \frac{9}{8}$$

33.
$$4 - 3d = 5d - 4$$
$$4 - 3d - 5d = 5d - 5d - 4$$
$$4 - 8d = -4$$
$$4 - 4 - 8d = -4 - 4$$
$$-8d = -8$$
$$\frac{-8d}{-8} = \frac{-8}{-8}$$
$$d = 1$$

35.
$$12p = 3p + 21$$
$$12p - 3p = 3p - 3p + 21$$
$$9p = 21$$
$$\frac{9p}{9} = \frac{21}{9}$$
$$p = \frac{7}{3}$$

37.
$$-z - 2 = -2z$$
$$-z + z - 2 = -2z + z$$
$$-2 = -z$$
$$\frac{-2}{-1} = \frac{-1z}{-1}$$
$$2 = z$$

39.
$$1 + \frac{1}{4}p = 2 + \frac{3}{4}p$$
$$1 + \frac{1}{4}p - \frac{1}{4}p = 2 + \frac{3}{4}p - \frac{1}{4}p$$
$$1 = 2 + \frac{1}{2}p$$
$$1 - 2 = 2 - 2 + \frac{1}{2}p$$
$$-1 = \frac{1}{2}p$$
$$2(-1) = 2\left(\frac{1}{2}p\right)$$
$$-2 = p$$

41.
$$4 + 2a - 7 = 3a + a + 3$$
$$4 - 7 + 2a = 3a + a + 3$$
$$-3 + 2a = 4a + 3$$
$$-3 + 2a - 4a = 4a - 4a + 3$$
$$-3 - 2a = 3$$
$$-3 + 3 - 2a = 3 + 3$$
$$-2a = 6$$
$$\frac{-2a}{-2} = \frac{6}{-2}$$
$$a = -3$$

43.
$$-8w + 8 + 3w = 2 - 6w + 2$$
$$-8w + 3w + 8 = -6w + 2 + 2$$
$$-5w + 8 = -6w + 4$$
$$-5w + 6w + 8 = -6w + 6w + 4$$
$$w + 8 = 4$$
$$w + 8 - 8 = 4 - 8$$
$$w = -4$$

45.
$$6y + 2y - 2 = 14 + 3y - 12$$
$$6y + 2y - 2 = 3y + 14 - 12$$
$$8y - 2 = 3y + 2$$
$$8y - 3y - 2 = 3y - 3y + 2$$
$$5y - 2 = 2$$
$$5y - 2 + 2 = 2 + 2$$
$$5y = 4$$
$$\frac{5y}{5} = \frac{4}{5}$$
$$y = \frac{4}{5}$$

47. $3n - 4(n-1) = 16$
$3n - 4n + 4 = 16$
$-n + 4 = 16$
$-n + 4 - 4 = 16 - 4$
$-n = 12$
$\dfrac{-1n}{-1} = \dfrac{12}{-1}$
$n = -12$

49. $9q - 5(q-3) = 5q$
$9q - 5q + 15 = 5q$
$4q + 15 = 5q$
$4q - 4q + 15 = 5q - 4q$
$15 = q$

51. $2(1-m) = 5 - 3m$
$2 - 2m = 5 - 3m$
$2 - 2m + 3m = 5 - 3m + 3m$
$2 + m = 5$
$2 - 2 + m = 5 - 2$
$m = 3$

53. $-4(k-2) + 14 = 3k - 20$
$-4k + 8 + 14 = 3k - 20$
$-4k + 22 = 3k - 20$
$-4k - 3k + 22 = 3k - 3k - 20$
$-7k + 22 = -20$
$-7k + 22 - 22 = -20 - 22$
$-7k = -42$
$\dfrac{-7k}{-7} = \dfrac{-42}{-7}$
$k = 6$

55. $3z - 9 = 3(5z - 1)$
$3z - 9 = 15z - 3$
$3z - 3z - 9 = 15z - 3z - 3$
$-9 = 12z - 3$
$-9 + 3 = 12z - 3 + 3$
$-6 = 12z$
$\dfrac{-6}{12} = \dfrac{12z}{12}$
$-\dfrac{1}{2} = z$

57. $6w + 2(w-1) = 14 - (3w + 1)$
$6w + 2w - 2 = 14 - 3w - 1$
$6w + 2w - 2 = 14 - 1 - 3w$
$8w - 2 = 13 - 3w$
$8w + 3w - 2 = 13 - 3w + 3w$
$11w - 2 = 13$
$11w - 2 + 2 = 13 + 2$
$11w = 15$
$\dfrac{11w}{11} = \dfrac{15}{11}$
$w = \dfrac{15}{11}$

59. $6(u-1) + 5u + 1 = 5(u+6) - u$
$6u - 6 + 5u + 1 = 5u + 30 - u$
$6u + 5u - 6 + 1 = 5u - u + 30$
$11u - 5 = 4u + 30$
$11u - 4u - 5 = 4u - 4u + 30$
$7u - 5 = 30$
$7u - 5 + 5 = 30 + 5$
$7u = 35$
$\dfrac{7u}{7} = \dfrac{35}{7}$
$u = 5$

Problem Recognition Exercises: Solving Equations

1. Equation

3. Expression

5. Equation

7. $5t = 20$
$\dfrac{1}{5}(5t) = \dfrac{1}{5}(20)$
$t = 4$

9. $5 + t = 20$
$5 - 5 + t = 20 - 5$
$t = 15$

11.
$$5(t-3)=20$$
$$5t-15=20$$
$$5t-15+15=20+15$$
$$5t=35$$
$$\frac{1}{5}(5t)=\frac{1}{5}(35)$$
$$t=7$$

13.
$$5x-3=20$$
$$5x-3+3=20+3$$
$$5x=23$$
$$\frac{1}{5}(5x)=\frac{1}{5}(23)$$
$$x=\frac{23}{5}$$

15.
$$5+3p-2=0$$
$$3p+5-2=0$$
$$3p+3=0$$
$$3p+3-3=0-3$$
$$3p=-3$$
$$\frac{3p}{3}=\frac{-3}{3}$$
$$p=-1$$

17.
$$0=2x+5x+1$$
$$0=7x+1$$
$$0-1=7x+1-1$$
$$-1=7x$$
$$\frac{-1}{7}=\frac{7x}{7}$$
$$-\frac{1}{7}=x$$

19.
$$-\frac{2}{3}p-\frac{1}{6}=-\frac{2}{3}$$
$$-\frac{2}{3}p-\frac{1}{6}+\frac{1}{6}=-\frac{2}{3}+\frac{1}{6}$$
$$-\frac{2}{3}p=-\frac{4}{6}+\frac{1}{6}$$
$$-\frac{2}{3}p=-\frac{3}{6}$$
$$\left(-\frac{3}{2}\right)\left(-\frac{2}{3}p\right)=\left(-\frac{3}{2}\right)\left(-\frac{1}{2}\right)$$
$$p=\frac{3}{4}$$

21.
$$-14=\frac{r}{6}-12$$
$$-14+12=\frac{r}{6}-12+12$$
$$-2=\frac{r}{6}$$
$$6(-2)=6\left(\frac{r}{6}\right)$$
$$-12=r$$

23.
$$2.3u+0.2=-1.2u+7.2$$
$$2.3u+1.2u+0.2=-1.2u+1.2u+7.2$$
$$3.5u+0.2=7.2$$
$$3.5u+0.2-0.2=7.2-0.2$$
$$3.5u=7$$
$$\frac{3.5u}{3.5}=\frac{7}{3.5}$$
$$u=2$$

25.
$$6a+3a-21=4-5a-1$$
$$9a-21=-5a+4-1$$
$$9a-21=-5a+3$$
$$9a+5a-21=-5a+5a+3$$
$$14a-21=3$$
$$14a-21+21=3+21$$
$$14a=24$$
$$\frac{14a}{14}=\frac{24}{14}$$
$$a=\frac{12}{7}$$

27.
$$-2(x-3)+14=10-(x+4)$$
$$-2x+6+14=10-x-4$$
$$-2x+20=-x+6$$
$$-2x+x+20=-x+x+6$$
$$-x+20=6$$
$$-x+20-20=6-20$$
$$-x=-14$$
$$\frac{-x}{-1}=\frac{-14}{-1}$$
$$x=14$$

29.
$$2-3(y+1)=-4y+7$$
$$2-3y-3=-4y+7$$
$$-3y+2-3=-4y+7$$
$$-3y-1=-4y+7$$
$$-3y+4y-1=-4y+4y+7$$
$$y-1=7$$
$$y-1+1=7+1$$
$$y=8$$

Section 11.6   Applications and Problem Solving

Section 11.6   Practice Exercises

1. Answers will vary.

3.

$$\frac{b}{5} - 5 = -14$$

$$\frac{b}{5} - 5 + 5 = -14 + 5$$

$$\frac{1}{5}b = -9$$

$$5\left(\frac{1}{5}b\right) = 5(-9)$$

$$b = -45$$

5.

$$4(r+4) - 12 = 18 - r$$

$$4r + 16 - 12 = 18 - r$$

$$4r + 4 = 18 - r$$

$$4r + r + 4 = 18 - r + r$$

$$5r + 4 = 18$$

$$5r + 4 - 4 = 18 - 4$$

$$5r = 14$$

$$\frac{5r}{5} = \frac{14}{5}$$

$$r = \frac{14}{5}$$

7.

$$4.4p - 2.6 = 1.2p - 5$$

$$4.4p - 1.2p - 2.6 = 1.2p - 1.2p - 5$$

$$3.2p - 2.6 = -5$$

$$3.2p - 2.6 + 2.6 = -5 + 2.6$$

$$3.2p = -2.4$$

$$\frac{3.2p}{3.2} = \frac{-2.4}{3.2}$$

$$p = -0.75$$

9. (a) Let x represent the number. The quotient of a number and 3 is -8.

$$\frac{x}{3} = -8$$

(b)

$$\frac{x}{3} = -8$$

$$3\left(\frac{x}{3}\right) = 3(-8)$$

$$x = -24$$

The number is -24.

11. (a) Let x represent the number. A number subtracted from -30 results in 42.

$$-30 - x = 42$$

(b)

$$-30 - x = 42$$

$$-30 + 30 - x = 42 + 30$$

$$-x = 72$$

$$-1(-x) = -1(72)$$

$$x = -72$$

The number is -72.

13. (a) Let x represent the number. A total of 30 and a number is 13.

$$30 + x = 13$$

(b)

$$30 + x = 13$$

$$30 - 30 + x = 13 - 30$$

$$x = -17$$

The number is -17.

15. (a) Let x represent the number. Five less than the quotient of a number and 4 is equal to -12.

$$\frac{x}{4} - 5 = -12$$

(b)

$$\frac{x}{4} - 5 = -12$$

$$\frac{x}{4} - 5 + 5 = -12 + 5$$

$$\frac{1}{4}x = -7$$

$$4\left(\frac{1}{4}x\right) = 4(-7)$$

$$x = -28$$

The number is -28.

17. (a) Let x represent the number. One-half increased by a number is 4.

$$\frac{1}{2} + x = 4$$

(b)
$$\frac{1}{2} + x = 4$$
$$\frac{1}{2} - \frac{1}{2} + x = 4 - \frac{1}{2}$$
$$x = \frac{8}{2} - \frac{1}{2}$$
$$x = \frac{7}{2} \text{ or } 3\frac{1}{2}$$

The number is $\frac{7}{2}$ or $3\frac{1}{2}$.

19. (a) Let x represent the number. The product of -12 and a number is the same as the sum of the number and 26.

$$-12x = x + 26$$

(b)
$$-12x = x + 26$$
$$-12x - x = x - x + 26$$
$$-13x = 26$$
$$\frac{-13x}{-13} = \frac{26}{-13}$$
$$x = -2$$

The number is -2.

21. (a) Let x represent the number. Ten times the total of a number and 5.1 is 56.

$$10(x + 5.1) = 56$$

(b)
$$10(x + 5.1) = 56$$
$$10x + 51 = 56$$
$$10x + 51 - 51 = 56 - 51$$
$$10x = 5$$
$$\frac{10x}{10} = \frac{5}{10}$$
$$x = 0.5$$

The number is 0.5.

23. (a) Let x represent the number. The product of 3 and a number is the same as 10 less than twice the number.

$$3x = 2x - 10$$

(b)
$$3x = 2x - 10$$
$$3x - 2x = 2x - 2x - 10$$
$$x = -10$$

The number is -10.

25. Let x represent the length of the shorter piece. Then $3x$ represents the length of the other piece.

$$\begin{pmatrix} \text{Length of} \\ \text{one piece} \end{pmatrix} + \begin{pmatrix} \text{Length of} \\ \text{other piece} \end{pmatrix} = \begin{pmatrix} \text{Total} \\ \text{length} \\ \text{of wire} \end{pmatrix}$$

$$x + 3x = 12$$
$$4x = 12$$
$$\frac{4x}{4} = \frac{12}{4}$$
$$x = 3$$
$$3x = 3(3) = 9$$

The pieces are 3ft and 9ft long.

27. Let x represent the number of hits for Boyz II Men. Then $x - 6$ represents the number of hits for Metallica.

$$\begin{pmatrix} \text{Boyz II Men} \\ \text{hits} \end{pmatrix} + \begin{pmatrix} \text{Metallica} \\ \text{hits} \end{pmatrix} = \begin{pmatrix} \text{Total} \\ \text{number} \\ \text{of hits} \end{pmatrix}$$

$$x + x - 6 = 26$$
$$2x - 6 = 26$$
$$2x - 6 + 6 = 26 + 6$$
$$2x = 32$$
$$\frac{2x}{2} = \frac{32}{2}$$
$$x = 16$$
$$x - 6 = 16 - 6 = 10$$

Boyz II Men had 16 hits while Metallica had 10 hits.

29. Let w represent the width. Then $w + 30$ represents the length.

$$P = 2l + 2w$$
$$460 = 2(w + 30) + 2w$$
$$460 = 2w + 60 + 2w$$
$$460 = 4w + 60$$
$$460 - 60 = 4w + 60 - 60$$
$$400 = 4w$$
$$\frac{400}{4} = \frac{4w}{4}$$
$$100 = w$$
$$x + 30 = 100 + 30 = 130$$

The soccer field is 100 yd by 130 yd.

31. Let x represent the number of minutes over 500. Then $0.25x$ represents the cost for x minutes over 500.

$$\begin{pmatrix} \text{Monthly} \\ \text{fee} \end{pmatrix} + \begin{pmatrix} \text{Cost of} \\ \text{additional} \\ \text{minutes} \end{pmatrix} = \begin{pmatrix} \text{Total} \\ \text{cost} \end{pmatrix}$$
$$49.95 + 0.25x = 62.45$$
$$49.95 - 49.95 + 0.25x = 62.45 - 49.95$$
$$0.25x = 12.50$$
$$\frac{0.25x}{0.25} = \frac{12.50}{0.25}$$
$$x = 50$$

Jim used 50 min over the 500 min.

33. Let x represent the length of the shorter piece. Then $2x$ represents the length of the other piece.

$$\begin{pmatrix} \text{Length of} \\ \text{one piece} \end{pmatrix} + \begin{pmatrix} \text{Length of} \\ \text{other piece} \end{pmatrix} = \begin{pmatrix} \text{Total} \\ \text{length} \\ \text{of ribbon} \end{pmatrix}$$
$$x + 2x = 4$$
$$3x = 4$$
$$\frac{3x}{3} = \frac{4}{3}$$
$$x = 1\frac{1}{3}$$

The pieces are $1\frac{1}{3}$ ft and

$$2\left(1\frac{1}{3}\right) = 2\left(\frac{4}{3}\right) = \frac{8}{3} = 2\frac{2}{3} \text{ ft long.}$$

35. Let x represent the number of points for Oakland. Then $2x + 6$ represents the number of points for Tampa Bay.

$$\begin{pmatrix} \text{Oakland's} \\ \text{points} \end{pmatrix} + \begin{pmatrix} \text{Tampa Bay's} \\ \text{points} \end{pmatrix} = \begin{pmatrix} \text{Total} \\ \text{points} \\ \text{scored} \end{pmatrix}$$
$$x + 2x + 6 = 69$$
$$3x + 6 = 69$$
$$3x + 6 - 6 = 69 - 6$$
$$3x = 63$$
$$\frac{3x}{3} = \frac{63}{3}$$
$$x = 21$$

Oakland scored 21 points and Tampa Bay scored $2(21) + 6 = 42 + 6 = 48$ points.

37. Let x represent the rent. Then $x - 350$ represents the security deposit.

$$(\text{rent}) + \begin{pmatrix} \text{security} \\ \text{deposit} \end{pmatrix} = \begin{pmatrix} \text{1st month's} \\ \text{payment} \end{pmatrix}$$
$$x + x - 350 = 950$$
$$2x - 350 = 950$$
$$2x - 350 + 350 = 950 + 350$$
$$2x = 1300$$
$$\frac{2x}{2} = \frac{1300}{2}$$
$$x = 650$$

Charlene's rent is $650 a month with a security deposit of $650 − $350 = $300.

39. Let x represent the number of hours of overtime. Then $18x$ represents the earnings for working overtime.

$$(\text{Salary}) + (\text{overtime pay}) = \begin{pmatrix} \text{paycheck} \\ \text{amount} \end{pmatrix}$$
$$480 + 18x = 588$$
$$480 - 480 + 18x = 588 - 480$$
$$18x = 108$$
$$\frac{18x}{18} = \frac{108}{18}$$
$$x = 6$$

Stefan worked 6 hr of overtime.

41. Let x represent the hours for the fall. Then $x + 4$ represents the hours for the spring.

$$\left(\begin{array}{c}\text{Fall}\\\text{hours}\end{array}\right) + \left(\begin{array}{c}\text{Spring}\\\text{hours}\end{array}\right) = \left(\begin{array}{c}\text{Total}\\\text{hours}\end{array}\right)$$

$$x + x + 4 = 28$$
$$2x + 4 = 28$$
$$2x + 4 - 4 = 28 - 4$$
$$2x = 24$$
$$\frac{2x}{2} = \frac{24}{2}$$
$$x = 12$$

Raul took 12 hr in the fall and signed up for $12 + 4 = 16$ hr in the spring.

43. Let x represent the amount of merchandise to sell. Then $0.24x$ represents the commission earned.

$$\left(\begin{array}{c}\text{Earnings from}\\\text{commission}\end{array}\right) = (\text{Salary})$$

$$0.24x = 300$$
$$\frac{0.24x}{0.24} = \frac{300}{0.24}$$
$$x = 1250$$

Ann-Marie must sell $1250 of merchandise each week.

Chapter 11   Review Exercises

Section 11.1

1. (a) $a + 8$

    (b) $a + 8 = (\ ) + 8 = 35 + 8 = 43$ years old

3. $-2(x + y)^2 = -2[(\ ) + (\ )]^2$
$$= -2[(6) + (-9)]^2 = -2(-3)^2$$
$$= -2(9) = -18$$

5. (a) $t - 5 = t + (-5) = -5 + t$

    (b) $h \cdot 3 = 3h$

7. $3(2b + 5) = 3(2b) + 3(5) = 6b + 15$

Section 11.2

9. $3a^2$ is a variable term with coefficient 3; $-5a$ is a variable term with coefficient -5; 12 is a constant term with coefficient 12.

11. Unlike terms

13. *Like* terms

15. $6y + 8x - 2y - 2x + 10$
$$= 8x - 2x + 6y - 2y + 10$$
$$= 6x + 4y + 10$$

17. $4(u - 3v) - 5u + v = 4u - 4(3v) - 5u + v$
$$= 4u - 12v - 5u + v$$
$$= 4u - 5u - 12v + v$$
$$= -u - 11v$$

Section 11.3

19. $\quad 5x + 10 = -5$
$$5(-3) + 10 \ \blacklozenge\ -5$$
$$-15 + 10 \ \blacklozenge\ -5$$
$$-5 = -5$$

-3 is a solution.

21. If a constant is being added to the variable term, use the subtraction property. If a constant is being subtracted from a variable term, use the addition property.

23. $\quad k - 3 = -15$
$$k - 3 + 3 = -15 + 3$$
$$k = -12$$

25. $\quad 21 = q - 3$
$$21 + 3 = q - 3 + 3$$
$$24 = q$$

27. $\quad -3.1 + n = 1.9$
$$-3.1 + 3.1 + n = 1.9 + 3.1$$
$$n = 5$$

29.
$$b + \frac{1}{5} = -\frac{9}{10}$$
$$b + \frac{1}{5} - \frac{1}{5} = -\frac{9}{10} - \frac{1}{5}$$
$$b = -\frac{9}{10} - \frac{2}{10}$$
$$b = -\frac{11}{10}$$

Section 11.4

31. $4d = -28$
$$\frac{4d}{4} = \frac{-28}{4}$$
$$d = -7$$

33.
$$\frac{t}{2} = -13$$
$$\frac{1}{2}t = -13$$
$$2\left(\frac{1}{2}t\right) = 2(-13)$$
$$t = -26$$

35.
$$-\frac{4}{5}y = -16$$
$$-\frac{5}{4}\left(-\frac{4}{5}y\right) = -\frac{5}{4}\left(-\frac{16}{1}\right)$$
$$y = 20$$

37. $1.4 = -0.7m$
$$\frac{1.4}{-0.7} = \frac{-0.7m}{-0.7}$$
$$-2 = m$$

39.
$$\frac{1}{3}w = \frac{3}{7}$$
$$3\left(\frac{1}{3}w\right) = 3\left(\frac{3}{7}\right)$$
$$w = \frac{9}{7}$$

41. $-42 = -7p$
$$\frac{-42}{-7} = \frac{-7p}{-7}$$
$$6 = p$$

43.
$$9x + 7 = -2$$
$$9x + 7 - 7 = -2 - 7$$
$$9x = -9$$
$$\frac{9x}{9} = \frac{-9}{9}$$
$$x = -1$$

45.
$$45 = 6m - 3$$
$$45 + 3 = 6m - 3 + 3$$
$$48 = 6m$$
$$\frac{48}{6} = \frac{6m}{6}$$
$$8 = m$$

47.
$$4 = \frac{3}{5}m - 2$$
$$4 + 2 = \frac{3}{5}m - 2 + 2$$
$$6 = \frac{3}{5}m$$
$$\frac{5}{3}\left(\frac{6}{1}\right) = \frac{5}{3}\left(\frac{3}{5}m\right)$$
$$10 = m$$

49.
$$5x + 12 = 4x - 16$$
$$5x - 4x + 12 = 4x - 4x - 16$$
$$x + 12 = -16$$
$$x + 12 - 12 = -16 - 12$$
$$x = -28$$

51. $6(w - 2) + 15 = 3(w + 3) - 2$
$$6w - 12 + 15 = 3w + 9 - 2$$
$$6w + 3 = 3w + 7$$
$$6w - 3w + 3 = 3w - 3w + 7$$
$$3w + 3 = 7$$
$$3w + 3 - 3 = 7 - 3$$
$$3w = 4$$
$$\frac{3w}{3} = \frac{4}{3}$$
$$w = \frac{4}{3}$$

53. $-(5a+3)-3(a-2)=24-a$

$-5a-3-3a+6=24-a$

$-5a-3a-3+6=24-a$

$-8a+3=24-a$

$-8a+a+3=24-a+a$

$-7a+3=24$

$-7a+3-3=24-3$

$-7a=21$

$\dfrac{-7a}{-7}=\dfrac{21}{-7}$

$a=-3$

Section 11.6

55. (a) Let x represent the number.

$-6x=x+2$

(b) $-6x=x+2$

$-6x-x=x-x+2$

$-7x=2$

$\dfrac{-7x}{-7}=\dfrac{2}{-7}$

$x=-\dfrac{2}{7}$

The number is $-\dfrac{2}{7}$.

57. (a) Let x represent the number.

$\dfrac{1}{3}-x=2$

(b) $\dfrac{1}{3}-x=2$

$\dfrac{1}{3}-\dfrac{1}{3}-x=2-\dfrac{1}{3}$

$-x=\dfrac{6}{3}-\dfrac{1}{3}$

$-1x=\dfrac{5}{3}$

$-1(-1x)=-1\left(\dfrac{5}{3}\right)$

$x=-\dfrac{5}{3}$

The number is $-\dfrac{5}{3}$.

59. Let x represent the number of films Tom Hanks has starred in. Then $x-10$ represents the number of films Johnny Depp has starred in.

$$\left(\begin{matrix}\text{Tom Hanks}\\\text{films}\end{matrix}\right)+\left(\begin{matrix}\text{Johnny Depp}\\\text{films}\end{matrix}\right)=\left(\begin{matrix}\text{Total}\\\text{films}\end{matrix}\right)$$

$x+x-10=104$

$2x-10=104$

$2x-10+10=104+10$

$2x=114$

$\dfrac{2x}{2}=\dfrac{114}{2}$

$x=57$

$x-10=47$

Tom Hanks has starred in 57 films and Johnny Depp has starred in 47 films.

61. Let w represent the width. Then $w+32$ represents the length.

$P=2l+2w$

$224=2(w+32)+2w$

$224=2w+64+2w$

$224=2w+2w+64$

$224=4w+64$

$224-64=4w+64-64$

$160=4w$

$\dfrac{160}{4}=\dfrac{4w}{4}$

$40=w$

$w+32=40+32=72$

The width is 40 in. and the length is 72 in.

Chapter 11   Test

1. $19.95 m$

3. $A = 2lw + 2lh + 2wh$
$= 2(3)(2.5) + 2(3)(1.75) + 2(2.5)(1.75)$
$= 15 + 10.5 + 8.75 = 34.25 \text{ ft}^2$

5. Commutative property of addition

7. Distributive property of multiplication over addition

9. $4(a + 9) - 12 = 4a + 36 - 12 = 4a + 24$

11. $14y + 2(y - 9) + 21 = 14y + 2y - 18 + 21$
$= 16y + 3$

13. $-(3x - 5) + 2x - 1 = -3x + 5 + 2x - 1$
$= -3x + 2x + 5 - 1$
$= -x + 4$

15. An expression is a collection of terms. An equation has an equal sign that indicates that two expressions are equal.

17. $-6x = 12$
$\dfrac{-6x}{-6} = \dfrac{12}{-6}$
$x = -2$

19. $\dfrac{x}{-6} = 12$
$-\dfrac{1}{6}x = 12$
$-6\left(-\dfrac{1}{6}x\right) = -6(12)$
$x = -72$

21. $12 = -3p + 9$
$12 - 9 = -3p + 9 - 9$
$3 = -3p$
$\dfrac{3}{-3} = \dfrac{-3p}{-3}$
$-1 = p$

23. $p + \dfrac{1}{16} = -\dfrac{3}{4}$
$p + \dfrac{1}{16} - \dfrac{1}{16} = -\dfrac{3}{4} - \dfrac{1}{16}$
$p = -\dfrac{12}{16} - \dfrac{1}{16}$
$p = -\dfrac{13}{16}$

25. $5h - 2 = -h + 16$
$5h + h - 2 = -h + h + 16$
$6h - 2 = 16$
$6h - 2 + 2 = 16 + 2$
$6h = 18$
$\dfrac{6h}{6} = \dfrac{18}{6}$
$h = 3$

27. $-2(q - 5) = 6q + 10$
$-2q + 10 = 6q + 10$
$-2q + 2q + 10 = 6q + 2q + 10$
$10 = 8q + 10$
$10 - 10 = 8q + 10 - 10$
$0 = 8q$
$\dfrac{0}{8} = \dfrac{8q}{8}$
$0 = q$

29. Let x represent the number.
$-2x = 15 + x$
$-2x - x = 15 + x - x$
$-3x = 15$
$\dfrac{-3x}{-3} = \dfrac{15}{-3}$
$x = -5$
The number is -5.

31. Let x represent the length of the shorter piece. Then $4x$ represents the length of the other piece.

$$\begin{pmatrix} \text{Length of} \\ \text{one piece} \end{pmatrix} + \begin{pmatrix} \text{Length of} \\ \text{other piece} \end{pmatrix} = \begin{pmatrix} \text{Total} \\ \text{length} \\ \text{of pipe} \end{pmatrix}$$

$$x + 4x = 300$$
$$5x = 300$$
$$\frac{5x}{5} = \frac{300}{5}$$
$$x = 60$$
$$4x = 4(60) = 240$$

The lengths are 60 cm and 240 cm.

Chapters 1–11   Cumulative Review Exercises

1. **(a)** Hundreds
   (b) Ten-thousands
   (c) Hundred-thousands

3. $1{,}285{,}000 \approx 1{,}290{,}000$

5.
$$\begin{array}{r} 851 \\ 46\overline{)39{,}190} \\ \underline{-36\ 8} \\ 2\ 39 \\ \underline{-2\ 30} \\ 90 \\ \underline{-46} \\ 44 \end{array}$$

dividend: 39,190
divisor: 46
quotient: 851
remainder: 44

7. $\dfrac{23}{8}$

9. $\dfrac{21}{10} \div \dfrac{75}{8} = \dfrac{\overset{7}{\cancel{21}}}{\underset{5}{\cancel{10}}} \cdot \dfrac{\overset{4}{\cancel{8}}}{\underset{25}{\cancel{75}}} = \dfrac{28}{125}$

11. $\dfrac{9}{25} - \dfrac{1}{10} + \dfrac{4}{15} = \dfrac{54}{150} - \dfrac{15}{150} + \dfrac{40}{150} = \dfrac{79}{150}$

13. $3\dfrac{7}{8} = \dfrac{3 \cdot 8 + 7}{8} = \dfrac{31}{8}$

15. $2\dfrac{1}{4} + 5\dfrac{5}{6} \cdot 1\dfrac{1}{10} = \dfrac{9}{4} + \dfrac{35}{6} \cdot \dfrac{11}{10}$

$$= \dfrac{9}{4} + \dfrac{385}{60} = \dfrac{135}{60} + \dfrac{385}{60}$$

$$= \dfrac{520}{60} = 8\dfrac{2}{3}$$

17. $0.16\overline{)0.08}$

$$\begin{array}{r} 0.5 \\ 16\overline{)8.0} \\ \underline{-8\ 0} \\ 0 \end{array}$$

19.
$$\begin{array}{r} 78.002 \\ -\ 34.250 \\ \hline 43.752 \end{array}$$

21. Total rooms $= 5 + 6 + 4 = 15$

$$\dfrac{\$300}{15} = \$20$$

Sarah makes $20 per room.

23.
$$\frac{n}{1\frac{1}{2}} = \frac{8}{15}$$

$$15n = \left(1\frac{1}{2}\right)(8)$$

$$15n = 12$$

$$\frac{15n}{15} = \frac{12}{15}$$

$$n = \frac{4}{5}$$

25. Kitty Treats: $\dfrac{\$1.84}{2 \text{ oz}} = \$0.92/\text{oz}$

Cat Goodies: $\dfrac{\$2.25}{2.5 \text{ oz}} = \$0.90/\text{oz}$

Kitty Treats costs $0.92 per oz. Cat Goodies costs $0.90 per oz. Cat Goodies is the better buy.

27. $\dfrac{1}{8} = 0.125$

$$\frac{1}{8} = \frac{1}{8} \times 100\% = \frac{100}{8}\% = 12.5\%$$

29. $\dfrac{2}{9} = 0.\overline{2}$

$$\frac{2}{9} = \frac{2}{9} \times 100\% = 22.\overline{2}\%$$

31. $1 \text{ gal} = \dfrac{1 \text{ gal}}{1} \cdot \dfrac{4 \text{ qt}}{1 \text{ gal}} \cdot \dfrac{2 \text{ pt}}{1 \text{ qt}} \cdot \dfrac{2 \text{ c}}{1 \text{ pt}} \cdot \dfrac{8 \text{ oz}}{1 \text{ c}}$

$$= 128 \text{ oz}$$

$$\frac{128 \text{ oz}}{6 \text{ oz}} = 21\frac{1}{3}$$

21 cups can be filled.

33. 680 cc = 680 mL = 0.68 L

35. 45 g = 4500 cg

37. $8\dfrac{1}{2}$ in. $- 2\left(\dfrac{1}{2}$ in.$\right) = 8\dfrac{1}{2}$ in. $- 1$ in. $= 7\dfrac{1}{2}$ in.

10 in. $- 2\left(\dfrac{1}{2}$ in.$\right) = 10$ in. $- 1$ in. $= 9$ in.

$A = lw = \left(7\dfrac{1}{2}$ in.$\right)(9 \text{ in.}) = 67.5 \text{ in.}^2$

The area is 67.5 in.^{2}.

39. $C = \pi d \approx (3.14)(6 \text{ yd}) = 18.84 \text{ yd}$

41.
```
   15
   12
    8
    5
    6
   12
   10
   20
    7
    5
    8
    9
   11
 + 12
 ────
  140
```
$\dfrac{140}{14} = 10$

43. 12

45. (a) 32% are enrolled in sports. Find 32% of 520.
(0.32)(520) = 166.4
Approximately 166 students enrolled in sports.

(b) 20% do not participate in activities Find 20% of 650.
(0.20)(650) = 130 students

47. $16 \div (-4) \cdot 3 = -4 \cdot 3 = -12$

49. $5 - 23 + 12 - 3 = -18 + 12 - 3 = -6 - 3 = -9$

51. $3y - (5y + 6) - 12 = 3y - 5y - 6 - 12$
$$= -2y - 18$$

53. $9(t - 1) - 7t + 2 = t - 15$
$$9t - 9 - 7t + 2 = t - 15$$
$$9t - 7t - 9 + 2 = t - 15$$
$$2t - 7 = t - 15$$
$$2t - t - 7 = t - t - 15$$
$$t - 7 = -15$$
$$t - 7 + 7 = -15 + 7$$
$$t = -8$$

55.

$$16 = 7 - (3x - 1)$$
$$16 = 7 - 3x + 1$$
$$16 = 7 + 1 - 3x$$
$$16 = 8 - 3x$$
$$16 - 8 = 8 - 8 - 3x$$
$$8 = -3x$$
$$\frac{8}{-3} = \frac{-3x}{-3}$$
$$-\frac{8}{3} = x$$

Additional Topics Appendix

Section A.1   Energy and Power

Section A.1   Practice Exercises

1. **(a)** **Energy** is the amount of work that a physical system is capable of performing.

   (b) A **Calorie** is a metric measurement of heat energy.

3. Energy = (5 ft)(3000 lb) = 15,000 ft · lb

5. $1.5 \text{ yd} = \dfrac{1.5 \text{ yd}}{1} \cdot \dfrac{3 \text{ ft}}{1 \text{ yd}} = 4.5 \text{ ft}$

   Energy = (4.5 ft)(50 lb) = 225 ft · lb

7. $1.5 \text{ tons} = \dfrac{1.5 \text{ tons}}{1} \cdot \dfrac{2000 \text{ lb}}{1 \text{ ton}} = 3000 \text{ lb}$

   Energy = (4 ft)(3000 lb) = 12,000 ft · lb

9. $3000 \text{ Btu} \approx \dfrac{3000 \text{ Btu}}{1} \cdot \dfrac{778 \text{ ft} \cdot \text{lb}}{1 \text{Btu}}$
   $= 2{,}334{,}000 \text{ ft} \cdot \text{lb}$

11. $8000 \text{ Btu} \approx \dfrac{8000 \text{ Btu}}{1} \cdot \dfrac{778 \text{ ft} \cdot \text{lb}}{1 \text{ Btu}}$
    $= 6{,}224{,}000 \text{ ft} \cdot \text{lb}$

13. $53{,}000 \text{ ft} \cdot \text{lb} \approx \dfrac{53{,}000 \text{ ft} \cdot \text{lb}}{1} \cdot \dfrac{1 \text{ Btu}}{778 \text{ ft} \cdot \text{lb}}$
    $\approx 68 \text{ Btu}$

15. $5{,}800{,}000 \text{ Btu} \approx \dfrac{5{,}800{,}000 \text{ Btu}}{1} \cdot \dfrac{778 \text{ ft} \cdot \text{lb}}{1 \text{ Btu}}$
    $= 4{,}512{,}400{,}000 \text{ ft} \cdot \text{lb}$

17. $1026 \text{ Btu} \approx \dfrac{1026 \text{ Btu}}{1} \cdot \dfrac{778 \text{ ft} \cdot \text{lb}}{1 \text{ Btu}}$
    $\approx 798{,}228 \text{ ft} \cdot \text{lb}$

19. $45 \text{ min} = \dfrac{45 \text{ min}}{1} \cdot \dfrac{1 \text{ hr}}{60 \text{ min}}$
    $= \dfrac{3}{4} \text{ hr or } 0.75 \text{ hr}$

21. $30 \text{ min} = \dfrac{30 \text{ min}}{1} \cdot \dfrac{1 \text{ hr}}{60 \text{ min}} = \dfrac{1}{2} \text{ hr or } 0.5 \text{ hr}$

    $2 \text{ hr } 30 \text{ min} = 2 + \dfrac{1}{2} \text{ hr} = \dfrac{5}{2} \text{ hr or } 2.5 \text{ hr}$

23. $6 \text{ min} = \dfrac{6 \text{ min}}{1} \cdot \dfrac{1 \text{ hr}}{60 \text{ min}} = \dfrac{1}{10} \text{ hr or } 0.1 \text{ hr}$

    $1 \text{ hr } 6 \text{ min} = 1 + \dfrac{1}{10} \text{ hr} = \dfrac{11}{10} \text{ hr or } 1.1 \text{ hr}$

25. $45 \text{ min} = 0.75 \text{ hr}$

    $\dfrac{590}{1} = \dfrac{x}{0.75}$
    $x = 590(0.75)$
    $\approx 443 \text{ Cal}$

27. $2 \text{ hr } 30 \text{ min} = 2.5 \text{ hr}$

    $\dfrac{500}{1} = \dfrac{x}{2.5}$
    $x = 500(2.5)$
    $= 1250 \text{ Cal}$

29. $1 \text{ hr } 40 \text{ min} = \dfrac{5}{3} \text{ hr}$

    $\dfrac{560}{1} = \dfrac{x}{\frac{5}{3}}$
    $x = \dfrac{5}{3}(560)$
    $\approx 933 \text{ Cal}$

31. $\text{Power} = \dfrac{(3 \text{ ft})(40 \text{ lb})}{2 \text{ sec}} = 60 \ \dfrac{\text{ft} \cdot \text{lb}}{\text{sec}}$

33. $1.5 \text{ yd} = \dfrac{1.5 \text{ yd}}{1} \cdot \dfrac{3 \text{ ft}}{1 \text{ yd}} = 4.5 \text{ ft}$

    $\text{Power} = \dfrac{(4.5 \text{ ft})(300 \text{ lb})}{10 \text{ sec}} = 135 \ \dfrac{\text{ft} \cdot \text{lb}}{\text{sec}}$

35. $1 \text{ ton} = 2000 \text{ lb}$

    $\text{Power} = \dfrac{(3 \text{ ft})(2000 \text{ lb})}{30 \text{ sec}} = 200 \ \dfrac{\text{ft} \cdot \text{lb}}{\text{sec}}$

37. $1100 \frac{ft \cdot lb}{sec} = \frac{1100 \frac{ft \cdot lb}{sec}}{1} \cdot \frac{1 \, hp}{550 \frac{ft \cdot lb}{sec}} = 2 \, hp$

39. $6050 \frac{ft \cdot lb}{sec} = \frac{6050 \frac{ft \cdot lb}{sec}}{1} \cdot \frac{1 \, hp}{550 \frac{ft \cdot lb}{sec}} = 11 \, hp$

41. $315 \, hp = \frac{315 \, hp}{1} \cdot \frac{550 \frac{ft \cdot lb}{sec}}{1 \, hp} = 173,250 \frac{ft \cdot lb}{sec}$

43. $215 \, hp = \frac{215 \, hp}{1} \cdot \frac{550 \frac{ft \cdot lb}{sec}}{1 \, hp} = 118,250 \frac{ft \cdot lb}{sec}$

45. (a) (3500 W)(6 hr/day)(30 days)
= 630,000 Wh

(b) 630,000 Wh = 630 kWh

(c) (630 kWh)($0.082) = $51.66

Section A.2　Scientific Notation

Section A.2　Practice Exercises

1. Scientific notation involves writing a number as the product of two factors. One factor is a number greater than or equal to 1 but less than 10. The other factor is a power of 10.

3. $100,000 = 10^5$

5. $1,000,000 = 10^6$

7. $0.01 = 10^{-2}$

9. $0.1 = 10^{-1}$

11. No; 82 > 10

13. Yes

15. Yes

17. No; 0.052 < 1

19. $\$13,247,000,000,000 = \$1.3247 \times 10^{13}$

21. $0.0625 \, in. = 6.25 \times 10^{-2} \, in.$

23. $5000 = 5 \times 10^3$

25. $62,000 = 6.2 \times 10^4$

27. $0.0009 = 9 \times 10^{-4}$

29. $0.58 = 5.8 \times 10^{-1}$

31. $25,500,000 = 2.55 \times 10^7$

33. $0.000116 = 1.16 \times 10^{-4}$

35. $0.15 = 1.5 \times 10^{-1}$

37. $13,400 = 1.34 \times 10^4$

39. $3 \times 10^4 = 30,000$

41. $2 \times 10^{-5} = 0.00002$

43. $2.1 \times 10^{-3} = 0.0021$

45. $5.5 \times 10^3 = 5500$

47. $6.13 \times 10^7 = 61,300,000$

49. $4.04 \times 10^{-4} = 0.000404$

51. $5.02 \times 10^{-5} = 0.0000502$

53. $7.07 \times 10^6 = 7,070,000$

55. Already in scientific notation

57. $0.64 \times 10^{24} = 6.4 \times 10^{23}$

59. $586.5 \times 10^{24} = 5.865 \times 10^{26}$

61. $102.4 \times 10^{24} = 1.024 \times 10^{26}$

Section A.3   Rectangular Coordinate System

Section A.3 Practice Exercises

1. (a) The **origin** is the point where the *x*-axis and *y*-axis intersect.
(b) The *x*- and *y*-axes divide the coordinate system into four regions called **quadrants.**
(c) A graph with two number lines drawn at right angles to each other is called a **rectangular coordinate system.**
(d) The ***x*-axis** is the horizontal number line in a rectangular coordinate system.
(e) The first number in an ordered pair is called the ***x*-coordinate.**
(f) The ***y*-axis** is the vertical number line in a rectangular coordinate system.
(g) The second number in an ordered pair is called the ***y*-coordinate.**
(h) Points graphed in a rectangular coordinate system are defined by two numbers as an **ordered pair.**

3–7.

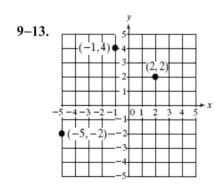

9–13.

15. First move to the left 1.8 units from the origin. Then go up 3.1 units. Place a dot at

the final location. The point is in Quadrant II.

17–21.

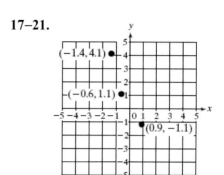

23–27.

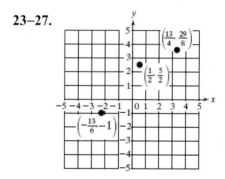

29–33.

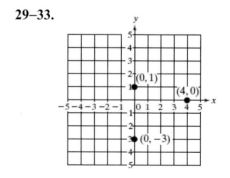

35. Quadrant IV

37. Quadrant III

39. *x*-axis

41. *y*-axis

43. Quadrant II

45. Quadrant I

211

47. (0, 3)

49. (2, 3)

51. (−5, −2)

53. (4, −2)

55. (−2, −5)

57. The ordered pairs are given by (1, −6), (2, −5), (3, 1), (4, 11), (5, 18), (6, 22), (7, 24), (8, 22), (9, 15), (10, 8), (11, 1), and (12, −4).

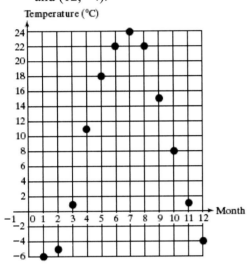

59. The ordered pairs are given by (1, 13,200), (2, 11,352), (3, 9649), (4, 8201), (5, 6971), and (6, 5925).

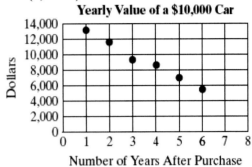

61. The ordered pairs are given by (30, 33), (40, 35), (50, 40), (60, 37), and (70, 30).

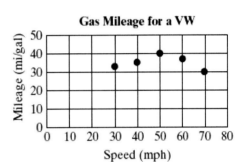

Notes

Notes

Notes

Notes